Bioinspired and Green Synthesis of Nanostructures

Scrivener Publishing
100 Cummings Center, Suite 541J
Beverly, MA 01915-6106

Publishers at Scrivener
Martin Scrivener (martin@scrivenerpublishing.com)
Phillip Carmical (pcarmical@scrivenerpublishing.com)

Bioinspired and Green Synthesis of Nanostructures

A Sustainable Approach

Edited by
Mousumi Sen
Department of Chemistry, Amity University, Noida, India
and
Monalisa Mukherjee
Amity Institute of Click Chemistry Research and Studies, Noida, India

This edition first published 2023 by John Wiley & Sons, Inc., 111 River Street, Hoboken, NJ 07030, USA and Scrivener Publishing LLC, 100 Cummings Center, Suite 541J, Beverly, MA 01915, USA
© 2023 Scrivener Publishing LLC

For more information about Scrivener publications please visit www.scrivenerpublishing.com.

All rights reserved. No part of this publication may be reproduced, stored in a retrieval system, or transmitted, in any form or by any means, electronic, mechanical, photocopying, recording, or otherwise, except as permitted by law. Advice on how to obtain permission to reuse material from this title is available at http://www.wiley.com/go/permissions.

Wiley Global Headquarters
111 River Street, Hoboken, NJ 07030, USA

For details of our global editorial offices, customer services, and more information about Wiley products visit us at www.wiley.com.

Limit of Liability/Disclaimer of Warranty
While the publisher and authors have used their best efforts in preparing this work, they make no representations or warranties with respect to the accuracy or completeness of the contents of this work and specifically disclaim all warranties, including without limitation any implied warranties of merchantability or fitness for a particular purpose. No warranty may be created or extended by sales representatives, written sales materials, or promotional statements for this work. The fact that an organization, website, or product is referred to in this work as a citation and/or potential source of further information does not mean that the publisher and authors endorse the information or services the organization, website, or product may provide or recommendations it may make. This work is sold with the understanding that the publisher is not engaged in rendering professional services. The advice and strategies contained herein may not be suitable for your situation. You should consult with a specialist where appropriate. Neither the publisher nor authors shall be liable for any loss of profit or any other commercial damages, including but not limited to special, incidental, consequential, or other damages. Further, readers should be aware that websites listed in this work may have changed or disappeared between when this work was written and when it is read.

Library of Congress Cataloging-in-Publication Data

ISBN 978-1-394-17446-1

Cover image: Pixabay.Com
Cover design by Russell Richardson

Set in size of 11pt and Minion Pro by Manila Typesetting Company, Makati, Philippines

Printed in the USA

10 9 8 7 6 5 4 3 2 1

Contents

Preface		**xv**
1	**Green Synthesis: Introduction, Mechanism, and Effective Parameters**	**1**
	Mousumi Sen	
	1.1 Introduction	2
	1.2 What Are Nanoparticles?	2
	1.3 Types of Nanoparticles	4
	1.3.1 Inorganic Nanoparticle	4
	1.3.1.1 Green Synthesis of Silver (Ag) Nanoparticles	4
	1.3.1.2 Green Synthesis of Gold (Au) Nanoparticles	7
	1.3.1.3 Green Synthesis of Copper (Cu) Nanoparticles	8
	1.3.1.4 Iron Oxide Nanoparticles	9
	1.3.2 Organic Nanoparticles	9
	1.3.2.1 Liposomes	10
	1.3.2.2 Micelles	10
	1.3.2.3 Dendrimers	10
	1.4 Approaches	10
	1.5 Conclusion	18
	References	19
2	**Greener Nanoscience: Proactive Approach to Advancing Nanotechnology Applications and Reducing Its Negative Consequences**	**25**
	Utkarsh Jain and Kirti Saxena	
	2.1 Introduction	26
	2.2 Why Do We Need Green Nanoscience Approaches?	27
	2.3 Green Nanotechnology	28
	2.4 Green Synthesis of Nanomaterials	29
	2.5 Advantages of Green Nanoscience	33
	2.5.1 Green Nanoscience in Industries	34

		2.5.2	Green Nanoscience in Automobiles	34
		2.5.3	Green Nanoelectronics	35
		2.5.4	Green Nanoscience in Food and Agriculture	35
		2.5.5	Green Nanoscience in Medicines	35
	2.6	Conclusion		36
		References		37
3	**Optimization of the Process Parameters to Develop Green-Synthesized Nanostructures with a Special Interest in Cancer Theranostics**			**43**
	Tathagata Adhikary, Chowdhury Mobaswar Hossain and Piyali Basak			
	3.1	Introduction		44
		3.1.1	Conventional Techniques in Nanoparticle Synthesis	44
		3.1.2	Green Nanotechnology	46
	3.2	Mechanism Underlying Green Synthesis		47
	3.3	Green Synthesized Nanoparticles in Cancer Theranostics		52
	3.4	Optimizing the Synthesis and Subsequent Characterizations		55
		3.4.1	Approaches to Achieve Optimization	55
		3.4.2	Characterization of Nanoparticles	57
		Acknowledgment		58
		References		59
4	**Sustainability: An Emerging Design Criterion in Nanoparticles Synthesis and Applications**			**65**
	Yashtika Raj Singh, Abhyavartin Selvam, P.E. Lokhande and Sandip Chakrabarti			
	4.1	Introduction		66
	4.2	Biotemplates		69
		4.2.1	Plant-Based Biotemplates	70
		4.2.2	Microorganism-Based Biotemplates	75
			4.2.2.1 Bacteria	75
			4.2.2.2 Fungi	79
			4.2.2.3 Yeast	79
			4.2.2.4 Algae	82
	4.3	Synthesis Routes		84
		4.3.1	Effect of pH	84
		4.3.2	Effect of Temperature	85
		4.3.3	Effect of Biomolecules	86
			4.3.3.1 Plant-Based	86
			4.3.3.2 Microorganism-Based	87

		4.4	Applications	88
			4.4.1 Biomedical Application	88
			4.4.1.1 Antimicrobial Activity	88
			4.4.1.2 Biomedication	90
			4.4.1.3 Vaccines	90
			4.4.1.4 Antidiabetic	91
			4.4.1.5 Diagnostic Applications	91
			4.4.2 Environmental Application	92
			4.4.2.1 Environmental Remediation	93
			4.4.2.2 Catalytic Removal of Textile Dyes	93
			4.4.2.3 Wastewater Treatment	94
			4.4.2.4 Agriculture	94
		4.5	Conclusion and Outlook	96
			References	98
5	**Green Conversion Methods to Prepare Nanoparticle**			**115**
	Pradip Kumar Sukul and Chirantan Kar			
		5.0	Introduction	116
		5.1	Bacteria	118
		5.2	Fungi	122
		5.3	Yeast	127
		5.4	Viruses	129
		5.5	Algae	132
		5.6	Plants	134
		5.7	Conclusion and Perspectives	135
			References	136
6	**Bioinspired Green Synthesis of Nanomaterials From Algae**			**141**
	Reetu, Monalisa Mukherjee and Monika Prakash Rai			
		6.1	Introduction	141
		6.2	Algal System-Mediated Nanomaterial Synthesis	143
		6.3	Factors Affecting the Green Synthesis of Nanomaterials	145
			6.3.1 Light	146
			6.3.2 Temperature	146
			6.3.3 Incubation Period	146
			6.3.4 pH	147
			6.3.5 Precursor Concentration and Bioactive Catalyst	147
		6.4	Applications of the Green Synthesized Nanomaterials	147
			6.4.1 Antimicrobial Agents	148
			6.4.2 Anticancerous	149
			6.4.3 Biosensing	149
			6.4.4 Bioremediation	149

	6.5	Future Perspectives	150
	6.6	Conclusion	150
		References	151

7 Interactions of Nanoparticles with Plants: Accumulation and Effects 157
Indrajit Roy

	7.1	Introduction	158
	7.2	Uptake and Translocation of Nanoparticles and Nanocarriers in Plants	160
	7.3	Nanoparticle-Mediated Sensing and Biosensing in Plants	164
	7.4	Tolerance Versus Toxicity of Nanoparticles in Plants	168
	7.5	Nanoparticle-Mediated Delivery of Fertilizers, Pesticides, Other Agrochemicals in Plants	173
	7.6	Nanoparticle-Mediated Non-Viral Gene Delivery in Plants	177
	7.7	Conclusions	181
		Acknowledgments	182
		References	183

8 A Clean Nano-Era: Green Synthesis and Its Progressive Applications 189
Susmita Das and Kajari Dutta

	8.1	Introduction	190
	8.2	Green Synthetic Approaches	190
		8.2.1 Microorganism-Induced Synthesis of Nanoparticles	190
		8.2.2 Biosynthesis of Nanoparticles Using Bacteria	191
		8.2.3 Biosynthesis of Nanoparticles Using Fungi	191
		8.2.4 Biosynthesis of Nanoparticles Using Actinomycetes	192
		8.2.5 Biosynthesis of Nanoparticles Using Algae	192
		8.2.6 Plant Extracts for Biosynthesis of Nanoparticles	193
	8.3	Nanoparticles Obtained Using Green Synthetic Approaches and Their Applications	193
		8.3.1 Synthesis of Silver (Ag) and Gold (Au)	193
		8.3.2 Synthesis of Palladium (Pd) Nanoparticles	195
		8.3.3 Synthesis of Copper (Cu) Nanoparticles	196
		8.3.4 Synthesis of Silver Oxide (Ag_2O) Nanoparticles	197
		8.3.5 Synthesis of Titanium Dioxide (TiO_2) Nanoparticles	197
		8.3.6 Synthesis of Zinc Oxide (ZnO) Nanoparticles	198
		8.3.7 Synthesis of Iron Oxide Nanoparticles	199
	8.4	Conclusion	200
		References	200

9	A Decade of Biomimetic and Bioinspired Nanostructures: Innovation Upheaval and Implementation		207
	Vishakha Sherawata, Anamika Saini, Priyanka Dalal and Deepika Sharma		
	9.1	Introduction	208
	9.2	Bioinspired Nanostructures	209
		9.2.1 Materials Inspired by Structural Properties of Natural Organism	210
	9.3	Biomimetic Structures	213
	9.4	Biomimetic Synthesis Processes and Products	214
	9.5	Application of Bioinspired and Biomimetic Structure	219
	9.6	Conclusion	223
	9.7	Future Outlook	224
		Acknowledgments	225
		References	225
10	A Feasibility Study of the Bioinspired Green Manufacturing of Nanocomposite Materials		231
	Arpita Bhattacharya		
	10.1	Introduction	232
	10.2	Biopolymers	233
		10.2.1 Cellulose	234
		10.2.2 Chitosan	234
		10.2.3 Starch	234
		10.2.4 Chitin	235
		10.2.5 Polyhydroxyalkanoates (PHA)	235
		10.2.6 Polylactic Acid (PLA)	235
	10.3	Different Types of Bioinspired Nanocomposites	236
		10.3.1 Polymer-HAp Nanoparticle Composites	236
		10.3.2 Nanowhisker-Based Bionanocomposites	237
		10.3.3 Clay-Polymer Nanocomposites	238
	10.4	Fabrication of Bionanocomposites	240
		10.4.1 Electrospinning	240
		10.4.2 Solvent Casting	240
		10.4.3 Melt Moulding	241
		10.4.4 Freeze Drying	242
		10.4.5 3D Printing	242
		10.4.6 Ball Milling Method	243
		10.4.7 Microwave-Assisted Method for Bionanocomposite Preparation	244
		10.4.8 Ultraviolet Irradiation Method	245

	10.5	Application of Bionanocomposites	246
		10.5.1 Orthopedics	246
		10.5.2 Dental Applications	248
		10.5.3 Tissue Engineering	251
	10.6	Conclusion	252
		References	252
11	**Bioinspiration as Tools for the Design of Innovative Materials and Systems Bioinspired Piezoelectric Materials: Design, Synthesis, and Biomedical Applications**		**263**
	Santu Bera		
	11.1	Bioinspiration and Sophisticated Materials Design	264
		11.1.1 Piezoelectricity in Natural Bulk Materials	266
		11.1.2 Piezoelectricity in Proteins	267
		11.1.3 Piezoelectric Ultra-Short Peptides	270
		11.1.4 Single Amino Acid Assembly and Coassembly-Based Piezoelectric Materials	273
	11.2	Biomedical Applications	276
		11.2.1 Piezoelectric Sensors	276
		11.2.2 Tissue Regeneration	279
	11.3	Conclusion and Future Perspectives	281
		Acknowledgment	282
		References	282
12	**Protein Cages and their Potential Application in Therapeutics**		**291**
	Chiging Tupe and Soumyananda Chakraborti		
	12.1	Introduction	292
	12.2	Different Methods of Cage Modifications and Cargo Loading	295
	12.3	Applications of Protein Cages in Biotechnology and Therapeutics	298
		12.3.1 Protein Cage as Targeted Delivery Vehicles for Therapeutic Protein	298
		12.3.2 Protein Cage-Based Encapsulation and Targeting of Anticancer Drugs	299
		12.3.3 Protein Cage-Based Immune-Therapy	300
	12.4	Future Perspective	301
	12.5	Conclusion	301
		Acknowledgment	301
		References	302

13 Green Nanostructures: Biomedical Applications and Toxicity Studies — 307
Radhika Chaurasia, Omnarayan Agrawal, Rupesh, Shweta Bansal and Monalisa Mukherjee

- 13.1 Introduction — 308
- 13.2 Moving Toward Green Nanostructures — 309
- 13.3 Methods of Nanoparticle Synthesis — 309
- 13.4 Plant-Mediated Synthesis of Green Nanostructures — 310
  - 13.4.1 Silver Nanoparticles — 310
  - 13.4.2 Gold Nanoparticles — 311
  - 13.4.3 Zinc Oxide Nanoparticles — 313
  - 13.4.4 Selenium Nanoparticles — 314
- 13.5 Microbe-Based Synthesis — 314
  - 13.5.1 Bacteria-Mediated Synthesis of NPs — 315
  - 13.5.2 Fungus-Mediated Synthesis of NPs — 316
  - 13.5.3 Actinomycete-Mediated Synthesis of NPs — 317
- 13.6 Toxicity of Nanostructures — 318
- 13.7 Conclusion — 319
- References — 319

14 Future Challenges for Designing Industry-Relevant Bioinspired Materials — 325
Warren Rosario and Nidhi Chauhan

- 14.1 Introduction — 326
- 14.2 Bioinspired Materials — 327
- 14.3 Applications of Bioinspired Materials and Their Industrial Relevance — 327
- 14.4 Bioinspired Materials in Optics — 328
  - 14.4.1 Applications in Optics — 328
  - 14.4.2 Bioinspired Materials in Energy — 329
  - 14.4.3 Applications in Energy — 331
  - 14.4.4 Bioinspired Materials in Medicine — 333
- 14.5 Applications in Medicine — 333
- 14.6 Future Challenges for Industrial Relevance — 336
- 14.7 Optics-Specific Challenges — 341
- 14.8 Energy-Specific Challenges — 342
- 14.9 Medicine-Specific Challenges — 342
- 14.10 Conclusion — 343
- References — 344

15 Biomimetic and Bioinspired Nanostructures: Recent Developments and Applications — 353
Sreemoyee Chakraborty, Debabrata Bera, Lakshmishri Roy and Chandan Kumar Ghosh

- 15.1 Introduction — 354
- 15.2 Designing Bioinspired and Bioimitating Structures and Pathways — 357
- 15.3 Nanobiomimicry—Confluence of Nanotechnology and Bioengineering — 359
- 15.4 Biofunctionalization of Inorganic Nanoparticles — 361
  - 15.4.1 Strategies to Develop Biofunctionalized Nanoparticles — 361
  - 15.4.2 Fate of Biofunctionalized Nanoparticles — 362
  - 15.4.3 Biofunctionalization Nanoparticles with Different Organic Compounds — 363
    - 15.4.3.1 Carbohydrates — 363
    - 15.4.3.2 Nucleic Acid — 363
    - 15.4.3.3 Peptides — 364
    - 15.4.3.4 DNA — 364
    - 15.4.3.5 Antibody — 364
    - 15.4.3.6 Enzyme — 365
    - 15.4.3.7 Stability of Biofunctionalized Nanoparticles — 365
    - 15.4.3.8 Applications of Biofunctionalized Nanoparticles — 365
- 15.5 Multifarious Applications of Biomimicked/Bioinspired Novel Nanomaterials — 367
  - 15.5.1 Implementation of Nanobiomimicry for Sustainable Development — 367
  - 15.5.2 Bioinspired Nanomaterials for Biomedical and Therapeutic Applications — 370
  - 15.5.3 Nanomaterial-Based Biosensors for Environmental Monitoring — 376
    - 15.5.3.1 Nanosensor Design — 378
    - 15.5.3.2 Operation of a Biomimetic Sensor — 380
    - 15.5.3.3 Applications in Environmental Monitoring — 381
  - 15.5.4 Biomimetic Nanostructure for Advancement of Agriculture and Bioprocess Engineering — 383

		15.5.5 Nanobiomimetics as the Future of Food	
		Process Engineering	387
	15.6	Emerging Trends and Future Developments	
		in Bioinspired Nanotechnology	389
	15.7	Conclusion	390
		References	391

Index   **405**

Preface

This book focuses on the recent developments and novel applications of bioinspired and biomimetic nanostructures as functionally advanced biomolecules with huge prospects for research, development, and engineering industries. The population explosion, automation and urbanization have had numerous harsh environmental effects that have ultimately led to climate change. Therefore, the future of the world depends on immediately investing our time and effort into advancing ideas on ways to restrict the use of hazardous chemicals, thereby arresting further environmental degradation. To achieve this goal, nanotechnology has been an indispensable arena which has extended its wings into every aspect of modernization. For example, green synthetic protocols are being extensively researched to inhibit the harmful effects of chemical residues and reduce chemical wastes. This involves the study of nanotechnology for artful engineering at the molecular level across multiple disciplines. In recent years, nanotechnology has ventured away from the confines of the laboratory and has been able to conquer new domains to help us live better lives.

The green synthetic techniques produce nanostructures that generally possess unique properties that set them apart from those produced using physicochemical techniques. In addition to being eco-friendly, economic, and appropriate for mass production, these nanostructures possess diverse chemical, optical, mechanical, and magnetic properties as compared to bulk materials because of the increase in the surface area. An influential tenet of nanotechnology is the fabrication of nanoscale materials as well as their controlled morphology and dimensions. Learning from nature has given us different ways to address problems that arise when developing novel materials, which are known as biologically inspired and biomimetic strategies. These strategies, which rely on learning from surrounding entities, have experienced an unprecedented surge in the last decade, spurred on by advances in nanoscience and technology. Globally, the scientific community has recognized the prospects of an environmental catastrophe, and equitably providing clean air, food, water, and sustainable sources of energy is a matter of major concern. In the next 30 years, the desire for sustainable green alternatives is anticipated to double; therefore, the interdisciplinary holistic approaches pushing the idea of turning waste into profit require special emphasis.

In pursuit of a sustainable and eco-friendly abode, research focusing on the green synthesis of materials has revolutionized the design, development, and application of chemical products. Meticulous efforts for minimal waste products, synthesis of recyclable materials, and energy conservation have led to the research and discovery of ingenious strategies. Currently, green nanostructure synthesis is becoming extremely prevalent because it is safe and works well with living things. Green synthesis is merely a simplification of so-called logic that surpasses the fundamental concepts and techniques of synthesis. Therefore, the significance of green nanostructure synthesis must be examined in terms of how it is produced, its quality, and potential applications. The application of nanotechnology has enabled us to develop bioinspired materials using unique structures which can result in desired properties. With our increasing awareness of the scarcity of resources and surging pollution, there is a growing push towards the development of more bioinspired materials with better sustainability. As a result, they are growing in popularity, which makes studying them, their properties, and their fabrication techniques extremely important. Many bioinspired materials have already been developed that show great promise in solving many of our problems. But on the road to mass production, there are still some obstacles that are yet to be overcome. This book provides detailed coverage of the chemistry of each major class of synthesis of bioinspired nanostructures and their multiple functionalities. In addition, it reviews the new findings currently being introduced, and analyzes the various green synthetic approaches for developing nanostructures, their distinctive characteristics, and their applications.

Chapter 1 focuses on the synthesis and application of the nanostructures categorized as reliable, eco-friendly, and sustainable that have sparked a drive to develop environmentally acceptable methods. Hence, greener ways of identifying the biomolecules present in plants that mediate the formation of nanostructures, along with their production, testing and applications are also discussed.

Next, Chapter 2 discusses the limitations of existing nanotechnology-based methods to produce nanostructure and why we need the green nanoscience approach to overcome these limitations. The advantages of greener nanoscience have been described together with the processes for green nanostructure synthesis and the design and optimization of green processes to reduce or eliminate environmental and health hazards.

Chapter 3 gives the reader an insight into the mechanisms underlying different green nanofabrication techniques and the effect of various factors in the fabrication process. Statistical models and other *in-silico* approaches are frequently employed along with experimental data to ease the optimization. Although these techniques remain valid in optimizing the green synthesis of any nanomaterial, this chapter attempts to review the related reports and recent advancements in the field of cancer theranostics.

Chapter 4 evaluates the emerging nanomaterials possessing copious applications due to their nature and biological compatibility, high synthesis rate, stability, selectivity, sensitivity, and so on. Along the same lines, the practicality of biogenically developed nanostructure for biomedical applications, which has been recently ameliorated, is explored. This chapter also recounts sustainable approaches to effectively engineer nanostructures biogenically to be applied in demanding situations and applications.

A green synthesis strategy furnishes safe, clean and environment-friendly methodology to produce metallic nanoparticles. There is great demand for developing new protocols to enable the cost-effective and high-yield production of nanoparticle comparable to conventional methods. A significant step toward this would be improving eco-friendly processes for the creation of metallic nanoparticles. Thus, Chapter 5 is designed to explain the method of green synthesis, and the effects of various parameters on the size, morphology, and amount of metal nanoparticles produced.

The goal of Chapter 6 is to provide a brief overview of the variety of algal strains used in this booming field and the factors affecting them, along with the disparate nanocomposites synthesized.

The objective of Chapter 7 is to frame extensive guidelines and regulations based on the knowledge already available in the area of bioinspired "green" nanoparticles and implemented for the safe and efficient use of nanoparticles in farming, agriculture, and other botanical practices, aimed at the restoration of the delicate balance between living organisms and the environment.

Biogenic reduction of metal salts generally results in nanostructures possessing unique properties compared to those produced using physicochemical techniques. Thus, green synthetic techniques are eco-friendly, economic and appropriate for mass production. Chapter 8 provides a detailed review and analysis of the various green synthetic approaches for developing nanostructures, their distinctive characteristics and their applications. It also highlights the applications and improved properties of the nanostructures obtained using green synthesis.

Chapter 9 attempts to explain the advances in biomimetic and bioinspired nanostructures and present them as promising solutions to many unresolved problems in the biomedical field. Biomimetic nanostructures regulate the cell behavior reported in *in-vitro* studies, where they play an important role in cell nuclear alignment, cell spreading, cell differentiation, phagocytosis, and viability. Here, recent developments in the preparation of bioinspired and biomimetic nanostructures through different routes of synthesis are presented. The different templates used for the synthesis of nanostructures and binding the template with other useful materials to enhance the therapeutic efficacy are also discussed.

The recent trends in nano-functional materials and renewable materials for the preparation of bioinspired nanocomposites especially used in the agricultural, biomedical and healthcare sectors are discussed in Chapter 10.

Chapter 11 systematically discusses the recent development of bio-piezoelectric materials based on natural or nature-inspired biomolecules, with an emphasis on the design strategy, synthesis, integration into bio-piezoelectric platforms and finally their deployment in the latest biomedical applications.

Chapter 12 provides various ideas for designing nanoscale structures with targeted delivery ability which can be used in various applications, including therapeutics that may sound like science fiction.

Various green synthesis techniques and the contribution of green nanostructures in a variety of applications are highlighted in Chapter 13. The goal of this chapter is to provide a brief overview of the different green nanostructures used in this emerging field.

In Chapter 14, the industrial relevance of bioinspired materials is highlighted by focusing on the fields of optics, energy and medicine. Also discussed are the bioinspired materials that have found use in sensing, construction, adhesive manufacturing, communication, thermoregulation and many other fields.

Finally, Chapter 15 presents the broader concept of recent developments and novel applications of bioinspired and biomimetic nanomaterials. Biomimetics (biomimicry) is the development of novel biomaterials which not only mimic the composition of natural systems but also copy their structure, morphology and functionality. These bioinspired materials generally have their origins in nature and are designed by studying and imitating the remarkable biological processes of organisms and pathways of occurrence of different natural phenomena. The core technology of bio-inspiration is built upon deciphering how biological materials are constructed, and understanding the interactions that cause their unique properties.

In conclusion, we would like to express our gratitude to the many contributors for the hard work they put into this book. We would also like to thank all the authors for sharing their insightful research and information with us. We are very much thankful for Aarushi Sen for her unending encouragement and support throughout the making of this book. Her help was much appreciated. We are also most grateful for the efforts of Martin Scrivener of Scrivener Publishing, whose help made this book possible. We thank him for his patience and consistent support throughout the journey.

Mousumi Sen
Monalisa Mukherjee
April 2023

1
Green Synthesis: Introduction, Mechanism, and Effective Parameters

Mousumi Sen

Department of Chemistry, Amity Institute of Applied Sciences, Amity University, Uttar Pradesh, India

Abstract

Nanoparticles are synthesized by different methods, such as physical methods, chemical methods, and biological methods. The need is the greener pathway and method for the synthesis of nanoparticles so that the process in nontoxic and do not harm the environment. In a number of industries, including medicine, pharmaceuticals, and agriculture, nanoparticles are used. Nanobiotechnology when combined with green technology benefits the industries, environment, and human health. Green chemistry plays a vital role in generating the plant extract-derived nanoparticles (especially gold and silver). Plant extracts contain the biomolecules that help in reducing metal ions and create the nanoparticles, and this can be achieved via a single-step synthesis process. In addition to plant species extracts, there are lots of diverse range of plant species that helped in the production of nanoparticles. Microbial synthesis or biological methods are applied for the synthesis in a greener fashion using different microorganisms, such as algae, bacteria, fungus, etc. Silver nanoparticles being antibacterial in nature are of great interest. They are used to treat cancer, tumors, and used in drug delivery process and many more countless applications.

Synthesis has been categorized as a reliable, eco-friendly, and sustainable way of synthesizing nanoparticles that contain substances, like metal oxides and others. Hence, greener way which are involved in the formation of the nanoparticles using plant extracts and types of nanoparticles with their production, testing and applications of these nanoparticles as well have been highlighted.

Keywords: Green synthesis, nanoparticles, sustainable, inorganic nanoparticles, organic nanoparticles, phytochemical screening

---

Email: mousumi1976@gmail.com

1.1 Introduction

In this fast-growing and technology-oriented world, advancements in almost every field are going on, and science is no exception. Research in every possible field is currently going to improve the present situation, methods, problems. Researchers are working hard in various fields. One of the filed in which there is much research going on in present times is nanoparticles. Nanotechnology is the branch of science that deals with dimensions of approximately 1 to 100 nm [1]. Due to their size, orientation, and chemical and physical characteristics, they are widely used. All the properties (chemical, physical, and biological) of individual atoms/molecules and their related bulk vary in fundamental ways within this size range of particles. This plays a vital role in many technologies, such as nanoparticles in optics, electronics, and medicinal science industries [2]. They are of particular importance due to their high surface-to-volume ratio and extremely small size, which, when compared to the majority of the same chemical compositions, causes both chemical and physical alterations in their properties. Massive advancements in the technologies had ushered forth new eras. This comprises creating nanoscale materials and then studying or utilizing their intriguing physicochemical and optoelectronic features. The methods to produce nanoparticles using plants extracts are stable, bio-degradable, environmentally safe, and cost-effective [3, 4]. These particles tend to form enormous clusters that result in deposition, diminishing their effectiveness, although they are independent of shape and size and instead suggest the stability of particles. These can be formed from larger molecules or created from the ground up, for example, by nucleating and growing particles from low concentration levels in the liquid or gaseous phase. Functionalization via conjugation to bioactive compounds is another method of synthesis. Since the early days of nanoscience, the synthesis of high-yielding, low-cost nanomaterials have been a major issue. The ability to produce particles with diverse forms, monodispersed, chemical content, and size is critical for the use of nanoparticles in medicine.

1.2 What Are Nanoparticles?

The name "nanoparticles" derives from the Greek word "nanos," which means dwarf or extremely small. Nanoparticles are also called "nanomaterials." Particles between 1 and 100 nm in size are referred to as nanoparticles. When these particles are compared to atoms, they are larger than only one atom. Atom clusters are defined as particles smaller than 1 nm.

As we know particles are small, and they cannot be detected by the human eye and microscopes. They are particles, smaller in size, so they can pass through candles or filters. The smaller the size of the particle, the larger will be the surface area. The main characteristic of nanoparticles is that they have a large surface area to volume ratio. Their sizes may vary as they are very small. They differ in their physical and chemical characteristics. The term nanoparticle can also be used for bigger particles and having a range up to 500 nm. They are differentiated from three particles: (i) microparticles—particles have ranged between 1 and 100 μm; (ii) Fine, particles have ranged between 100 and 2500 nm; (iii) Coarse- particles have ranged between 2500 and 10,000 nm.

They are environmentally friendly. These particles do not need any high temperature, energy, pressure, or any other kind of toxic chemicals [5]. The advantage of the nanoparticle is by reducing microorganisms and their cultures. These particles appear in nature and are mostly used in medicines, industries, laboratories, etc.

Richard Feynman, an 11-year-old American physicist and Nobel Prize winner, introduced the idea of nanotechnology for the first time in 1959. His goal was to use machines to create even smaller, molecular-level devices. He is also called as father of modern nanotechnology. The term "nanotechnology" was first used and defined by a Japanese scientist named Norio Taniguchi in 1974, which was 15 years later. Nanotechnology, according to him, is the separation, distortion, and consolidation of material by one atom or one molecule during processing. Following this, scientists began to find the field of nanotechnology to be very interesting, and research in this area began. For the potential synthesis of nanoparticles, two paths or approaches—top-down approach and bottom-up approach—have been established. Both have pros and cons in terms of price, quality, turnaround time, and speed. Nanotechnology and nanoparticles are not brand-new fields. Even today, there are indications that nanotechnology has existed and been used in the past [6, 7]. The Lycurgus Cup is an illustration of ancient nanotechnology. It is a cup constructed from dichroic glass. It is a 4-century Roman glass, which has a unique property of changing color. When light is shining on the outside, it appears olive green, and when light is shining on the inside, it turns ruby red. By utilizing transmission electron microscopy (TEM) to examine the cup in 1990, scientists were able to determine the cause of the colour change: the existence of nanoparticles with a diameter between 50 and 100 nm. After additional X-ray research, it was discovered that the nanoparticles detected in 21 of the cups were made of a silver-gold alloy with a silver to gold content of 7:3 and roughly 10% copper scattered throughout the glass matrix.

1.3 Types of Nanoparticles

Nanostructures (nanomaterials) are made up of organic polymers and inorganic polymers. so, these are classified into two types: (1) inorganic nanoparticles, (2) organic nanoparticles, both materials are of two dimensions or more than that. These are also of the same size as these nanoparticles are (1–100 nm) [8]. The organic nanomaterials are composed of liposomes, micelles, and polymer nanoparticles mostly used for drug delivery systems. A liposome is a lipid-based particle that includes a core that is surrounded by a thick layer called the phospholipid layer, whereas the inorganic nanoparticles are used for industrial, therapeutic purposes. These particles include Au, Ag, ZnO, CuO, and other metals and their oxides [9, 10].

1.3.1 Inorganic Nanoparticle

1.3.1.1 Green Synthesis of Silver (Ag) Nanoparticles

Silver (Ag) nanoparticles have a range between 1 and 100 nm in size. Due to the higher surface area of silver atoms, these particles are composed of silver oxide, as well as particles containing a solution of silver metal ions and a reducing agent. These have different physical, chemical, and biological properties. The ability to absorb water in Ag particles is high. The procedures of stabilization and reduction are used to create silver nanoparticles [11]. These are the simplest techniques. Stabilization can be achieved by the breakdown of a molecule, such as vitamins, proteins, etc. Extraction of silver particles is done by plants, such as *Aloe vera, Saccharum officinarum*, etc. (plants must be medicinal because Ag particles are very fundamental to biomedical applications). The most used shapes of Ag nanoparticles are diamond, spherical, etc. Ag plays a crucial role in electrochemical sensor platforms or biomedical applications [12].

The method for the synthesis of nanoparticle involves the following:

The target plant component is meticulously twice washed with tap water after being obtained from various locations in order to remove both epiphytes and necrotic plants. After that, any related material is removed using sterile distilled water [13]. Before being powdered in a home blender, the clean sources are dried for 10 to 15 days in the shade. Boil around 10 g of the dry powder in 100 mL of deionized distilled water to make the plant broth (hot percolation method). The infusion is then filtered until the soup contains no more insoluble particles. Pure Ag(I) ions are converted into Ag(0) when a small amount of plant extract is introduced to a 103-M AgNO$_3$ solution [14]. This process may be observed by periodically analyzing the solution's UV-visible spectra.

The preparation of Ag nanoparticles required a large portion of the flora. Different plants, as well as their various parts, are tested. Using the Alternanthera dentate aqueous extract, green fast production of spherical shaped Ag nanoparticles with diameters of 50–100 nm were observed [15, 16]. This extract's conversion of silver ions to silver nanoparticles took only ten minutes. Extracellular Ag nanoparticles are synthesized using aqueous leaf extract, which is a fast, easy, and cost-effective method that are comparable to chemical and microbiological methods. Pseudomonas aeruginosa, Escherichia coli, Klebsiella 3 pneumoniae, and Enterococcus faecalis are all susceptible to silver nanoparticles. Acorus calamus was also utilized to make silver nanoparticles in order to test its antioxidant, antibacterial, and anticancer properties. A plant extract called boerhaavia diffusa was employed as the reducing agent in the green production of silver nanoparticles. According to XRD and TEM analysis, Ag nanoparticles with a face-centered cubic structure and spherical shape have an average particle size of 25 nm [17]. These nanoparticles were tested for their antibacterial potency against *Pseudomonas fluorescens*, *Aeromonas hydrophila*, and *Flavobacterium branchiophilum*, three fish bacterial infections. The most sensitive bacterium was *F. branchiophilum*, when compared to the other two. The reducing agents are present in relatively high concentrations in steroid, sapogenin, carbohydrates, and flavonoids, whereas phyto-constituents serve as capping agents and stabilise silver nanoparticles. The nanoparticles produced were determined to be spherical in shape, with an average size of the 7 to 17 nm. These nanoparticles have crystalline structure with a face cantered cubic (FCC) shape, as demonstrated by the XRD technique. Tea was employed as a capping agent to make crystalline silver nanoparticles with 20 diameters ranging from 20 to 90 nm. The amount of tea extract employed and the temperature of the reaction have an effect on the efficiency of production and also on the rate of nanoparticle formation. Ag nanoparticles with a spherical shape range in size from 5 to 20 nm, according to TEM. With callus extracted, the salt marsh plant *Sesuvium portulacastrum* L., Ag nanoparticles indicated a gradual shift in the colouration of the extracts to yellow-brown as the intensity of the extract rose over time [18]. A dried fruit body 41 *Tribulus terrestris* L. extract was combined to produce Ag nanoparticles with Ag nitrate This extract was utilized to create Ag nanoparticles with a spherical shape and a size range of 16 to 28 nm using the Kirby-Bauer process, which demonstrated antibacterial efficacy against multidrug-resistant bacteria, such as *Streptococcus pyogens, Escherichia coli, Bacillus subtilis, Pseudomonas aeruginosa,* and *Staphylococcus aureus.* By combining ethyl acetate and methanol with tree extracts from 13 Cocousnucifera, a silver nanoparticle with a diameter of 22 nm was created. *Salmonella paratyphi, Klebsiella pneumoniae, Bacillus subtilis,* and *Pseudomonas aeruginosa* have all been found to exhibit antibiotic action. The Abutilon indicum extract was used to create a stable and spherical Ag nanoparticle. Those nanoparticles are having high antimicrobial action against the *S. typhi, E. coli, S. aureus,* and also, *B. substilus* bacteria. Silver nanoparticles are also made from Ziziphoratenuior leaves, and they were described utilizing a variety of methods. These nanoparticles were spherical in shape and uniformly distributed, according to the FTIR spectroscopic approach, with diameters ranging from 8 to 40 nm. Biomolecules with primary amine groups, carbonyl groups, hydroxyl groups, and other stabilizing functional groups were used to functionalize them [19]. In a recent work, these nanoparticles were also created using an aqueous mixture of Ficuscarica leaf extract and irradiation. Silver nanoparticles were created using aqueous solution of 5 mM silver nitrate after three hours of incubation at 37°C. *P. aeruginosa, P. mirabilis, E. coli, Shigella flexaneri, Shigella somenei,* and also *Klebsiella pneumonia* were all killed by the native fragrant plant *Cymbopogan citratus* (DC) stapf (commonly known as lemon grass) from India, which is widely cultivated in other tropical and subtropical nations. Krishnaraj *et al.* employed Acalypha indica leaf extract to make silver nanoparticles that developed in less than 30 minutes. When stable silver nanoparticles are generated at various $AgNO_3$ concentrations, these particles are generally spherical with sizes ranging from 15 to 50 nm. TEM imaging revealed spherical particles with a size range of 3 to 12 nm that were well distributed. Dwivedi *et al.* showed how to make silver nanoparticles from the noxious plant Chenopodium album in a simple and quick way. Silver and gold nanoparticles with diameters ranging from 10 to 30 nm were effectively synthesized using the leaf extract. The spherical nanoparticles were discovered at increasing concentrations of the leaf extract, according to TEM imaging [20]. The growth kinetics of the Ag nanoparticles with the diameters of 10–35 nm were explored by Prathna *et al.*, who made Ag nanoparticles by reducing silver nitrate solution with an aqueous extract of *Azadirachta indica* leaves. In a straightforward green approach, thermal treatment of the aq. silver nitrate solutions and the natural rubber latex obtained from *Hevea brasiliensis* resulted in colloidal silver nanoparticles.

1.3.1.2 Green Synthesis of Gold (Au) Nanoparticles

The size of gold (Au) particles can vary from 1 to 100 nm, and they have various visual and physical characteristics. The most crucial physical property is the Tyndall effect, i.e., scattering of light, perhaps the most crucial optical property is based on their structure, like size, shape, etc. and another optical property is Plasmon Resonance. Au consists of a core and protective coating, this protective coating shields the core of gold particle and hence prevent Gold Nanoparticle [21]. Au nanoparticles are toxic-free, environment-friendly, simple, and economical. The particle is used in various applications as a diagnostic tool, biosensors, drug delivery, etc. These are more favorable with peptides, proteins, and antibodies. The stability of Au particle is higher than Ag particle due to Au sulfur bond. The synthesis of Au nanoparticles can be done by three methods: biologically, physically, and chemically [22].

1. Chemically, through the reduction method of HAuCl4 with a solution of Thiolate Chitosan.
2. Physically, achieved via gamma–radiation technique.
3. Biologically, it can be achieved by reduction of $HAuCl_4$.

Gold nanoparticles have gotten a lot of interest due to their one-of-a-kind potential for 28 usage in medicine and biology. They have a more biocompatible nature, tunable surface plasmon resonance, minimal toxicity, high scattering and absorption, easy surface functionalization, and simple synthesis processes, among other things [23]. When creating gold nanoparticles, reducing agents from biogenic complexes with a variety of chemical molecular composition are used. reacting with gold metal ions to produce reduction and nanoparticle formation. Various studies were also established that biomolecules contained in plant extracts, such as favonoids, phenols, protein, and others, play a great role in metal ion reduction and gold nanoparticle topping. Shankar and his colleagues were the first to use geranium leaf extract as a reducing and capping agent in the production of gold nanoparticles, which they did in 2003. The terpenoids in the leaf extract which are 39 responsible for the reduction of the gold ions to gold nanoparticles, which took 48 hours to complete [24]. According to morphological investigations, these nanoparticles are triangular, spherical, decahedral, and icosahedral in shape. They also used Azadirachta indica leaf extract to make gold nanoparticles in 2.5 hours. The neem extract's terpenoids and favanones were probably absorbed on the nanoparticles' surface and controlled their stability for four weeks. According to morphological studies, the nanoparticles are spherical and mainly planar, with the majority being triangular and some being hexagonal [25].

Aloe vera leaf extract was used by Chandran *et al.* to modify the size and form of gold nanoparticles. The amount of leaf extract used determines the size and form of the triangles, which range in size from 50 to 350 nm [26]. Less leaf extract was used to create larger nano-gold triangles in the HAuCl4 solution, but more leaf extract produced more spherical nanoparticles, which decreased the ratio of nano-triangle to nano-spherical particles. Using a modest extract quantity of 35 mushroom extract, some anisotropic gold nanoparticles were produced, with a maximum of triangles and prisms and a very small number of hexagons and spheres. When the amount of mushroom extract was increased, the nanoparticles' shape changed to hexagons and spheres, reduced considerably, and the number of nanotriangles also shrank. The nanoparticles generated when the extracted quantity was increased to its greatest concentration were 25 nm in size [27]. Temperature had a great effect on the nanoparticles, which was clarified by receiving hexagons at 313 K at the greatest extract quantity, while nanoparticles in dendrites shapes were obtained at 353 Singh *et al.* observed temperature effects in the production of gold nanoparticles by Diopyros kaki and Magnolia kobus leaf extracts [28]. The nanoparticles were generated in the 10- to 35-nm range, according to morphological characterization using transmission electron microscopy [29].

1.3.1.3 *Green Synthesis of Copper (Cu) Nanoparticles*

Copper nanoparticles are made by reducing aqueous copper ions with various plant extracts, such as Aloe vera plant extract. A 578-nm signal on a UV–Visible spectrometer verified the making of Cu nanoparticles with an average size of 40 nm [30]. $Cu/GO/MnO_2$ nanocomposite was synthesized using a leaf extract from Cuscuta refexa, which is high in antioxidant phytochemicals as Myricetin, Myricetin glucoside, Kaempferol 3-Oglucoside (Astragalin), Kaempferol-3-O-galactoside, Kaempferol, Quercetin, Quercetin-3-O-glucoside, Quercetin 3-O. The ingredients listed above are responsible for converting plant extract into an antioxidant-rich feedstock for nanoparticle production [31–33]. Cu nanoparticles were fixed on surface of graphene oxide/MnO2 nanocomposites after Cu^{+2} ions were reduced to Cu nanoparticles using Cuscuta refexa leaf extract. For reduction of the rhodamine B, congo red, methylene blue, methyl orange, 4-nitro phenol, and 2,4-DNPH by $NaBH_4$ in an aqueous solution, these nanocomposites with Cu nanoparticles were used as the heterogeneous and recoverable catalyst. Cheirmadurai and his colleagues used henna leaves extract as a reductant to make copper nanoparticles on a massive scale. Cu nanoparticles and collagen fibres left over from the leather industry were used to create nanobiocomposites conducting film. The film can be used in a wide variety of 15 17 electronic devices. Tamarind and lemon juice were also used to make large-scale Cu nanoparticles with sizes that are ranging from 20 to 50 nm. Using barberry fruit extract as a stabilizing and reducing agent, Cu nanoparticles were created in situ on reduced graphene oxide/Fe3O4 and were found to be useful as an active catalyst for the reaction of phenol with aryl halides to produce O-arylation of phenol under ligand-free circumstances. Additionally, it is recoverable and can be reused repeatedly without losing its catalytic properties [34].

1.3.1.4 Iron Oxide Nanoparticles

These nanoparticles contain iron oxide particles with sizes varying from one to one hundred nanometers. Magnetite (Fe_3O_4) and its oxidized counterpart maghemite (Fe_2O_3) are the two primary types [35]. They have piqued the interest of many people due to their superparamagnetic capabilities and prospective and are used in various industries. Due to their magnetic property, small size, and wide surface area, iron oxide nanoparticles are desirable for the elimination of heavy metal contamination from water, indicating their promise in metal-ion detection. It can be demonstrated that the produced iron oxide nanoparticles have a broad range of applications and are in high demand. Iron oxide nanoparticles exhibit a variety of magnetic behaviors and qualities, including high magnetic perceptivity and superparamagnetic activity. Magnetic iron oxide nanoparticles, such as magnetite and maghemite, are known for their biocompatibility and low toxicity [36]. This form of nanoparticle possesses potential to be a major source of concern for researchers working on bio-applications, data storage, and catalysis. The surface-to-volume ratio of these nanoparticles is extremely high, necessitating large surface energies. As a result, they can combine to minimize surface energy.

1.3.2 Organic Nanoparticles

Organic nanoparticles are two-dimensional materials ranges 1 to 100 nm in size. Nanoparticles have distinct size-dependent physical and chemical characteristics, such as optical, magnetic, catalytic, thermodynamic, and electrochemical capabilities. Liposomes, micelles, protein/peptide-based carriers, and dendrimers are the four basic types of organic Nano Particles [37]. Dendrimers are operating in multiple (15 nm) synthetic polymers having layered topologies with various terminal groups that regulate the dendrimer's characteristics, a core structure, and an inner region.

1.3.2.1 Liposomes

Liposomes are circular vesicles consisting of one (or more) phospholipid bilayers. These mediums present themselves as an alluring delivery framework due to their physicochemical properties and biochemical nature permitting them to be effectively manipulated. Liposomes have an interesting capacity to encase lipophilic and hydrophilic compounds, making them appropriate carriers for a run of drugs. Other preferences, incorporate their capacity to self-assemble, capacity to carry expansive drug loads, and biocompatibility [38]. Being composed of characteristics phospholipids makes them "pharmacologically inactive", meaning they show negligible harmfulness. Liposomes can be categorized into four fundamental types:

1. Conventional, 2. Theragnostic, 3. PEGylated, 4. Ligand-targeted

1.3.2.2 Micelles

A Micelle is a loosely bonded aggregation of 100 to 1000 atoms, ions, or macromolecules that constitute a colloidal particle—that is, one of several ultramicroscopic particles scattered through some continuous media. Micelles are significant in surface chemistry; for example, the ability of soap solutions to dissipate organic molecules that are insoluble or only weakly dissolve in water is described as a micelle.

1.3.2.3 Dendrimers

These macromolecules are highly branched, and globular [39]. These are used to encapsulate small individual drug molecules which can also be served as "hubs" which further results in huge numbers of drug particles and can be attached using covalent bonds. Example – 5-fluorouracil to poly amino amine dendrimers.

1.4 Approaches

Two types of Approaches help in producing Nanoparticles.

1. Top-down approach
2. Bottom-up Approach

Top-down Approach: This is the approach that starts with the bulk material and cuts the large material into smaller pieces to produce Nanoparticles. Examples: Ball Milling method. The Ball Milling method is a process that usually consists of Balls and a chamber called a mill chamber and stainless steel (balls must be of Iron, silicon Carbide, or Tungsten Carbide). they are made to rotate inside a Mill. this method helps in the production of nanoparticles. These mills are highly equipped with grinding media that contain tungsten carbide or steel. These mills rotate on the horizontal axis and are partially filled with the substance to be processed as well as the grinding medium [40]. The balls spin at high speeds inside a container before colliding with the solid and crushing it into nano crystallites due to gravity.

In a top-to-bottom technique, nanoparticle synthesis is typically accomplished by evaporation–condensation in the tube furnace at air pressure. The foundation material is vaporized into a carrier gas in this procedure, which takes place within a boat and is centred at the furnace. The evaporation/condensation approach was previously used to make Ag, Au, PbS, and fullerene nanoparticles [41]. The tube furnace has a number of drawbacks, including taking up a lot of area and using a lot of energy to raise the temperature in the area around source material, as well as taking a long time to reach thermal stability. A typical tube furnace also requires several kilowatts of power and several tens of minutes of pre-heating time to reach a stable operating temperature [42]. One of the method's 10 key shortcomings is that it causes defects in the product's surface structure, plus further physical characteristics of nanoparticles are highly dependent on the surface structure in terms of surface chemistry. In general, chemical techniques, regardless of the method used, have some limitations, either with the help of chemical contamination during nanoparticle synthesis or in subsequent applications [43]. Their ever-increasing use in everyday life, however, cannot be underestimated. For example, "the noble silver nanoparticles" are pursuing cutting edge applications in every sphere of science and technology, including medicine, and hence cannot be dismissed solely as a result of their source of production. Ag 7 nanoparticles are used in over a total of 200 consumer goods, including apparel, medications, and cosmetics, due to their medical and antibacterial characteristics. Their expanding uses are bringing together chemists, physicists, material scientists, biologists, and doctors/pharmacologists to maintain their most recent foundations. As a result, it is now an obligation of every researcher to create a synthetic alternative that is not only economically effective but also environmentally beneficial Green synthesis is establishing itself as an important method and displaying its potential at the top in terms of aesthetics [44, 45]. Green production of nanoparticles has various compared to chemical and physical methods, including being environmentally friendly, cost-effective, and easily scaled up for large-scale nanoparticle creation. High temperatures, pressures, or energy, as well as hazardous substances, are not required.

A large number of literature has been published to date on biological production of microorganisms that use silver nanoparticles include fungi, bacteria, and plants due to their antioxidant or the reducing properties, which are typically responsible in the lowering of metallic compounds in their respective nanoparticles. Because the need for exceptionally aseptic conditions and their upkeep, microbe-mediated synthesis, one of numerous biological approaches for silver nanoparticle creation, is not industrially viable. As a result, plant extracts may be preferable to micro-organisms for this purpose due to their simplicity of improvement, lower biohazard, and more complicated cell culture maintenance method. It is the finest platform for nanoparticle syntheses since it is free of hazardous compounds and provides natural capping agents to stabilise silver nanoparticles. Furthermore, using plant extracts decreases the cost of culture media and microbe separation, making microbial nanoparticle manufacturing more cost-effective [46]. As a result, a report had been published that details bio-inspired silver nanoparticle syntheses that are more eco-friendly, cost-effective, and effective in the various types of applications, 16 particularly bactericidal activities, than physical and chemical methodologies.

The Technologies Based on the top-down approach are:

a. Evaporation/Condensation
   The procedure is evaporation, in which liquid converts into vapors. The evaporation of the metal usually forms thin films, which are performed, under a high vacuum. this heat is produced by electrical resistance. This technique is based upon the pressure, which is created by vaporizing the metal, whose ability to evaporate is determined by its chemical strength. this is because of the evaporation via heating and then condensing the vapor to obtain nanoparticles. Steam generates through radiation and heating through the oxidation of Fe, Ni, Co, Cu, Pd, and Pt. The heating is required for oxygen and refractory metals. in the surroundings, if some metallic elements are reactive, then nanoparticles will oxidize [47]. the main disadvantage of this method is the lack of control over the nanomaterial size. The particles are synthesized by quickly condensing the escaped gas which is ensuring us the development of many Nanoparticles and other

methods. this technique is most useful in the factory for manufacturing metal and ceramic nanoparticles. in the Air, nanoparticles are vulnerable to overheating and exploding.

b. Laser Ablation

This method makes use of a strong laser beam to generate nanoparticles. A high-energy laser focused on a target in a solvent in a top-down process. Pulses of light are being emitted by the laser and are sufficient to vaporize small patches of the metal target, which condenses as a nanoparticle in the solvent [48]. This procedure is typically used to create noble metallic nanoparticles such as gold, silver, and platinum. However, it may be easily extended to produce other nanomaterials such as metal alloy nanoparticles, which is a distinct advantage of this approach. A pulsed laser, a set of focusing optics, and a container carrying a metal target are typical components of a typical setup. The tank is filled with a solution (for example, water or ethanol), into which the metal object is placed. The metal object is placed near the laser's focal point. The pulse duration, wavelength, and strength of the laser pulses can help to adjust the size and dispersion of the nanoparticles. Laser pulses are often measured in femtoseconds (1/1,000,000,000,000,000 of a second), picoseconds (1/1,000,000,000,000 of a second), or nanoseconds (1/1,000,000,000 of a second). Lasers of visible and near-infrared wavelengths are common in laser ablation equipment. Because the action is largely physical, this method can be utilized in multiple materials and solvents. Another distinct feature of this technology is the capacity to create nanoparticles with extremely high purity. Because there are no by-products or leftover compounds produced by this procedure.

c. Mechanical Milling

A mechanical milling method is a top-down approach to manufacturing nanomaterials, its simplicity, versatility of processes (suitable for manufacturing many types of nanomaterials), expandability of processes., And are popular at low cost. There is. Most ultra-fine bulk materials are milled to the nano-range by strong mechanical shear forces in the milling process [49]. Three distinct types of grinding equipment are more frequently used than others. Shaker mill, planetary ball mill, allocation mill. As the name implies, in a shaker mill, the material to be ground is set up in a small bottle containing a "crushing ball", which is a spherical ball of hard material. Then attach the sample firmly to the shaker and shake it vigorously back and forth for thousands of cycles per minute. During this shaking process, the crushed balls collide with one another and bounce off the walls of the vial. High shear and impact forces as a result crush and mix the solids. The planetary ball mill is named after the movement of the ampoules in the device. These vials are attached to a rotating disk that rotates about its axis, and each vial also rotates about its axis, but in the opposite direction of the main rotating disk. The entire system rotates at thousands of revolutions per minute, and strong friction and impact forces shred the material into smaller sizes [50]. Planetary ball mills are popular with many because they the ability to grind hundreds of materials at once. The attribution mill is similar to a ball mill, where the grinding balls are placed in a horizontal cylinder and rotated to complete the grinding process. However, in attraction mills, the vertical drum is secured internally via a string of carefully placed impellers. These impellers are fixed so that they are at right angles to each other. Unlike ball mills, attritor mills rotate at high speeds with the inner impeller running [51]. This can result in very high impact and shear forces that cannot be achieved with an attribution mill.

Advantages of Top-down Approach:
(i) manufacturing on a large scale (ii) Decontamination is not Required.

Disadvantages of Top-down Approach:
(i) Expensive Technique (ii) Broad Size Distribution (iii) Controlling over parameters is difficult to achieve (iv) do not have control over the Material dimensions.

Bottom-up Approach: This is the Approach that combines small atoms and molecules into larger structures. Example: - Colloidal Dispersion.

Colloidal Dispersion Method: This is the method in which the Dispersed phase is mixed with dispersion medium and then these are shaken well to Obtain the form of suspension. the suspension has to pass through a colloid mill. This Simplest type of colloid mill called disc mill barely consists of 2 metals discs that are nearly touching each other and are rotating in opposite directions and these rotations take place at high speed. This dispersion goes through these circular discs and is subjected to a tremendous shearing force, resulting in colloidal particles [52].

Advantages of Bottom-up Approach:
- Cheaper Technique.
- Narrow Size distribution is possible.
- Ultrafine particles and nanotubes can be prepared.
- More control over material dimensions.

Disadvantages of Bottom-Up Approach:
- Large-Scale fabrication is difficult.
- Decontamination is Required.

The Technologies based on the bottom-up approach are:
The bottom-up strategy involves the development of both physical and chemical processes that function at the nanoscale to integrate major tiny components into larger structures. The strategy offers an acceptable completion to the top-down approach with a reduction in unit size. This method is motivated by biological systems, in which natural forces of life harness their chemical equivalents to form structures [53]. Scientists on the reverse side, are striving to create the same forces that nature uses to self-integrate into larger structures. When allowing undesirable deterioration, gold-palladium alloy nanoparticles based on carbon treated with acid and breaking down hydrogen peroxide are created from the conjunction of white hydrogen and red oxygen. This method has been utilized to create nanoparticles via condensation to the coalescence of atomic vapors.

Chemical techniques
a. Sol-Gel techniques
This method is based on Chemise douche chemistry, which provides quick and diverse ways for synthesizing extremely homogeneous, cost-effective nanomaterials with high purity into final oxidic structure networks. The size and morphological properties can be fine-tuned by varying the concentration ratio of capping agent to precursor and thus selecting a suitable reducing agent, as well as other reaction conditions, such as (1) generation of an appropriate precursor solution; (2) deposition of the solution onto a substrate, followed by hydrolysis and condensation; and (3) thermal treatment of their precursor and its conversion into the oxide nanoparticles. This enables the manufacturing of nanoparticles from alkoxides or colloidal solutions. They can be monoliths, crystallized nano pigments, or

thin layers [54]. The sol-gel method is relating to the formation of chemical compounds in a solution. Following interactions between the species and the liquid, the sol transforms into a three-sided channel enlarged in the solvent. The process is then frozen, and the sol-gels are converted to deformed solid material via liquid removal from the air-gel or evaporation [55]. One advantage of the method is the ability to control the size and uniformity of the material spread out. They oversee large-scale production.

b. Aerosol-Based Process

Aerosol-based technologies are a prominent way of producing nanoparticles in the industrial environment. Aerosols are characterized as solid or liquid particles in a gas phase, with particles ranging in size from molecules to 100 m. Long before the basic science and engineering of aerosols were understood, aerosols were used in industrial manufacture. Carbon black particles used in pigments and reinforced car tires, for example, are produced by hydrocarbon combustion; titania pigment used in paints and plastics is produced by the oxidation of titanium tetrachloride, titania and fumed silica is produced by flame pyrolysis from their respective tetrachloride's, and optical fibers are produced similarly. Spraying is generally used to both dry wet materials or deposit coatings. Spraying the precursor chemicals onto a heated surface or into a hot environment causes precursor pyrolysis and particle production [56]. At Oxford University, for example, a room temperature electro-spraying technique was devised to generate nanoparticles of compound semiconductors and certain metals. CDs nanoparticles were created by creating aerosol micro-droplets containing Cd salt in an environment containing hydrogen sulfide.

c. Chemical Vapor Deposition (CVD)

CVD is the process of initiating a chemical reaction between the surface of a substrate and a gaseous precursor. Temperature (thermal CVD) or plasma is used to activate the material (plasma-enhanced chemical vapor deposition [PECVD]). The fundamental advantage of this technology is that it is non-directive [57]. When compared to the thermal CVD technique, plasma allows for a substantial reduction in process temperature. Carbon nanotubes are commonly manufactured via CVD.

Phyto Nanotechnology

Phyto nanotechnology is the process of creating nanoparticles from fresh plants or plant extracts. Plant-derived nanoparticles are made from readily available plant materials, and their non-toxic nature makes them ideal for meeting the high demand for nanoparticles with applications in biomedicine and the environment [58]. Recently, nanoparticles of gold and silver were successfully synthesized by using extracts of leaf and root from the herbal plant Panax Ginseng. Furthermore, plant parts such as leaves, fruits, stems, roots, and extracts have been employed in the creation of metal nanoparticles. Proteins, amino acids, organic acids, vitamins, and secondary metabolites, such as flavonoids, alkaloids, polyphenols, terpenoids, heterocyclic compounds, and polysaccharides, have been proposed to play a vital role in the reduction of metal salt and to act as stabilizing agents for synthesized nanoparticles. Officinalis extract, for example, could aid in the formation and stabilization of gold nanoparticles. According to reports, plant species have diverse processes for generating nanoparticles. Notably, dicot plants possess many metabolites that could be used to synthesize nanoparticles. Various components, such as emodin, ditech, and a purgative resin, contain quinone compounds, and these quinone compounds can easily be found in xerophytes plants (mostly found in low-level water conditions or near-desert areas) are responsible for silver nanoparticle formation. Ditech quinone is very beneficial for metal nanoparticle synthesis through mesophytic plants (i.e., terrestrial plants which are neither acclimated to a particularly dry nor a particularly moist environment) [59]. The primary terpenoid of cinnamomum, i.e., eugenol was discovered to play a vital role in the creation of gold and silver nanoparticles.

Phytochemical Screening
It is the process of extracting, screening, and identifying medicinally active compounds present in plants. Flavonoids, alkaloids, carotenoids, tannin, antioxidants, and some of the phenolic compounds are like bioactive molecules that can be procured from plants [60]. The existence of these phytochemicals is indicated by color changes or precipitation production.

Tests For Phytochemical Screening
Tests for Flavonoids: 1 ml of the extracted sample and add a few drops of dilute NaOH. after adding NaOH, there is an appearance of the yellow color while adding dilute HCl it becomes colorless. So, the Formation of yellow color to colorless indicates the presence of Flavonoids [61].
Tests for Terpenoids: 0.5 gm of extract in the test tube and add 2 ml of chloroform. After adding chloroform add 5 ml of Conc. H_2SO_4, the results form a layer, and the formed layer is of reddish-brown color, which indicates the presence of terpenoids.

Tests for Carbohydrates
The carbohydrates test can be achieved by benedict's reagent test. for this test, take a few ml of benedict's solution which is blue and contains Cu^{2+} ions, take the sample from the extract and then mix both extract and benedict's solution and heat them. The solution gives brick-red color and also shows the amount/presence of carbohydrates.

Advantages of Phyto Nanotechnology
- Helps in the production of nanoparticles
- Environmentally safe
- Cost-efficient
- Less time consuming
- Long-term safety
- Easy method

Applications of Nanoparticles
- ❖ Nanoparticles produced by the different technologies have been employed in various vitro diagnostic applications.
- ❖ Both gold and silver nanoparticles have been proven to have antibacterial action against human and animal pathogens.
- ❖ Antimicrobial silver nanoparticles have become often used in commercial medical and consumer applications.
- ❖ Zinc oxide nanoparticles are used in antimicrobial food packaging, also Crop protection and agriculture are two areas where nanoparticles are being used.

1.5 Conclusion

Nature has created the most efficient tiny functional materials in lovely and innovative ways. A growing interest in green chemistry and also in the application of green synthesis methods have sparked a drive to develop environmentally acceptable methods. The advantage of synthesizing Ag nanoparticles with plant extracts is that it is cost-effective, provides healthier work environments and communities, protects human health and the environment, and results in less waste and safer goods [62]. Green silver nanoparticle production has major characteristics of nanotechnology through unrivalled applications. Plants may be preferable to other biological entities for nanoparticle synthesis because they can avoid the time-consuming procedure of employing bacteria and maintaining their culture, which can lead to the loss of their capacity for nanoparticle synthesis.

As a result, the utilization of the plant extract for synthesis could have a huge impact in the next decades.

Increasing demand for nanotechnology has pushed the use of synthesis methods to produce nanomaterials via plants, microbes, and others over the last few decades [63, 64]. There is a need for a commercially feasible, cost-effective, and environmentally friendly method of determining the capacity of the natural reducing constituents to create silver nanoparticles that needs to be investigated. When plant extracts from the same species are gathered from different parts of the world, there is a considerable difference in chemical makeup, which might lead to the different results in the different laboratories [65]. One of the biggest drawbacks of employing the plant extracts as reducing and stabilizing agents to the production of Ag nanoparticles is that they are difficult to control. Many studies have been conducted on the synthesis of plant extract-mediated nanoparticles, and their prospective is used in various forms of disciplines because of their low cost, nontoxic approach, ease of availability, and environmental friendliness. Furthermore, nanoparticles have major importance in our daily life, and each particle has its importance as they are produced using plants extracts. These (different plants) have several uses in catalysis, medicine, water treatment, dye degradation, textile engineering, bioengineering sciences, sensors, imaging, biotechnology, electronics, optics, and other biomedical domains [66]. Plants also contain chemicals that help in the synthesis of particles and thereby also improve nanoparticles. These plants are also used in the production of nanoparticles, which is the most fascinating feature of nanotechnology that seems to have a huge impact on the environment in terms of their growth and sustainability. Future outcomes from the production of nanoparticles are that these will grow rapidly, but these have a threat to humans, as well as animals and can be accumulated in the environment. However, identification of biomolecules present in plants that mediate nanoparticle formation enabling a speedy single-step protocol to address the challenge could give green silver nanoparticle syntheses a new lease on life.

References

1. Mekonnen, G., Review on application of nanotechnology in animal health and production. *J. Nanomed. Nanotechnol.*, 12, 559, 2021.
2. Sriram, P. and Suttee, A., Nanotechnology advances, benefits, and applications in daily life, in: *Nanotechnology*, pp. 23–44, CRC Press, Boca Raton, Florida, USA, 2020.

3. Modena, M.M., Rühle, B., Burg, T.P., Wuttke, S., Nanoparticle characterization: What to measure? *Adv. Mater.*, 31, 32, 19015, 2019.
4. Nobahar, A., Carlier, J.D., Miguel, M.G., Costa, M.C., A review of plant metabolites with metal interaction capacity: A green approach for industrial applications. *BioMetals*, 34, 4, 761–793, 2021.
5. Wang, Q., Qu, J., Wang, B., Wang, P., Yang, T., Green technology innovation development in China in 1990–2015. *Sci. Total Environ.*, 696, 134008, 2019.
6. Abbas, J. and Sağsan, M., Impact of knowledge management practices on green innovation and corporate sustainable development: A structural analysis. *J. Cleaner Prod.*, 229, 611–620, 2019.
7. Singh, C.S., Green construction: Analysis on green and sustainable building techniques. *Civ. Eng. Res. J.*, 4, 3, 555638, 2018.
8. AlMasoud, N., Alhaik, H., Almutairi, M., Houjak, A., Hazazi, K., Alhayek, F., Aljanoubi, S. *et al.*, Green nanotechnology synthesized silver nanoparticles: Characterization and testing its antibacterial activity. *Green Process. Synth.*, 10, 1, 518–528, 2021.
9. Palit, S. and Hussain, C.M., Recent advances in green nanotechnology and the vision for the future, in: *Green Metal Nanoparticles: Synthesis, Characterization and Their Applications*, pp. 1–21, 2018.
10. Nargund, V.B., Vinay, J.U., Basavesha, K.N., Chikkanna, S., Jahagirdar, S., Patil, R.R., Green nanotechnology and its application in plant disease management, in: *Emerging Trends in Plant Pathology*, pp. 591–609, Springer, Singapore, 2021.
11. El Shafey, A.M., Green synthesis of metal and metal oxide nanoparticles from plant leaf extracts and their applications: A review. *Green Process. Synth.*, 9, 1, 304–339, 2020.
12. Shah, M., Fawcett, D., Sharma, S., Tripathy, S.K., Poinern, G.E.J., Green synthesis of metallic nanoparticles via biological entities. *Materials*, 8, 11, 7278–7308, 2015.
13. Kalishwaralal, K., Deepak, V., Ram Kumar Pandian, S., Kottaisamy, M., BarathmaniKanth, S., Kartikeyan, B., Gurunathan, S., Biosynthesis of silver and gold nanoparticles using Brevibacterium casei. *Colloids Surf. B Biointerfaces*, 77, 2, 257–62, 2010 Jun 1.
14. Virkutyte, J. and Varma, R.S., Green synthesis of nanomaterials: Environmental aspects, in: *Sustainable Nanotechnology and the Environment: Advances and Achievements*, pp. 11–39, American Chemical Society, Washington, DC, 20036, 2013.
15. Pal, G., Rai, P., Pandey, A., Green synthesis of nanoparticles: A greener approach for a cleaner future, in: *Green Synthesis, Characterization and Applications of Nanoparticles*, pp. 1–26, Elsevier, Amsterdam, Netherlands, 2019.
16. Shah, R., Oza, G., Pandey, S., Sharon, M., Biogenic fabrication of gold nanoparticles using Halomonas salina. *J. Microbiol. Biotechnol. Res.*, 2, 4, 485–492, 2012.
17. Raheman, F., Deshmukh, S., Ingle, A., Gade, A., Rai, M., Silver nanoparticles: Novel antimicrobial agent synthesized from a endophytic fungus Pestalotia

sp. isolated from leaves of Syzygium cumini (L.). *Nano Biomed. Eng.*, 3, 3, 174–178, 2011.
18. Shakibaie, M., Forootanfar, H., Mollazadeh-Moghaddam, K., Bagherzadeh, Z., Nafissi-Varcheh, N., Shahverdi, A.R., Faramarzi, M.A., Green synthesis of gold nanoparticles by the marine microalga Tetraselmis suecica. *Biotechnol. Appl. Biochem.*, 57, 2, 71–75, 2010.
19. Bindhu, M.R. and Umadevi, M., Antibacterial activities of green synthesized gold nanoparticles. *Mater. Lett.*, 120, 122–125, 2014.
20. Kaur, H., Kaur, S., Singh, J., Rawat, M., Kumar, S., Expanding horizon: Green synthesis of TiO2 nanoparticles using Carica papaya leaves for photocatalysis application. *Mater. Res. Express*, 6, 9, 095034, 2019.
21. Agarwal, H. and Shanmugam, V., A review on anti-inflammatory activity of green synthesized zinc oxide nanoparticle: Mechanism-based approach. *Bioorg. Chem.*, 94, 103423, 2020.
22. Parashar, M., Shukla, V.K., Singh, R., Metal oxides nanoparticles via sol–gel method: A review on synthesis, characterization and applications. *J. Mater. Sci.: Mater. Electron.*, 31, 5, 3729–3749, 2020.
23. Parveen, K., Banse, V., Ledwani, L., Green synthesis of nanoparticles: Their advantages and disadvantages. *AIP Conf. Proc.*, 1724, 1, 020048, 2016.
24. El Shafey, A.M., Green synthesis of metal and metal oxide nanoparticles from plant leaf extracts and their applications: A review. *Green Process. Synth.*, 9, 1, 304–339, 2020.
25. Murty, B.S., Shankar, P., Raj, B. et al., Applications of nanomaterials, in: *Textbook of Nanoscience and Nanotechnology*, Springer, Heidelberg, Berlin, 2013.
26. Hussain, I., Singh, N.B., Singh, A. et al., Green synthesis of nanoparticles and its potential application. *Biotechnol. Lett.*, 38, 4, 545–560, 2016.
27. Noruzi, M., Biosynthesis of gold nanoparticles using plant extracts. *Bioprocess Biosyst. Eng.*, 38, 1, 1–14, 2015.
28. Singh, P., Kim, Y.J., Yang, D.C., A strategic approach for rapid synthesis of gold and silver nanoparticles by Panax ginseng leaves. *Artif. Cells Nanomed. Biotechnol.*, 44, 8, 1949–1957, 2016.
29. Alomar, T.S., AlMasoud, N., Awad, M.A. et al., An eco-friendly plant-mediated synthesis of silver nanoparticles: Characterization, pharmaceutical and biomedical applications. *Mater. Chem. Phys.*, 249, 123007, 2020.
30. Singh, P., Pandit, S., Garnæs, J. et al., Green synthesis of gold and silver nanoparticles from Cannabis sativa (industrial hemp) and their capacity for biofilm inhibition. *Int. J. Nanomed.*, 13, 3571, 2018.
31. Rani, P., Kumar, V., Singh, P.P. et al., Highly stable AgNPs prepared via a novel green approach for catalytic and photocatalytic removal of biological and non-biological pollutants. *Environ. Int.*, 143, 105924, 2020.
32. Taran, M., Rad, M., Alavi, M., Biosynthesis of TiO2 and ZnO nanoparticles by Halomonas elongata IBRC-M 10214 in different conditions of medium. *BioImpacts: BI*, 8, 2, 81, 2018.

33. Ahmad, S., Munir, S., Zeb, N. et al., Green nanotechnology: A review on green synthesis of silver nanoparticles—An ecofriendly approach. *Int. J. Nanomed.*, 14, 5087, 2019.
34. Kaur, K. and Thombre, R., Nanobiotechnology: Methods, applications, and future prospects, in: *Nanobiotechnology*, pp. 1–20, Elsevier, Netherlands, 2021.
35. Mandhata, C.P., Sahoo, C.R., Padhy, R.N., Biomedical applications of biosynthesized gold nanoparticles from cyanobacteria: An overview. *Biol. Trace Elem. Res.*, 200, 5307–5327, 2022, https://doi.org/10.1007/s12011-021-03078-2.
36. Fahimmunisha, B.A., Ishwarya, R., AlSalhi, M.S. et al., Green fabrication, characterization and antibacterial potential of zinc oxide nanoparticles using Aloe socotrina leaf extract: A novel drug delivery approach. *J. Drug Delivery Sci. Technol.*, 55, 101465, 2020.
37. Saravanan, A., Kumar, P.S., Karishma, S. et al., A review on biosynthesis of metal nanoparticles and its environmental applications. *Chemosphere*, 264, 128580, 2021.
38. Ahmed, S.F., Mofijur, M., Parisa, T.A. et al., Progress and challenges of contaminate removal from wastewater using microalgae biomass. *Chemosphere*, 286, 131656, 2022.
39. Ahmed, S.F., Mofijur, M., Rafa, N. et al., Green approaches in synthesizing nanomaterials for environmental nanobioremediation: Technological advancements, applications, benefits and challenges. *Environ. Res.*, 204, 111967, 2022.
40. Kumar, B.R., Mathimani, T., Sudhakar, M.P., Rajendran, K., Nizami, A.S., Brindhadevi, K., Pugazhendhi, A., A state of the art review on the cultivation of algae for energy and other valuable products: Application, challenges, and opportunities. *Renewable Sustainable Energy Rev.*, 138, 110649, 2021, https://doi.org/10.1016/j.rser.2020.110649.
41. Palem, R.R., Ganesh, S.D., Kronekova, Z., Sláviková, M., Saha, N., Saha, P., Green synthesis of silver nanoparticles and biopolymer nanocomposites: A comparative study on physico-chemical, antimicrobial and anticancer activity. *Bull. Mater. Sci.*, 41:55, 2018, https://doi.org/10.1007/s12034-018-1567-5.
42. Dahoumane, S.A., Mechouet, M., Wijesekera, K., Filipe, C.D.M., Sicard, C., Bazylinski, D.A., Jeffryes, C., Algae-mediated biosynthesis of inorganic nanomaterials as a promising route in nanobiotechnology-a review. *Green Chem.*, 19, 552-587, 2017, https://doi.org/10.1039/c6gc02346k.
43. Kumar, N., Balamurugan, A., Balakrishnan, P., Vishwakarma, K., Shanmugam, K., Biogenic nanomaterials: Synthesis and its applications for sustainable development, in: *Biogenic Nano-Particles and Their Use in Agro-Ecosystems*, 2020, https://doi.org/10.1007/978-981-15-2985-6_7.
44. Nagajyothi, P.C., Prabhakar Vattikuti, S.V., Devarayapalli, K.C., Yoo, K., Shim, J., Sreekanth, T.V.M., Green synthesis: Photocatalytic degradation of textile dyes using metal and metal oxide nanoparticles-latest trends and

advancements. *Crit. Rev. Environ. Sci. Technol.*, 50, 2617-2723, 2020, https://doi.org/10.1080/10643389.2019.1705103.
45. Yılmaz Öztürk, B., Yenice Gürsu, B., Dağ, İ., Antibiofilm and antimicrobial activities of green synthesized silver nanoparticles using marine red algae Gelidium corneum. *Process Biochem.*, 89, 208–219, 2020, https://doi.org/10.1016/J.PROCBIO.2019.10.027.
46. Abdel-Raouf, N., Al-Enazi, N.M., Ibraheem, I.B.M., Green biosynthesis of gold nanoparticles using Galaxaura elongata and characterization of their antibacterial activity. *Arabian J. Chem.*, 10, S3029–S3039, 2017, https://doi.org/10.1016/J.ARABJC.2013.11.044.
47. Hussein, A.K., Applications of nanotechnology in renewable energies—A comprehensive overview and understanding. *Renewable Sustainable Energy Rev.*, 42, 460, 2015.
48. Bundschuh, M., Filser, J., Lüderwald, S., McKee, M.S., Metreveli, G., Schaumann, G.E., Schulz, R., Wagner, S., Nanoparticles in the environment: Where do we come from, where do we go to? *Environ. Sci. Eur.*, 30, 6, 2018.
49. Dinesh, R., Anandaraj, M., Srinivasan, V., Hamza, S., Engineered nanoparticles in the soil and their potential implications to microbial activity. *Geoderma*, 173–174, 19, 2012.
50. Tripathi, D.K., Shweta, Singh, S., Singh, S., Pandey, R., Singh, V.P., Sharma, N.C., Prasad, S.M., Dubey, N.K., Chauhan, D.K., An overview on manufactured nanoparticles in plants: Uptake, translocation, accumulation and phytotoxicity. *Plant Physiol. Biochem.*, 110, 2, 2017.
51. Kumar, A., Choudhary, A., Kaur, H., Mehta, S., Husen, A., Smart nanomaterial and nanocomposite with advanced agrochemical activities. *Nanoscale Res. Lett.*, 16, 156, 2021.
52. Troy, E., Tilbury, M.A., Power, A.M., Wall, J.G., Nature-based biomaterials and their application in biomedicine. *Polym. (Basel)*, 13, 19, 3321, 2021.
53. Harish, V., Tewari, D., Gaur, M., Yadav, A.B., Swaroop, S., Bechelany, M. et al., Review on nanoparticles and nanostructured materials: Bioimaging, biosensing, drug delivery, tissue engineering, antimicrobial, and agro-food applications. *Nanomaterials*, 12, 3, 457, 2022
54. Sanchez, C., Arribart, H., Guille, M.M.G., Biomimetism and bioinspiration as tools for the design of innovative materials and systems. *Nat. Mater.*, 4, 277, 2005.
55. Das, S., Ahn, B.K., Martinez-Rodriguez, N.R., Biomimicry and bioinspiration as tools for the design of innovative materials and systems. *Appl. Bionics Biomech.*, 6103537, 1–9, 2018.
56. Katiyar, N.K., Goel, G., Hawi, S., Goel, S., Nature-inspired materials: Emerging trends and prospects. *NPG Asia Mater.*, 13, 56, 2021.
57. Andre, R., Tahir, M.N., Natalio, F., Tremel, W., Bioinspired synthesis of multifunctional inorganic and bio-organic hybrid materials. *FEBS J.*, 279, 1737–1749, 2012, https://doi.org/10.1111/j.1742-4658.2012.08584.x.

58. Lepora, N.F., Verschure, P., Prescott, T.J., The state of the art in biomimetics. *Bioinspiration Biomimetics*, 8, 013001, 2013.
59. Katiyar, N.K., Goel, G., Hawi, S. et al., Nature-inspired materials: Emerging trends and prospects. *NPG Asia Mater.*, 13, 56, 2021.
60. Wang, Y., Naleway, S.E., Wang, B., Biological and bioinspired materials: Structure leading to functional and mechanical performance. *Bioact. Mater.*, 5, 745–757, 2020.
61. Sahoo, B., Sahu, S.K., Pramanik, P., A novel method for the immobilization of urease on phosphonate grafted iron oxide nanoparticle. *J. Mol. Catal. B Enzym.*, 69, 95–102, 2011.
62. Huang, J., Lin, L., Sun, D., Chen, H., Yang, D., Li, Q., Bio-inspired synthesis of metal nanomaterials and applications. *Chem. Soc. Rev.*, 44, 6330–6374, 2015.
63. Kumar, A., Sharma, G., Naushad, M., Al-Muhtaseb, A.H., García-Peñas, A., Mola, G.T. et al., Bio-inspired and biomaterials-based hybrid photocatalysts for environmental detoxification: A review. *Chem*, 382, [122937], 2020
64. Li, M., Gou, H., Al-Ogaidi, I., Wu, N., Nanostructured sensors for detection of heavy metals: A review. *ACS Sustain. Chem. Eng.*, 1, 713–23, 2013.
65. Prasad, R., Bhattacharyya, A., Nguyen, Q.D., Nanotechnology in sustainable agriculture: Recent developments, challenges, and perspectives. *Front. Microbiol.*, 8, 1014, 2017.
66. Katiyar, N.K., Goel, G., Hawi, S. et al., Nature-inspired materials: Emerging trends and prospects. *NPG Asia Mater.*, 13, 56, 2021.

2

Greener Nanoscience: Proactive Approach to Advancing Nanotechnology Applications and Reducing Its Negative Consequences

Utkarsh Jain[1]* and Kirti Saxena[2]

[1]*School of Health Sciences & Technology (SoHST), University of Petroleum and Energy Studies (UPES), Bidholi, Dehradun, India*
[2]*Amity Institute of Nanotechnology (AINT), Amity University Uttar Pradesh (AUUP), Noida, India*

Abstract

Although nanotechnology continuously offers new materials to be utilized in various applications beneficial to society, some nanomaterial hazards as end products have also been observed by researchers in hundreds of studies. This observation indicates the necessity of research on improvement in present nanomaterial complexities through the implication of new methods for nanomaterial's synthesis. In this context, green nanoscience is a rapidly growing approach to the production and utilization of nanomaterials developed by biological methods. Green nanoscience is beneficial to society and the environment, in several ways through the implementation of designing rules to synthesize safe nanomaterials without the production of any toxic end products. In this chapter, we have discussed the limitation of existing nanotechnology-based methods to produce nanomaterials and why we need the green nanoscience approach to overcome these limitations. The advantages of greener nanoscience have been described together with the processes for green nanomaterials synthesis. Green nanotechnology focuses on product designing and optimization of processes to reduce or eliminate environmental and health hazards.

Keywords: Nanotechnology, greener nanoscience, green technology, nanomaterials, green synthesis, nanoengineered materials, nanoemulsions

---

Corresponding author: utkarsh.jain@ddn.upes.ac.in

Mousumi Sen and Monalisa Mukherjee (eds.) *Bioinspired and Green Synthesis of Nanostructures: A Sustainable Approach*, (25–42) © 2023 Scrivener Publishing LLC

2.1 Introduction

The interdisciplinary nanotechnology field emerges as a promising arena to provide a nanoworld by miniaturization of molecules for generating new applications in biology, physics, and chemistry's fields [1, 2]. A renowned physicist, Dr. Richard Feynman first gave the idea about nanotechnology in his talk in a conference on "there is plenty of room at the bottom." The nanotechnology term was coined in 1974 by Norio Taniguchi [3]. Basically, nanotechnology deals with the manipulation of materials on the scale of atomic, molecular as well as supramolecular. Nanotechnology depends on the nanostructured materials known as nanoparticles with sizes between 1 and 100 nm that exhibit improved properties as compared to their bulk materials [4, 5]. These properties have attracted researchers to synthesize bulk amounts of nanoparticles by locating novel techniques. The defined abilities of nanoparticles have lightened up the path of numerous possibilities in various fields, such as clothing, security, environmental, medical, fuel synthesis, fuel cells, food industries, and sports as well [6–8]. It shows tremendous growth in the past few years by employing new technological approaches through chemical, physical as well as biological routes. These routes are explored to synthesize nanoparticles of desired size, shape, and properties. However, physical and chemical methods required toxic chemicals which leads to adverse effects on the environment [9, 10]. As a result, concerns develop about human health and environment's safety by nanotechnology was raised due to hazardous waste streams produced during nanoparticle manufacturing. However, nanoparticle presence is not itself a danger but it can be risky in certain conditions like their increased reactivity and their mobility [11]. The expert society made a point that if nanostructured materials are showing new properties then additional size-dependent dangers are likely to emerge [12]. The societies addressed research needs to develop new approaches for understanding the risk of nanoengineered materials and developed a framework in this direction. This framework involves management of potential risk during synthesis processes of nanomaterials and understanding of the interaction of nanostructured materials with the environment and humans, their route of exposure, and the development of characterization techniques for their detection [13–15]. Now, in the present time, researchers are moving toward green technology to develop eco-friendly processes.

2.2 Why Do We Need Green Nanoscience Approaches?

Due to some serious societal concerns with nanoparticles, it is necessary to develop alternative methods for nanoparticle synthesis without any harm to our environment. Therefore, researchers now focused on green technology for nanoparticle synthesis by using plants. Green technology focuses on contaminant reduction and harmless synthesis processes by altering the root cause conditions of these threats. Green technology does not disturb natural resources conserved by our environment due to their way of production and utilization [16]. Green synthesis of nanomaterials gains much attention to reduce toxic effects on the surrounding environment. This approach helps to synthesize biocompatible nanoparticles with safe properties and it is reliable and eco-friendly too [17]. The plant-based green approach is economic and also offers proteins as natural capping agents [18]. Chemical toxicity of the environment can be regulated by meticulous biological synthesis of nanoparticles like metals and metal oxides by using plant extract. Green synthesis routes also involve the use of microorganisms like bacteria, viruses, fungi, and actinomycetes (Figure 2.1).

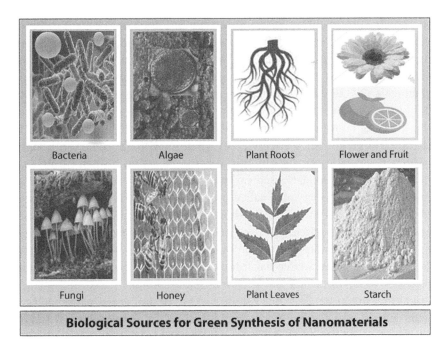

Figure 2.1 Representation of some biological sources used in the green synthesis of nanomaterials.

Microorganisms have the capability to synthesize inorganic materials and are explored extensively in the green synthesis of nanomaterials [19, 20].

2.3 Green Nanotechnology

Green technology is a continuously evolving group of processes that deal with techniques to generate energy with nontoxic end products [21, 22]. The basic reason for global warming is rapid urbanization and industrialization which is responsible for the release of toxic elements into the environment [23]. Over the past five decades, the rise in global temperature has been noticed at the fastest rate. Moreover, global warming affects living beings in several aspects [24]. Thus, in view of this, green technology would help to reduce the effect of global warming by developing new trends for energy generations [25]. For example, a solar cell converts light energy into electrical energy by following the photovoltaics technique. The use of solar cells ensures reduced pollution by less consumption of fossil fuels and prevents greenhouse gas emissions. Green technology also changes the way of production, as well as the elimination of waste without causing any harm to the environment [26]. The aim of green technology is to provide sustainability, source reduction, and the development of innovative synthesis methods to generate energy efficiently as an alternative fuel. It also involves the synthesis of green chemistry products with the elimination of less hazardous end products by avoiding the use of toxic chemicals [27, 28]. Recently, the green building term has also come into light which involves the synthesis of nontoxic building materials, as well as the choice of building location [29, 30]. One of the green technology parts is known as green nanotechnology, which deals with the green synthesis of nanomaterials without causing any adverse effects on human health and the environment. As discussed earlier, nanoparticles are the fundamental unit of nanotechnology also called building blocks [31]. The synthesis of nanoparticles exploits several methods like physical, chemical, biological, and hybrid. Nanoparticles are used in bottom-up approaches for the development of nanomaterials and devices [32]. However, sometimes, it may produce some unavoidable hazardous end products that directly or indirectly harm environment and known as nanopollutants. To overcome this limitation, green nanoscience can be used to design green nanomaterials with defined structure, purity and composition by developing a synthetic strategy. Developing new and innovative design rule(s) are essential criteria in green nanotechnology [33]. Designing rules should be more focused on innovation and commercialization of new class of nanomaterials with high degree of safety and desirable properties. Green nanotechnology applies the famous green chemistry rules for designing, production, and utilization of nanomaterials. Like green chemistry, green nanotechnology focuses on product designing and optimization of process to reduce or eliminate environmental and health hazards. Some basic criteria to be optimized and understand for green synthesis of nanomaterials are as follows: mechanism, bio, or eco testing methods, characterization strategies and tools and synthetic methods. By using these criteria, it would be easy to design and develop well-defined, reliable, efficient, and novel nanomaterials to replace hazardous materials [34–36]. Green nanotechnology is also useful to estimate the possible positioning of environmental impacts in the production chain just like life cycle approaches [37]. Comparing the strengths and weaknesses of competing technologies also requires some complementary strategies, such as analysis of decision support tools. Thus, green nanotechnology can be beneficial in terms of innovation (exploration of novel nanomaterials and their properties), enhanced possibilities of commercialization (by increased safety and more production approaches), and providing protection of investment from the uncertain risk of nanotechnology to consumers [38]. Apart from some well-known applications like solar cells, fuel cells, biofuels, etc., green nanotechnology can also be used in clean manufacturing processes, where sunlight and recycled industrial waste can be used to synthesize nanoparticles [39–42]. Furthermore, utilization of true green nanotechnology (fully plant-grown nanomaterials) on an industrial basis is still limited due to limited production as they will never reach up to the required scale. However, green nanotechnology is the safest way to advance nanotechnology in a sustainable world. Green nanotechnology also offers protection from adverse effects even before they occur [43–45].

2.4 Green Synthesis of Nanomaterials

The nanomaterials and their interaction with environmental or biological entities are really complex to understand. There are so many factors involved that influence nanomaterials toxicity such as nanomaterials aggregation in a biological medium, type of biological assay, and used quantity (Figure 2.2). The structure and tunable composition of nanomaterials make them complex; however, they can be produced in distinct shapes, sizes, compositions, surface areas, and functionalities. In nanomaterial synthesis, one should check the impurity profiles of materials used by characterization techniques, as well as analysis of purification methods

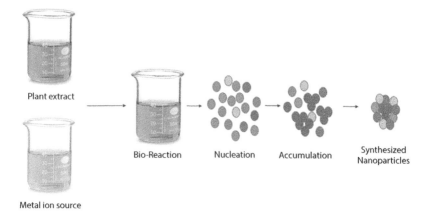

Figure 2.2 Schematic representation of green nanomaterial synthesis.

because impurities influence the properties, reactivity, as well as toxicity of nanomaterials. The aim of greener synthesis of nanomaterials is to provide nanomaterials with higher performance and lower toxicity. Green production methods avoid the exploitation of hazardous materials, thus minimizing the chances of hazards in end products also. Some factors like improvement in material efficiency, enhanced throughput, product control, and reduction in waste materials should be focused on at the time of nanomaterial synthesis. Green synthesis of nanomaterials involved the use of plant products or extracts [46–48]. A summary of some plant materials involved in green synthesis of nanomaterials has given below in a tabular form in Table 2.1.

In this context, some lipid nanoparticles, polymeric nanoparticles, such as nanoemulsions, nanospheres, and nanocapsules, liposomes, can be prepared by using green synthesis approaches. These green synthesis-derived nanoparticles showed enhanced bioavailability, solubility stability, and improved pharmaceutical properties. Apart from these, they also exhibit sustained delivery and defense against easy degradation. The green synthesis approach is widely accepted for silver and gold nanoparticles synthesis [49]. Gold nanoparticles are used extensively in sensing and theranostic applications due to their shape, controllable size, and surface properties. Some specific plant materials have been employed for gold nanoparticle synthesis based on green chemistry approaches. These methods are environmental-friendly and non-toxic. It was also observed that by adjusting pH of the reaction medium, distinct shapes of nanoparticles can be achieved, such as hexagonal, icosahedral, decahedral, rod-shaped, and irregular. In some studies, eucalyptus leaf extract was used for spherical

Table 2.1 Summary of plant materials used for green synthesis of various nanoparticles.

Plant material	Synthesized nanoparticles	Size of nanoparticle	Shape of nanoparticle
Eucalyptus leaf extract, cashew nut, mahogany, neem, *Aloe vera*, pear	Gold, silver	20–80 nm	Spherical, hexagonal, icosahedral, decahedral, rod-shaped, and irregular
Magnolia leaves, clove extract	Copper and copper oxide	40–100 nm	Spherical to granular
Cinnamon bark, Custard apple peel extract, tea, coffee extracts, Persimmon leaf extract, Red pine lignin extract	Palladium and platinum	75–85 nm, 20–60 nm	Spherical, crystalline
Annona squamosa peel, *Nyctanthes arbor-tristis* leaf, *Eclipta prostrata* leaf, *Psidium guajava*, and *Catharanthus roseus* leaf, *Aloe vera*	Titanium dioxide nanoparticles	5–50 nm	Irregular
sorghum bran, neem, lemon grass, *Tridax procumbens*, *Datura innoxia*, *Tinospora cordifolia*, *Euphorbia milii*, and *Calotropis procera*	Iron oxide	100 nm	Spherical
Jatropha curcas latex	Lead	>100 nm, 10–12 nm	-
Citrus reticulata peel	Selenium	>100 nm, 70 nm	-

gold nanoparticle synthesis. Nanoparticle sizes can be obtained in the 20- to 80-nm range from green synthesis approaches [50]. Moreover, some other plants like cashew nut, mahogany, neem, Aloe vera, pear, etc are also used to synthesize silver and gold nanoparticles. Some studies reported the synthesis of bi-metallic nanoparticles also like gold-silver apart from pure metal nanoparticles [51].

Furthermore, copper nanoparticles and copper oxide nanoparticles are also used widely in various applications due to their distinguishable properties. Several approaches have been reported so far for the synthesis of copper and copper oxides from plant materials or extracts. For example; leaves of magnolia were utilized to develop highly stable nanoparticles ranging in size from 40 to 100 nm. Likewise, clove extract also showed the potential to develop 40 nm-sized copper nanoparticles. Some biocompatible nanoparticles of palladium and platinum were also produced by using green chemistry-based techniques [52]. The well-known medicinal plant cinnamon was employed for palladium nanoparticle synthesis. The bark of cinnamon is used not only as a spice but also as a traditional medicine for thousands of years. The concentration, temperature, and pH can affect the nanoparticle's size and morphology [53, 54]. Palladium nanoparticles (PdNPs) of sizes 75 to 85 nm and 20 to 60 nm were prepared by using custard apple peel extract and tea, as well as coffee extracts, respectively [55, 56]. A green method for platinum nanoparticle synthesis was established by using persimmon leaf extract. Apart from that, red pine lignin extract has also been reported for the synthesis of platinum and palladium nanoparticles [57, 58]. Some metal oxide nanoparticles, such as titanium dioxide, zinc oxide, indium oxide, and iron oxide, have also been produced using plant materials. *Annona squamosa* peel, *Nyctanthes arbor-tristis* leaf, *Eclipta prostrata* leaf, *Psidium guajava*, and *Catharanthus roseus* leaf were exploited in titanium dioxide synthesis. Titanium dioxide nanoparticles are known for their antibacterial and antioxidant properties against several kinds of pathogens. *Aloe vera* plant is used in the treatment of several diseases and has a superior place in Ayurveda due to its medicinal properties. Leaf extract from *Aloe vera* was used to develop 5 to 50 nm sized spherical nanoparticles of indium oxide. Iron nanoparticle is very important in environmental remediation techniques. Thus, various research has been carried out for iron nanoparticle synthesis based on green nanotechnology. For example, plant extracts from sorghum bran, neem, lemon grass, *Tridax procumbens, Datura innoxia, Tinospora cordifolia, Euphorbia milii,* and *Calotropis procera* were recently utilized for metal oxide nanoparticles synthesis. Furthermore, the nanoparticles like lead and selenium were synthesized within 100 nm size by using Jatropha curcas latex and citrus reticulata peel, respectively. Plant extracts are rich in proteins, alkaloids, sugars, terpenoids phenolic acid, and polyphenols. These bioactive compounds play an important role in reducing and stabilizing metal ions during the process of nanomaterial synthesis. Plant materials are also capable in hyper-accumulation of ions. Plant extract concentration, composition, and interaction with metal ions determine nanoparticle shape and size. Green synthesis of nanoparticles involved mixing plant extracts with metal salt solutions. Biochemical reduction of metal salts is achieved instantly by the plant extract indicated by a color change in the solution. The reduction process consists of the conversion of a metal ion from a monovalent or divalent state to a zero valent state to achieve nucleation of the reduced metal atom. The process also involved the integration of smaller particles to give rise to thermally stable large nanoparticles. In this way, one can produce nanoparticles of different shapes, sizes, stable morphologies, and improved quality [59].

2.5 Advantages of Green Nanoscience

Greener nanoscience followed the basic principle of green chemistry for designing, synthesis, and use of nanostructured materials. It strives to reduce as well as eliminate health and environmental hazards through synthesis optimization and product designing. Green nanoscience can spur innovation by exploring new materials and their properties. It can also enable commercialization by providing assurance of material safety and more reliable manufacturing methods. Green synthesis may also protect investors from threats of consumers' fear regarding the uncertain danger of nanotechnology. Green synthesis help to improve nanomaterials properties and their functionalization towards various applications in distinct fields of physics, chemistry, and biology. Green nanoscience can lead to a reduction in non-renewable resource uses thus these resources can be conserved for the future. In this way, green nanoscience help in cost savings in a long term. With the help of green technology, living beings can spend a quality life in a healthier environment with less pollution. Therefore, the green nanoscience approach leads toward better and healthier as well as more conductive life and environmental conditions to survive within. A summary of applications of green nanoscience is given below (Figure 2.3).

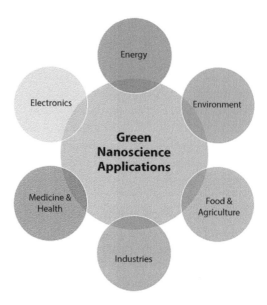

Figure 2.3 Applications of green nanoscience.

2.5.1 Green Nanoscience in Industries

Nanotechnology has great potential for multiple concrete applications, thus research focus is now shifted to the commercialization of innovations produced by green nanotechnology. In view of the present scenario of global warming and energy scarcity, this is necessary to pay much attention to the use of clean and green technology in industries. In industries, production methods can be modified for existing materials with newly developed green materials. Some Indian corporations are following green technology at present time. Developed countries have already been encouraging green technology for manufacturing in industries. For example; biomass plants and green wind projects. A few examples of international industries are Dell, Cisco, Nokia, Intel, etc. Green nanomanufacturing involves the synthesis of low toxic nanomaterials with reduced environmental and health impacts. Biological synthesis of nanomaterials like nanofibers, and nanotubes, are encouraged widely to produce contaminants free, large quantities of nanoparticles [60, 61].

2.5.2 Green Nanoscience in Automobiles

The use of green synthesis for fuel production can match the strict emission norms of managing authorities. Nanomaterial-derived products can redefine energy as well as applications of materials. They also have the potential to replace expensive materials like platinum in fuel cells to develop environmentally friendly vehicles. Green machines also came into light recently to benefit the environment for example; hybrids and hydrogen vehicles. Carbon nanotubes are extensively used for manufacturing fuel cells that proved to increase reactivity and store hydrogen [62].

2.5.3 Green Nanoelectronics

In the present time, everyone uses electronic gadgets in day-to-day life. Therefore, they must be human health and environment friendly due to direct exposure to these electronic machines. Electronic gadgets manufactured with green technology result in low cost, degradable without producing any toxic end products, and compatibility with humans and the environment [63].

2.5.4 Green Nanoscience in Food and Agriculture

The rising global population has reduced natural resources per capita. A number of challenges are faced by the entire world related to agriculture and food security management. The green nanomaterials uses have the potential to reduce the harmful environmental effects, cost of purchasing nutrients, and increase yields of crops. Green technology can reduce the production of greenhouse gases like carbon dioxide, methane, and nitrous oxide in the atmosphere. Nanotechnology-based sensors can help to monitor food quality as well as food packaging to keep food products safe from microorganisms. Green nanoscience also facilitates the on-site monitoring of water, soil, and environmental pollutants like pesticides and pathogens very quickly [64, 65]. Green nanomaterials used in the agriculture field can reduce the loss of nutrients, improve the delivery of nutrients, optimize water and nutrient levels and increase crop yield.

2.5.5 Green Nanoscience in Medicines

Some ground-breaking developments have been seen in recent years in the biomedical field based on nanotechnology. Green nanoscience can help in fighting against several life-threatening diseases and better health conditions. It can help in sensing disease-related biomarkers and the presence of bacteria and viruses to check infections caused by them [66–68]. Nanomaterial-based site-specific drug delivery and tissue engineering applications have also been used to provide therapy against several

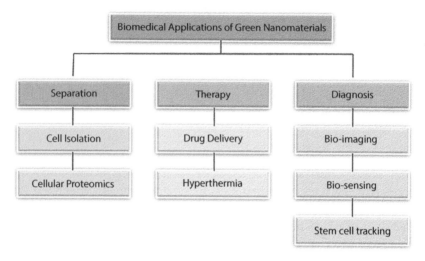

Figure 2.4 Schematic representation of green nanomaterials-based biomedical applications.

diseases. Machines for imaging diseases are also utilized nanoparticles like quantum dots for clarity (Figure 2.4).

2.6 Conclusion

Green nanoscience is a boon for the modern industrialized world. Adaptation of nanotechnological methods for improvement in traditional methods can help to reduce the environmental pollution and scarcity of natural resources. Green nanoscience can also improve the day-to-day life of human beings in several ways. Several industries of developed countries already embraced the changes by encouraging green nanoscience to avoid the negative consequences of toxic materials. The exploitation of nanoparticles in biomedical applications required properties like improved biodegradability, biocompatibility, and functionalization; however, it remains a challenge for researchers to develop such kind of nanomaterials. Till now, nanoparticles, like paramagnetic, quantum dots, carbon nanotubes, and nanoshells, are extensively used and examined for various applications, for example, robotics, electronics, tissue engineering, drug targeting, imaging, and diagnosis. The focus on green synthesis of nanomaterials will help to develop new approaches for the determination and implementation of designing new and safer nanomaterials with more efficient processes. Green nanoscience will help in the sustainable development of various fields with innovations, elimination of toxic end products, and reusability in a more successful manner.

References

1. Mekonnen, G., Review on application of nanotechnology in animal health and production. *J. Nanomed. Nanotechnol.*, 12, 559, 2021.
2. Nasrollahzadeh, M., Mohammad Sajadi, S., Sajjadi, M., Issaabadi, Z., An introduction to nanotechnology, in: *Interface Science and Technology*, vol. 28, pp. 1–27, Elsevier, Netherland, 2019.
3. Jain, U. and Saxena, K., Smart nanobiosensors, in: *Nanosensors for Smart Manufacturing*, pp. 231–245, Elsevier, Amsterdam, 2021.
4. Bayda, S., Adeel, M., Tuccinardi, T., Cordani, M., Rizzolio, F., The history of nanoscience and nanotechnology: From chemical–physical applications to nanomedicine. *Molecules*, 25, 1, 112, 2019.
5. Aithal, P.S. and Aithal, S., Ideal technology concept & its realization opportunity using nanotechnology. *Int. J. Appl. Innov. Eng. Manage. (IJAIEM)*, 4, 2, 153–164, 2015.
6. Sriram, P. and Suttee, A., Nanotechnology advances, benefits, and applications in daily life, in: *Nanotechnology*, pp. 23–44, CRC Press, Boca Raton, 2020.
7. Mazari, S.A., Ali, E., Abro, R., Khan, F.S.A., Ahmed, I., Ahmed, M., Nizamuddin, S. et al., Nanomaterials: Applications, waste-handling, environmental toxicities, and future challenges–A review. *J. Environ. Chem. Eng.*, 9, 2, 105028, 2021.
8. Bhargava, C. and Amit S., (eds),. Nanotechnology: Advances and real-life applications. CRC Press, Boca Raton, London, New York, 2020.
9. Hadef, F., An introduction to nanomaterials, in: *Environmental Nanotechnology*, pp. 1–58, Springer, Cham, 2018.
10. Hussain, C.M., *Handbook of functionalized nanomaterials for industrial applications*, Elsevier Science, Elsevier, Netherland, 2020.
11. Brewer, A., Dror, I., Berkowitz, B., The mobility of plastic nanoparticles in aqueous and soil environments: A critical review. *ACS ES&T Water*, 1, 1, 48–57, 2020.
12. Bundschuh, M., Filser, J., Lüderwald, S., McKee, M.S., Metreveli, G., Schaumann, G.E., Schulz, R., Wagner, S., Nanoparticles in the environment: Where do we come from, where do we go to? *Environ. Sci. Eur.*, 30, 1, 1–17, 2018.
13. Sajid, M., Ilyas, M., Basheer, C., Tariq, M., Daud, M., Baig, N., Shehzad, F., Impact of nanoparticles on human and environment: Review of toxicity factors, exposures, control strategies, and future prospects. *Environ. Sci. Pollut. Res.*, 22, 6, 4122–4143, 2015.
14. Savage, D.T., Zach Hilt, J., Dziubla, T.D., *In vitro* methods for assessing nanoparticle toxicity, in: *Nanotoxicity*, pp. 1–29, Humana Press, New York, NY, 2019.

15. Modena, M.M., Rühle, B., Burg, T.P., Wuttke, S., Nanoparticle characterization: What to measure? *Adv. Mater.*, 31, 32, 19015, 2019.
16. Palit, S. and Hussain, C.M., Recent advances in green nanotechnology and the vision for the future, in: *Green Metal Nanoparticles: Synthesis, Characterization and their Applications*, pp. 1–21, 2018, 5. Do We Consume Too Much?
17. Ahmed, S.F., Mofijur, M., Rafa, N., Chowdhury, A.T., Chowdhury, S., Nahrin, M., Saiful Islam, A.B.M., Ong, H.C., Green approaches in synthesising nanomaterials for environmental nanobioremediation: Technological advancements, applications, benefits and challenges. *Environ. Res.*, 204, 111967, 2022.
18. Ahmed, S., Ahmad, M., Swami, B.L., Ikram, S., A review on plants extract mediated synthesis of silver nanoparticles for antimicrobial applications: A green expertise. *J. Adv. Res.*, 7, 1, 17–28, 2016.
19. Aslam, M., Abdullah, A.Z., Rafatullah, M., Recent development in the green synthesis of titanium dioxide nanoparticles using plant-based biomolecules for environmental and antimicrobial applications. *J. Ind. Eng. Chem.*, 98, 1–16, 2021.
20. Nobahar, A., Carlier, J.D., Miguel, M.G., Costa, M.C., A review of plant metabolites with metal interaction capacity: A green approach for industrial applications. *BioMetals*, 34, 4, 761–793, 2021.
21. Soni, G.D., Advantages of green technology. Social issues and environmental problems. *Int. J. Res. Granthaalayah*, 3, 9, 97–100, 2015.
22. Wang, Q., Qu, J., Wang, B., Wang, P., Yang, T., Green technology innovation development in China in 1990–2015. *Sci. Total Environ.*, 696, 134008, 2019.
23. Appannagari, R.R., Environmental pollution causes and consequences: A study. *North Asian Int. Res. J. Soc. Sci. Humanit.*, 3, 8, 151–161, 2017.
24. Singh, R.L. and Singh, P.K., Global environmental problems, in: *Principles and Applications of Environmental Biotechnology for a Sustainable Future*, pp. 13–41, Springer, Singapore, 2017.
25. Owusu, P.A. and Asumadu-Sarkodie, S., A review of renewable energy sources, sustainability issues and climate change mitigation. *Cogent Eng.*, 3, 1, 1167990, 2016.
26. Lai, C.S., Jia, Y., Lai, L.L., Xu, Z., McCulloch, M.D., Wong, K.P., A comprehensive review on large-scale photovoltaic system with applications of electrical energy storage. *Renew. Sustain. Energy Rev.*, 78, 439–451, 2017.
27. Abbas, J. and Sağsan, M., Impact of knowledge management practices on green innovation and corporate sustainable development: A structural analysis. *J. Cleaner Prod.*, 229, 611–620, 2019.
28. Jiang, W., Chai, H., Shao, J., Feng, T., Green entrepreneurial orientation for enhancing firm performance: A dynamic capability perspective. *J. Cleaner Prod.*, 198, 1311–1323, 2018.
29. Gomaa Mayhoub, M.M., El Sayad, Z.M.T., Ali, A.A.M.M., Ibrahim, M.G., Assessment of green building materials' attributes to achieve sustainable building façades using AHP. *Buildings*, 11, 10, 474, 2021.

30. Singh, C.S., Green construction: Analysis on green and sustainable building techniques. *Civ. Eng. Res. J.*, 4, 3, 555638, 2018.
31. Verma, A., Gautam, S.P., Bansal, K.K., Prabhakar, N., Rosenholm, J.M., Green nanotechnology: Advancement in phytoformulation research. *Medicines*, 6, 1, 39, 2019.
32. Chauhan, N. and Saxena, K., Nanoscale interface techniques for standardized integration of nanosensors in current devices, in: *Nanosensors for Smart Manufacturing*, pp. 91–114, Elsevier, Netherland, 2021.
33. Khan, S.H., Green nanotechnology for the environment and sustainable development, in: *Green Materials for Wastewater Treatment*, pp. 13–46, Springer, Cham, 2020.
34. Ganachari, S.V., Yaradoddi, J.S., Somappa, S.B., Mogre, P., Tapaskar, R.P., Salimath, B., Venkataraman, A., Viswanath, V.J., Green nanotechnology for biomedical, food, and agricultural applications, (eds) *Handbook of Ecomaterials*. Springer, Cham. University of Edinburgh Research Explorer, 2019.
35. Maksimović, M. and Omanović-Mikličanin, E., Towards green nanotechnology: Maximizing benefits and minimizing harm, in: *CMBEBIH 2017*, pp. 164–170, Springer, Singapore, 2017.
36. Geraldes, A.N., da Silva, A.A., Leal, J., Estrada-Villegas, G.M., Lincopan, N., Katti, K.V., Lugão, A.B., Green nanotechnology from plant extracts: Synthesis and characterization of gold nanoparticles. *Adv. Nanopart.*, 5, 03, 176, 2016.
37. Benko, A., Truong, L.B., Medina-Cruz, D., Mostafavi, E., Cholula-Díaz, J.L., Webster, T.J., Green nanotechnology in cardiovascular tissue engineering, in: *Tissue Engineering*, pp. 237–281, Academic Press, Cambridge, Massachusetts, 2022.
38. Aithal, S. and Aithal, P.S., Green nanotechnology innovations to realize UN sustainable development goals 2030. *Int. J. Appl. Eng. Manage. Lett. (IJAEML)*, 5, 2, 96–105, 2021.
39. Rodríguez-Couto, S., Green nanotechnology for biofuel production, in: *Sustainable Approaches for Biofuels Production Technologies*, pp. 73–82, Springer, Cham, 2019.
40. AlMasoud, N., Alhaik, H., Almutairi, M., Houjak, A., Hazazi, K., Alhayek, F., Aljanoubi, S. et al., Green nanotechnology synthesized silver nanoparticles: Characterization and testing its antibacterial activity. *Green Process. Synth.*, 10, 1, 518–528, 2021.
41. Sharma, P., Guleria, P., Kumar, V., Green nanotechnology for bioactive compounds delivery, in: *Biotechnological Production of Bioactive Compounds*, pp. 391–407, Elsevier, Netherland, 2020.
42. Nazir, R., Ayub, Y., Tahir, L., Green-nanotechnology for precision and sustainable agriculture, in: *Biogenic Nano-Particles and their Use in Agro-Ecosystems*, pp. 317–357, Springer, Singapore, 2020.
43. Silva, L.P. and Bonatto, C.C., Green nanotechnology for sustained release of eco-friendly agrochemicals, in: *Sustainable Agrochemistry*, pp. 113–129, Springer, Cham, 2019.

44. Palit, S. and Hussain, C.M., Recent advances in green nanotechnology and the vision for the future, in: *Green Metal Nanoparticles: Synthesis, Characterization and their Applications*, pp. 1–21, 2018.
45. Nargund, V.B., Vinay, J.U., Basavesha, K.N., Chikkanna, S., Jahagirdar, S., Patil, R.R., Green nanotechnology and its application in plant disease management, in: *Emerging Trends in Plant Pathology*, pp. 591–609, Springer, Singapore, 2021.
46. Parveen, K., Banse, V., Ledwani, L., Green synthesis of nanoparticles: Their advantages and disadvantages. *AIP Conf. Proc.*, AIP Publishing LLC, 1724, 1, 020048, 2016.
47. Singh, J., Dutta, T., Kim, K.-H., Rawat, M., Samddar, P., Kumar, P., 'Green' synthesis of metals and their oxide nanoparticles: Applications for environmental remediation. *J. Nanobiotechnol.*, 16, 1, 1–24, 2018.
48. El Shafey, A.M., Green synthesis of metal and metal oxide nanoparticles from plant leaf extracts and their applications: A review. *Green Process. Synth.*, 9, 1, 304–339, 2020.
49. Castillo-Henríquez, L., Alfaro-Aguilar, K., Ugalde-Álvarez, J., Vega-Fernández, L., de Oca-Vásquez, G.M., Vega-Baudrit, J.R., Green synthesis of gold and silver nanoparticles from plant extracts and their possible applications as antimicrobial agents in the agricultural area. *Nanomaterials*, 10, 9, 1763, 2020.
50. Correia, M., Uusimäki, T., Philippe, A., Loeschner, K., Challenges in determining the size distribution of nanoparticles in consumer products by asymmetric flow field-flow fractionation coupled to inductively coupled plasma-mass spectrometry: The example of Al2O3, TiO2, and SiO2 nanoparticles in toothpaste. *Separations*, 5, 4, 56, 2018.
51. Shah, M., Fawcett, D., Sharma, S., Tripathy, S.K., Poinern, G.E.J., Green synthesis of metallic nanoparticles via biological entities. *Materials*, 8, 11, 7278–7308, 2015.
52. Salem, S.S. and Fouda, A., Green synthesis of metallic nanoparticles and their prospective biotechnological applications: An overview. *Biol. Trace Elem. Res.*, 199, 1, 344–370, 2021.
53. Abdelfatah, A.M., Fawzy, M., Eltaweil, A.S., El-Khouly, M.E., Green synthesis of nano-zero-valent iron using ricinus communis seeds extract: Characterization and application in the treatment of methylene blue-polluted water. *ACS Omega*, 6, 39, 25397–25411, 2021.
54. Bautista-Guzman, J., Gomez-Morales, R., Asmat-Campos, D., Checca, N.R., Influence of the alcoholic/ethanolic extract of mangifera indica residues on the green synthesis of FeO nanoparticles and their application for the remediation of agricultural soils. *Molecules*, 26, 24, 7633, 2021.
55. Kalishwaralal, K., Deepak, V., Ram Kumar Pandian, S., Kottaisamy, M., BarathmaniKanth, S., Kartikeyan, B., Gurunathan, S., Biosynthesis of silver and gold nanoparticles using Brevibacterium casei. *Colloids Surf. B Biointerfaces*, 77, 2, 257–262, 2010 Jun 1.

56. Narayanan, K.B. and Park, H.H., Antifungal activity of silver nanoparticles synthesized using turnip leaf extract (*Brassica rapa* L.) against wood rotting pathogens. *Eur. J. Plant Pathol.*, 140, 185–192, 2014.
57. Attar, A. and Yapaoz, M.A., Biosynthesis of palladium nanoparticles using Diospyros kaki leaf extract and determination of antibacterial efficacy. *Prep. Biochem. Biotechnol.*, 48, 7, 629–634, 2018.
58. Fahmy, S.A., Preis, E., Bakowsky, U., El-Said Azzazy, H.M., Platinum nanoparticles: Green synthesis and biomedical applications. *Molecules*, 25, 21, 4981, 2020.
59. Verma, A., Gautam, S.P., Bansal, K.K., Prabhakar, N., Rosenholm, J.M., Green nanotechnology: Advancement in phytoformulation research. *Medicines*, 6, 1, 39, 2019.
60. Huguet-Casquero, A., Gainza, E. and Pedraz, J.L., Towards green nanoscience: From extraction to nanoformulation. *Biotechnol. Adv.*, 46, 107657, 2021.
61. Kemp, R. and Never, B., Green transition, industrial policy, and economic development. *Oxford Rev. Econ. Policy*, 33, 1, 66–84, 2017.
62. Verma, G., Various areas of green chemistry and safer environment an overview. *Int. J. Res. Appl. Nat. Soc. Sci.*, 3, 3, 87–94, 2015.
63. Eisenstein, G. and Bimberg, D. (Eds.), *Green photonics and electronics*, Springer International Publishing - Midtown Manhattan, New York City, 2017.
64. Jain, U., Saxena, K., Hooda, V., Balayan, S., Singh, A.P., Tikadar, M., Chauhan, N., Emerging vistas on pesticides detection based on electrochemical biosensors–An update. *Food Chem.*, 371, 131126, 2022.
65. Singh, A.P., Balayan, S., Gupta, S., Jain, U., Sarin, R.K., Chauhan, N., Detection of pesticide residues utilizing enzyme-electrode interface via nano-patterning of TiO2 nanoparticles and molybdenum disulfide (MoS2) nanosheets. *Process Biochem.*, 108, 185–193, 2021.
66. Chauhan, N., Gupta, S., Avasthi, D.K., Adelung, R., Mishra, Y.K., Jain, U., Zinc oxide tetrapods based biohybrid interface for voltammetric sensing of helicobacter pylori. *ACS Appl. Mater. Interfaces*, 10, 36, 30631–30639, 2018.
67. Gupta, S., Jain, U., Murti, B.T., Putri, A.D., Tiwari, A., Chauhan, N., Nanohybrid-based immunosensor prepared for Helicobacter pylori BabA antigen detection through immobilized antibody assembly with @Pdnano/rGO/PEDOT sensing platform. *Sci. Rep.*, 10, 1, 1–14, 2020.
68. Balayan, S., Chauhan, N., Chandra, R., Jain, U., Electrochemical based C-reactive protein (CRP) sensing through molecularly imprinted polymer (MIP) pore structure coupled with bi-metallic tuned screen-printed electrode. *Biointerface Res. Appl. Chem.*, 6, 38, 2022.

3

Optimization of the Process Parameters to Develop Green-Synthesized Nanostructures with a Special Interest in Cancer Theranostics

Tathagata Adhikary[1], Chowdhury Mobaswar Hossain[2] and Piyali Basak[1*]

[1]*School of Bioscience and Engineering, Jadavpur University, Kolkata, India*
[2]*Department of Pharmaceutical Technology, Maulana Abul Kalam Azad University of Technology, West Bengal, Nadia, India*

Abstract

Green synthesis follows eco-friendly laboratory protocols to limit the use of synthetic chemicals. Several methods are adopted for green synthesizing nanocomposites with optimal balance between their efficacy and toxicity. Each technique involves different parameters that primarily influence the characteristics (e.g., in terms of its mean particle size, concentration, or polydispersity index) of the synthesized nanoparticles. Assessing bioactivities of the nanoparticles combined with physicochemical characterizations (such as dynamic light scattering, FTIR, UV-Vis spectroscopy, electron microscopy, etc.) helps a researcher to reach the optimized product. A reader gets an insight into the mechanisms underlying different green nanofabrication techniques and the effect of various factors in the fabrication process. Statistical models and other *in silico* approaches are frequently employed along with experimental data to ease the optimization. Although these techniques remain valid in optimizing the green synthesis of any nanomaterial, this article attempts to review the related reports and recent advancements in the field of cancer theranostics.

Keywords: Green synthesis, biocompatible, cancer, optimization, nanoparticles

---

*Corresponding author: piyalibasak@gmail.com

3.1 Introduction

The term "nano" is originated from Greek word "nános" which translates to "dwarf." It indicates one billionth of anything it is prefixed to. In 1974, Prof. Norio Taniguchi from Tokyo Science University coined the term "nanotechnology" to highlight precision and nanometer tolerances in manufacturing processes and later in 1986, Kim Eric Drexler titled his book as 'Engines of creation: The Coming Era of Nanotechnology' [1]. Based on the dimensions, nanomaterials can be classified as 0 D, 1 D, and 2 D. The term "dimensions" here denotes any axis of the nanostructure that can be stretched beyond 100 nm to a few micrometers. 0 D nanostructures exist in triangular, hexagonal, and spherical forms, and they are widely used in optical, electronic, and biomedical applications, such as in drug delivery systems and bioimaging. 1 D nanostructures comprise nanorods, nanowires, and nanotubes. 2D nanomaterials typically include graphene, nanofilms, and nanocoatings [2]. The fact that the properties of a bulk material show significant changes when it is brought down to its nanometric form due to the increase in surface area is exploited in recent advancements in material sciences. Silver nanoparticles show distinct variations in physicochemical properties, such as high electrical and thermal conductivity, surface-enhanced Raman scattering, and non-linear optical behavior [3]. Gold has a melting point of 1064°C, while gold nanoparticles have a lower melting point of around 300°C (for a dimension of 2.5 nm) [4].

3.1.1 Conventional Techniques in Nanoparticle Synthesis

Among the physical techniques, some can be listed as plasma arcing (plasma torch is used to deposit positively charged ions on the cathode in the form of nanosized materials), ball milling (locally generating high pressure due to the collision between rigid balls to form nano-powder), thermal evaporate, spray pyrolysis, developing ultra-thin films, laser ablation, lithographic techniques, vapor deposition, molecular beam epitasis, and diffusion flame synthesis of nanoparticles. Other methods to chemically synthesize nanoparticles can be electro-deposition, sol-gel process, vapor deposition, Langmuir Blodgett method, and hydrolysis co-precipitation. The use of organic solvents in the synthesis of nanoparticles carries the risk of neurobehavioral and reproductive disorders if not handled properly [5]. An unsafe working environment is created if the synthesis protocol involves high pressure and/or heat treatments. Additionally, the generation of volatile vapors and excessive carbon dioxide aids in the greenhouse effect, which must be restricted while considering industrial upscaling. The sol-gel process typically occurs in five stages. At first, precursors are hydrolyzed (by using either water or an organic solvent) and then the condensation step starts where linkages between the neighboring molecules are formed. Followed by the drying of the so-formed "gel" (either freeze-dried or thermally treated), calcination is done to drive off residual water or other residues [6]. In the vapor deposition method, a substrate is made in contact with volatile gases that react with the substrate (generally inside a vacuum chamber at elevated temperatures in the presence of nitrogen gas and catalysts) to synthesize nanostructures. The reaction temperature and the chemical composition of the reactants (i.e., substrate and precursor) can be varied to generate nanoparticles with diverse physicochemical properties, morphologies, and sizes. Carbon nanomaterials including graphene, fullerene, carbon nanotubes, and diamond-like carbon films are successfully synthesized by this process [2]. In hydrothermal synthesis or solvothermal synthesis, the material is solubilized (in water or an appropriate solvent along with some stabilizing agents) and kept (in a steel autoclave) at extremely high temperature and pressure to form its crystalline structures. One can regulate the morphological characteristics and size distribution of the nanoparticle by changing the precursors, operational temperature, and pH of the solution [7]. Ultrasonic treatment uses acoustic waves to cause cavitation, which initiates the chemical reactions involved in the growth of nanostructures. During this cavitation process, microbubbles store the ultrasonic energy, grow, and finally burst to release the stored energy into the surrounding environment, generating intense heat and pressure for a small fraction of time. Optimization of the nanostructures can be achieved by using different precursors, the solvent medium and the frequency of the ultrasonic waves. In laser ablation, the material is fragmented by laser pulses in an appropriate fluidic medium. The dimensions of these smaller fragments finally reach the nanoscale to complete the synthesis. Usually, the synthesized nanomaterial is collected on the surface of the gaseous/vacuum chamber as a thin film while colloidal structures are observed in a liquid medium. Optimized nanoparticles are synthesized by adjusting the intensity of laser pulses, exposure time and wavelength of the laser. Gold nanoparticles obtained by laser ablation are conjugated with specific oligonucleotides/aptamers and found their use in detecting human prostate cancer [8]. Inorganic nanoparticles tend to be toxic in biological systems. Antineoplastic therapies are making use of non-toxic organic compounds for the green production of biocompatible nanoparticles with antitumor activity [9].

3.1.2 Green Nanotechnology

The advent of nanotechnology in the field of medical sciences broadens the research area to address a biological problem statement in a sustainable way. Green synthesis of nanoparticles offers a better approach in terms of cost and eco-friendliness when compared to physical and chemical synthesis techniques. Moreover, the use of synthetic chemicals contributes to the toxicity of nanoparticles in biological systems. This toxicity can be countered by surface modification of nanosized materials. Hybrid methods are sometimes adopted that involve the combination of physical, chemical, and/or biological techniques of nanoparticle synthesis [10]. Integrating green nanotechnology in researches and industrial production attempts to bypass the limitations of conventional techniques involved in the synthesis of nanomaterials.

Plants are reported to cause the reduction of metal ions on their surface and other internal tissues remotely located from the ion penetration site. This observation leads to the concept of phytomining, which refers to the extraction of precious metals from the soil and their accumulation in plant parts. Then the metals are recovered by sintering and smelting processes. This bioaccumulation of metals usually occurs in nanometric form. Interestingly, the size and shape of these metal nanoparticles vary with their location, i.e., the tissue in which it is accumulated. As an example, a significant accumulation of silver nanostructures is observed in *Brassica juncea* and *Medicago sativa* when cultivated in the presence of silver nitrate. Similarly, gold and copper nanoparticles are reported in *M. sativa* and *Iris pseudacorus* respectively when grown in the presence of corresponding metal salts [11]. Later, *in vitro* experiments are performed to use plant extracts in reducing the metal ions to form nanoparticles. This technique provides rapid nanoparticle synthesis and can control its morphological characteristics by regulating the process parameters, such as pH, temperature (at which reaction occurs) and time. Gemcitabine hydrochloride loaded on the surface of gum karaya stabilized gold nanoparticles significantly inhibited the growth of human non-small cell lung cancer cells (A549) and showed an increase in the level of reactive oxygen species (when compared to the native drug) [12].

While biological methods often involve a one-step synthesis with low energy requirement, physical and chemical techniques make use of high concentrations of reductants, stabilizing agents and radiation, posing a threat to the environment [13]. Green nanobiotechnology has contributed to the stability and biocompatibility of the synthesized nanoparticles. Nanomedicine provides cutting-edge alternatives to the conventional treatment procedures of several diseases, including cancer. The nanosized materials ensure easy penetration through the capillaries which are then taken up by the cells and gets accumulated at the target sites with prolonged clearance time, thus increasing their bioavailability.

Nano-pollution is a new term that arises from nanotechnology and must be addressed cautiously. A slight alteration in the structure of nanoparticles will exhibit particle growth, a tendency to gelate, or changes in the drug-release profile that can cause cytotoxicity, alveolar inflammation, and other non-intended biological responses [14].

3.2 Mechanism Underlying Green Synthesis

Using green technology, the biosynthesized nanoparticles exhibited diverse biological properties, enhanced stability and specific shape and dimensions. The 12 principles of green chemistry in adopting a sustainable production process are followed worldwide to restrict the use of hazardous chemicals and by-products [15]. The principles are stated below [2]:

1. Prevention: Steps to avoid waste generation
2. Atom economy: Maximizing the incorporation of the materials used in the synthesis in the end product
3. Less hazardous chemical synthesis: Requiring materials that are safe to the environment
4. Alternative non-hazardous chemicals: Use of non-toxic chemicals without compromising efficiency
5. Safe solvents: Allows easy handling and limited use of organic solvents
6. Energetically efficient: Limited use of energy in the synthesis process
7. Use of renewable feedstocks: Steps to ensure a sustainable supply of raw materials
8. Reduce derivatives: Limited use of derivatives (e.g., blocking agents) to avoid additional waste generation
9. Catalysis: Use of catalysts for rapid synthesis
10. Easy degradation: Chemicals need to be able to degrade into non-hazardous forms after the completion of synthesis
11. Analysis of pollution and its prevention: Monitoring the generation of toxic pollutants in real-time
12. Inherently safer chemistry for accident prevention: Steps to prevent hazardous environment

The concept of green chemistry in synthesizing nanoparticles majorly involves phytochemicals (which are present in plants—nature's chemical factories), hence bridging the gap between nanoscience, plant biochemistry, and phytochemistry. The development of nanosized particles is achieved by the bioreduction of metal ions in the presence of biological entities as precursors. These biological entities can be plant metabolites (e.g., phenols, alkaloids, flavonoids, triterpenoids, and even proteins), bacteria, diatoms, fungi, etc. and their by-products [16]. As an example, the reduction of ionic Au (III) to Au (0) and Ag^+ to Ag^0 yields gold and silver nanostructures, respectively. Some of the green synthesized metallic nanostructures include cobalt, copper, silver, gold, palladium, platinum, cadmium sulfide, zinc oxide, zinc sulfide (sphalerite), lead (II) sulfide, iron sulfide, and magnetite [13, 17]. Two approaches that are encountered in the synthesis of nanoparticles are referred to as "top-down" and "bottom-up." The methods following the "top-down" approach tend to reduce the size of the bulk material to gradually generate particles on a nanometric scale. On the other hand, the "bottom-up" approach relies on the self-assembly of atoms to form nanoparticles.

Plants and their phytochemicals (such as extracts of turmeric, garlic, clove, etc.) are widely explored in the green synthesis of nanoparticles. Polyphenols, polyols, and other organic compounds possess antioxidant activity and take part in oxidation–reduction reactions, which in turn form and stabilize the nanoparticles. Monometallic silver and gold nanoparticles, as well as their bimetallic nanoalloys (a result of simultaneous and competitive reduction of Au^{III} and Ag^I ions), are synthesized by using the crude extract of the plant *Swietenia mahogany* (rich in polyhydroxy limonoids) in the absence of any stabilizers (such as surfactants and synthetic polymers) [18].

Green synthesis of nanoparticles using plant extracts comprises three key stages:

(i) Activation phase: Metal ion is reduced here and subsequent nucleation occurs
(ii) Growth phase: Small adjacent nanoparticles spontaneously coalesce to form larger but stable nanoparticles of various sizes and shapes (nanoprisms, nanospheres, nanohexahedrons, etc.)
(iii) Termination phase: Determines the final shape of the nanoparticles to achieve energetically favorable conformation by minimizing the Gibbs free energy.

Anthocyanins, isoflavonoids, flavonols, chalcones, flavones, and flavanones belong to the phytochemical class called flavonoids that has the ability to cause chelation and reduction of metal ions to form nanoparticles. It is reported that tautomerism in flavonoids from the enol-form to the keto-form donates a hydrogen ion that causes this bioreduction. Such transformations are observed in the case of luteolin and rosmarinic acid (flavonoids) in synthesizing silver nanoparticles [19]. Certain flavonoids cause the chelation of metal ions due to the presence of carbonyl groups or π-electrons. Quercetin (an example of flavonoid) is able to chelate strongly at three positions in its structure due to the presence of carbonyl, hydroxyl, and catechol groups. Purified flavonoids and flavonoid glycosides such as apigenin glycoside are used to green synthesize gold and silver nanoparticles. The FTIR data confirmed the attachment of apiin to the nanoparticles via carbonyl groups [20]. Among sugars, monosaccharides, e.g., glucose having an aldehyde group can directly induce reduction, whereas fructose having a keto-group tautomerize into aldehyde to exhibit antioxidant activity. The oxidation of the aldehyde group (present in sugars) into the carboxyl group is responsible for reducing metal ions and forming stable nanostructures. The reducing power of oligosaccharides relies on its constituent monosaccharide to open up the chain and make the metal ion available to its aldehyde group. Because of this, disaccharides maltose and lactose can act as reducing agents. On the contrary, sucrose does not have a reducing ability since its linkage between monomers (glucose and fructose) does not allow sucrose to exist in an open chain form. However, acidic hydrolysis of sucrose into its monomers can use tetrachloroauric and tetrachloroplatinic acids to yield gold and platinum nanoparticles respectively [11]. Among amino acids, lysine, cysteine, arginine, and methionine are reported to synthesize silver nanoparticles, aspartate is used to synthesize nanoparticles, whereas valine and lysine lack the reducing capability [21]. This reducing potential of amino acids comes from the presence of certain groups such as amino, carbonyl, carboxyl, thiol, thioether, and hydroxyl group [22]. Studies have shown the importance of amino acid sequences in regulating the synthesis of nanoparticles, their size and morphological characteristics. GASLWWSEKL is a synthetic peptide that is used for the rapid synthesis of nanoparticles of size under 10 nm. Replacing some of its amino acids (which forms a peptide SEKLWWGASL) slows down the reduction process and yields nanostructures of around 40 nm in size.

Identifying the bioactive compounds that are responsible for directing the size and morphology of the synthesized nanomaterials will help us to elucidate their function/mechanism in the synthesis, thus finding a way to manipulate the structures of nanoparticles by using their chemical properties.

Azadirachta indica extract reduces tetrachloroauric acid ($HAuCl_4$) to form 2-D nanostructures (triangular and hexagonal shaped) of size less than 100 nm. FTIR spectroscopy suggested that metabolites namely sugars, triterpenes and triterpenoids, flavonoids, polyphenols, alkaloids, and plant proteins take part in reducing metal ions and forming stable nanoparticles. The size and shape (e.g., triangular, hexagonal, pentagonal, cubical, spherical, ellipsoidal, and nanorods) of the synthesized nanoparticles are also governed by the interaction of plant metabolites with the nanoparticles [23, 24]. Adopting green nanotechnology, honey can be used as a catalyst in the reduction, stabilization, and formation of honey functionalized gold, silver, carbon, platinum, and palladium nanoparticles [25]. Microemulsion and nano-emulsion (acting as a template and nano-reactor to restrict the size of nanoparticles) are prepared to synthesize silver nanoparticles (ranging from 25 to 150 nm) by using geranium (*P. hortorum*) leaf extract as the reducing and capping agent. In the process, castor oil serves as the oily phase while Brij 96 V and 1,2-hexanediol are used as the surfactant and co-surfactant, respectively [26]. Graphene in nanometric form and bimetallic nanoalloys are also synthesized using green technology. Pomegranate juice (containing anthocyanins) is introduced as the reducing and capping agent to green synthesize graphene nanosheets from graphene oxide (obtained from the oxidation of graphite by the Hummer method) [27]. *O. ficus-indica* is utilized to synthesize stable metallic nanoparticles of silver, cadmium, copper, lead and titanium using their nitrates (as the precursors) under thermal treatment and magnetic stirring. The extract is also used to develop graphene sheets and carbon quantum dots and is finally characterized by transmission electron microscopy, Raman spectroscopy, and X-ray photoelectron spectroscopy [9].

In a study, silver nanoparticles are formed (extracellularly) by utilizing cell-free culture supernatants of various psychrophilic bacteria [28]. *Bacillus amyloliquefaciens* and *Bacillus subtilis* are used in the formation of biogenic spherical silver nanoparticles with an average size of less than 140 nm. Owing to its surface plasmon resonance, a peak at 418 and 414 nm is observed in the spectroscopic data. It exhibited antifungal properties and antibacterial activity against both gram-positive and gram-negative bacteria [29]. Secretion of elevated levels of proteins by fungi (when compared to bacteria) makes them a suitable candidate for the green synthesis of nanoparticles [30]. *F. oxysporum* in contact with silver nitrate releases nitrate reductase to form stable nanoparticles of silver oxide (exhibiting a distinct peak at 408 to 411 nm in UV-visible spectroscopy and size ranging from 21.3 to 37.3 nm) with antimicrobial property against a few multidrug-resistant bacteria and pathogenic yeasts [31]. A comparative evaluation of green synthesized gold nanoparticles using several microbial cell-free extracts is reported. The formation of nanoparticles is confirmed by the UV-visible spectrum with peaks at the wavelength of 538, 539, and 543 nm. These nanoparticles are crystalline in nature with either spherical or pseudo-spherical structures [32].

Enzymes with defined structure and high purity act as good catalysts in the bioreduction process. Immobilized enzymes functioning as biocatalysts witness enhanced stability (with less denaturation) at different pH, temperature and ionic strength [33]. Proper surface modification of nanoparticles (e.g., silanization, carbodiimide activation, using cross-linkers, and PEG or PVA spacing) is done to achieve enzyme immobilization with minimal fouling. An enzyme, α-NADPH-dependent sulphite reductase, obtained from *Escherichia coli* and purified by performing ion exchange chromatography, is utilized as a cell-free extract to synthesize gold nanoparticles with antifungal properties [34]. The enzymes, catalase, and glucose oxidase, are incorporated in a multi-layered poly(acrylic acid) modified polyvinylidene fluoride membrane to obtain the *in situ* synthesis of bimetallic Fe/Pd nanoparticles and develop a nanocomposite membrane for the degradation of pollutants [35]. An immunofluorescence sensor is designed for the detection of alpha-fetoprotein (a cancer biomarker) by enzymatically controlling the inhibition of the dissolution of the MnO_2 nanoflakes and the formation of fluorescent polydopamine nanoparticles as signal-generation tags. This study can be extended to detect different biomarkers of cancer by considering their corresponding antibodies [36].

Self-assembly of gold and platinum nanomaterials of varying shapes are obtained by using vitamin B_2 (riboflavin) in the absence of any other reducing, capping, or dispersing agents to adopt an eco-friendly production process [37]. Ascorbic acid as the reducing agent and sodium alginate as the stabilizing agent are used in green synthesizing spherical and cubic silver nanoparticles. The characteristics of the formed nanoparticles (reported by UV-visible spectroscopy, transmission electron microscopy and X-ray diffraction) are investigated with the change in pH, ascorbic acid concentration, and different ultrasonic treatments. These nanoparticles exhibited increased membrane permeability causing the loss of cell wall integrity and acted as antibacterial agents against *Staphylococcus aureus* and *Escherichia coli* [38]. Using a one-step synthesis process, vitamin C is used (as a reducer and stabilizer) to generate gold nanoparticles (with sizes ranging from 8 to 80 nm) and silver nanoparticles (of size 20 to 175 nm). The nanoparticles are then surface-modified by using either neutral, positively, or negatively charged water-soluble surfactants [39].

Copper and copper oxide nanomaterials have found diverse applications. In order to detect nitrite ions in any sample with good selectivity and reproducibility, an electrochemical biosensor with the modified electrode is developed by depositing a dispersion of carboxymethyl cellulose stabilized copper nanoparticles and multi-walled carbon nanotubes on glassy carbon [40]. Biogenic silver and zinc oxide nanoparticles are synthesized using *Prosophis fracta* and coffee. They are incorporated into bandages to enhance their antimicrobial property, thus aiding in wound healing [41]. Nanoparticles of cerium oxide are investigated as an anti-obesity agent to inhibit adipogenesis and prevent the build-up of triglycerides in pre-adipocytes. Results from *in vivo* experiments indicated its non-toxicity in Wistar rats while decreasing the concentration of insulin, leptin, glucose and triglycerides in their plasma [42]. Cerium oxide nanoparticles act as a catalyzer and fluorescent quencher. It has found its use in various colorimetric, electrochemical, and chemoluminescent biosensors for detecting organic molecules, proteins, and metal ions [43]. Cadmium sulfide quantum dots are green synthesized using various microorganisms such as *Escherichia coli, Fusarium oxysporum,* and *Klebsiella pneumonia* [44]. *Trichoderma harzianum* is used in synthesizing cadmium sulfide nanoparticles (having a peak at 332 nm in UV-visible spectroscopy) with excellent photocatalytic properties [45]. Silver nanoparticles are incorporated in wound dressing materials and implant coatings because of their antimicrobial and anti-inflammatory activities.

Among other major functional groups, the carboxylic acid groups of plant metabolites direct the green synthesis and stability of nanostructure. These groups attach to the nanoparticles' surface and contribute to their stabilization through electrostatic interactions. Carboxylate ions from carboxylic acids/dicarboxylic acids (e.g., formic acid, propionic acid, caprylic acid, nonanoic acid, oxalic acid, malic acid, gallic acid, acetic acid, decanoic acid, lauric acid, trifluoroacetic acid, benzoic acid, and glutaric acid) successfully synthesized nanoparticles of palladium, magnetite, gold, silver, selenium, titanium oxide, zinc oxide quantum dots, tungstite nanoplates and nanoflowers [46–50]. Carboxylic acid functionalized selenium nanoparticles formed by using acetic acid, pyruvic acid, and benzoic acid reported anticancer activity and can act as nanomedicine in treating Dalton's lymphoma cancer cells [51].

3.3 Green Synthesized Nanoparticles in Cancer Theranostics

Genetic and epigenetic changes cause the prognosis of cancer with abnormal cell proliferation and tissue growth, accounting for 13% of all deaths and this figure is expected to reach 25% by 2030 [52]. The disadvantages (e.g., cancer cells being resistant to ionizing radiation or drugs, inadequate bioavailability and nonspecific drug delivery, real-time monitoring, etc.) associated with surgery, radiotherapy, and chemotherapy in the cancer treatment limit their success rate. Nanotheranostics has emerged as a promising interdisciplinary approach to provide cutting-edge solutions to numerous heterogeneous diseases including cancers. Researchers are giving constant efforts to improve the biocompatibility and efficiency of theranostic systems. Nanoparticles with specific properties in terms of optical, electronic, magnetic, and catalytic activity are being developed to combine the concepts of therapeutics and diagnostics in the early detection of a disease or its treatment outcome. Controlled delivery of drug doses at a specific site in a time-dependent manner remains the main challenge in designing nanomedicines that specifically target cancerous tissues. This targeting or the intracellular release of therapeutic agents is greatly influenced by the knowledge of cell surface receptors. Nanocarrier and functionalized nanoparticles with modified surfaces get tagged with various bioconjugated moieties for site-specific drug delivery, thus minimizing its side effects [52].

Using the extracts of *Abutilon indicum*, *Cucurbita maxima*, *Moringa oleifera*, *Coriandrum sativum* and *Acorus calamus*, phytochemicals (such as flavonoids, tannins, etc.) act as the reducing and capping agent to yield silver nanoparticles that exhibited anticancer activity against human colon cancer cells, skin cancer cells, and human breast adenocarcinoma cell line [53–55]. Glutathione (a thiolated tripeptide), found in normal cells, is responsible for maintaining cellular redox homeostasis through its facile oxidation into glutathione disulphide. Higher levels of glutathione and glutathione disulphide are observed in cancer cells, and the balance between them is also not maintained (as cancer cells generate reactive oxygen species). Thiols and disulfides exhibited favorable interactions with silver, forming self-assembled monolayers on their surface. This indicates that glutathione complexation and simultaneous reduction to silver nanoparticles are higher in cancer cells, thus taking a step toward its theranostic applications [56]. *In vitro* experiments confirmed the intracellular synthesis of silver nanoclusters by cancer cells (with high near-infrared fluorescence emission) upon incubation with silver salt derivative due to the complex formation between glutathione and silver nitrate, thus aiding in *in vivo* cancer cell imaging with high precision. Such observation is not seen when non-cancer cells are incubated [57]. A similar study on bioimaging is performed using fluorescent gold nanoclusters on human hepato-carcinoma and leukemia cells [58].

Depending on the shape and size, gold nanoparticles possess unique optical and electrical properties and are widely used in bioimaging. (-)-Epigallocatechin-3-gallate, a polyphenol from green tea, is used in the rapid synthesis of gold nanoparticles that showed selective toxicity towards mammary adenocarcinoma by altering the redox status and Nrf2 (nuclear factor erythroid 2–related factor activation [59]. Gold nanoparticles of size under 50 nm synthesized using the extract of *Musa paradisiaca* showed a cytotoxic effect in human lung cancer cells at a low concentration of 100 μg/ml [60]. Gold nanoparticles synthesized from *Candida albicans* are conjugated with antibodies specific to liver cancer cells and exploited to distinguish between normal and cancer cells [61].

The optical and photoacoustic properties of quantum dots are often exploited in bioimaging and drug delivery systems to diagnose and treat diseases including cancer. Quantum dots attached to a particular peptide, antibody, or immunoglobulin can identify specific cancer biomarkers. The better fluorescence capability of quantum dots (in comparison to other common chromophores and contrast agents) and the photothermal characteristics of carbon nanotubes are often conjugated in the recent advancements in cancer theranostics [62]. Biomass sources stand out to be the sustainable and eco-friendly precursors in the synthesis of carbon quantum dots [63]. Multifunctional cadmium sulfide quantum dots (size ranging from 2 to 5 nm) are green synthesized using the extract of *Camellia sinensis*. It is then evaluated for its antibactericidal effect, in bioimaging and induction of apoptosis in lung cancer cells. The results portrayed high-contrast fluorescence images of cancer cells, while data from flow cytometry suggested the arrest of cancer cell growth at the S phase of the cell cycle [64]. Synthesis of carbon nanotags by the biogenic route of mass production from cyanobacteria is reported. It exhibited high solubility, improved photostability, and low cytotoxicity in normal ovary cells and kidney fibroblast-like cell lines. Conjugation of doxorubicin with these tags inhibits the growth of cancer cells (namely HepG2 and MCF-7), while its evaluation in *in vivo* mice model resulted in higher anticancer activity by the nanotags at 0.01 mg/ml [65].

Selenium nanoparticles have emerged as next-generation anticancer agents, cytotoxic in cancer cell lines but having no toxicity in noncancerous cells. It is mostly synthesized using microbes and is spherical in nature [66]. In a study, the bioconstituents of honey are used for the rapid green synthesis of selenium nanoparticles, which induces antiproliferation in human ovarian cancer cells in a concentration-dependent manner (by generating reactive oxygen species causing oxidative damage) with a low IC_{50} value of 60.95 μg/ml [67].

Nanoparticles of iron oxide exhibit good magnetic properties (superparamagnetism) which makes it useful as contrast probes in magnetic resonance imaging. Also, its biodegradability and the ability to get manipulated by magnetic fields are exploited in targeted drug delivery or in localized heat therapy [68]. Iron oxide nanoparticles are formed by reducing ferric/ferrous iron ions into crystalline nanostructures, often enclosed within a stabilizing layer. An *in vivo* study on its anti-tumor property uses magnetosomes in glioblastoma tumors. A series of alternating magnetic fields are applied to cause mild hyperthermia (about 41–45°C) that cures these tumors with no reported side effects [69]. In order to sidestep the physiological barriers, namely liver, kidneys, and spleen, the size of superparamagnetic iron oxide nanoparticles is tuned and surface modification is done for maximizing its blood half-life and easy targeting of genes, proteins, or anticancer drugs in the body [70].

3.4 Optimizing the Synthesis and Subsequent Characterizations

3.4.1 Approaches to Achieve Optimization

Optimization of the process parameters (such as temperature, time, pH, etc.) is crucial in regulating the shape and size of nanoparticles, which in turn determines their application (i.e., whether to be used in drug delivery systems, biosensing and imaging, or as excipients). The morphology of synthesized nanoparticles also depends on the nature of solvents and reducing agents used. Apart from the composition of metabolites present in the plant extract, other parameters, such as pH of the reaction environment, temperature, exposure time, and electrochemical potential of a metal ion affect the green synthesis process. pH determines the surface charge of the bioactives which in turn controls their binding and reducing capability. *Avena sativa* extract is reported to have a higher yield value of gold nanoparticles (of size ranging from 5 to 20 nm) when the reaction medium has a pH value between 3.0 and 4.0, while particle aggregation (forming nanoparticles of size between 25 and 85 nm) is seen at pH 2.0 [71]. Flat nanosized gold triangles and hexagons are synthesized using pear fruit extract at alkaline pH but acidic pH does not initiate nanoparticle formation [72].

Another important process parameter in green synthesis is temperature. Increasing temperature to a certain limit improves the rate of reaction/synthesis by elevating the nucleation rate. The physical characteristics are also affected by temperature. At higher temperatures, *Aloysia citrodora* extract produced more silver nanoparticles of crystalline nature. In synthesizing silver nanostructures using *Cassia fistula* extracts, it is observed that at room temperature silver nanoribbons are formed, while nanospheres are produced at temperatures above 60°C [11]. Altering the temperature changes the interaction of metabolites with the surface of nanomaterial.

The factors, e.g., concentration of extracts, nature of metal salt, pH, and proper surface contact determine the characteristics, stability and time required for the formation of nanoparticles. In a short duration of time, plasmonic silver nanoparticles having dye degradation properties are formed by using beet juice under microwave irradiation [73]. In the green synthesis of silver nanoparticles using the crude extract of *Azadirachta indica*, an analysis of the growth kinetics of nanoparticle formation (initially by recording the absorption maxima at a regular interval of thirty minutes till 4 hours, and subsequently studying data from X-ray diffraction, FT-IR, zeta potential, inductively coupled plasma-optical emission spectrometry, and transmission electron microscopy) revealed that stable nanoparticles are synthesized after 2 hours and tend to agglomerate after 4 hours of reaction time [74]. Optimization of the green synthesis of silver nanoparticles using aqueous extract of *Rosa damascena* petals is done to obtain biocompatible nanoparticles (checked against erythrocytes) with anticancer properties against human lung cancer cell line (A549). The parameters under consideration are the concentration of the plant extract and silver salt, the ratio of reactants in the composition, and the time allowed for their interaction. With the formation of nanoparticles, the color of the solution changes from pale yellow to brown yellow, showing a characteristic peak of silver nanoparticles at 420 nm. DLS predicted their size to be around 84.00±10.08 nm while FE-SEM and HR-TEM portrayed their structure as spherical. The peak around 3 keV in EDAX confirmed the presence of silver and its IC_{50} value of 80 µg/ml against A549 established its anticancer activity [75].

The composition of nutrient medium, temperature, pH, inoculum size, light, agitation rate, and buffer strength are reported to play a key role in the optimization of green synthesis using bacteria to achieve the desired stability, and morphological and physicochemical characteristics of biogenic nanomaterials. Silver nanoparticles are synthesized using the enzyme nitrate reductase to reduce Ag^+ while retaining the enzymatic activity. This enzymatic activity is optimized in the synthesis process and by employing the design of experiments the optimization of the fermentation medium (to produce nitrate reductase by *Bacillus licheniformis*) is done. In the optimized condition with percentages of glucose, peptone, yeast extract and

KNO_3 as 1.5, 1, 0.35% and 0.35%, respectively, the nitrate reductase activity is observed to be 452.206 U/ml which yielded nanoparticles of size 10 to 80 nm [76].

The response surface methodology (RSM) is a dynamic tool that is widely exploited in the Design of Experiment (DOE). RSM tries to establish a relationship between output variables(s) and its input(s)/process parameter(s) in order to maximize or minimize the response(s)/output(s) for a predicted set input parameter(s) [77, 78]. In green synthesizing gold nanoparticle (using extract of *Pancratium parvum*), RSM is employed to optimize several factors namely pH (varying between 7 and 11), temperature (varying between 50°C and 150°C, the concentration of chloroauric acid tetrahydrate (varying between 0.25 and 1.25 mM), and plant extract volume (varying between 0.25 and 1.25 ml). The nanoparticle formation is indicated by the change in color of the reaction medium from yellow to purple and the corresponding UV-visible spectrum (wavelength ranging from 400 to 800 nm) is recorded [79]. Optimization of biogenic spherical zinc oxide nanoparticles (synthesized using a fungi *Aspergillus niger* and showing absorption maxima at 320 nm) is done by using a three-level central composite design (a RSM model). It considers temperature and concentration of fungi as the input variables, while the responses/outputs include IC_{50} value (in breast cancer cell line MCF-7) and inhibition zone (against *E. coli*) [80].

3.4.2 Characterization of Nanoparticles

Various techniques involving spectroscopy and microscopy are employed to characterize the morphology and nature of the synthesized nanoparticles. These include X-ray diffraction (XRD), Fourier transmission infrared (FTIR) spectroscopy, dynamic light scattering (DLS), energy dispersive X-ray examination (EDAX), scanning electron microscopy (SEM), and transmission electron microscopy (TEM). XRD is a non-contact and non-destructive technique and X-ray diffractograms generated from it predict the crystallinity of nanoparticles. In order to perform elemental composition analysis of nanoparticles, EDAX is used which separates the characteristic X-rays of a particular element in an energy spectrum.

The property of surface plasmon resonance exhibited by nanoparticles is exploited in determining its formation by UV-visible spectroscopy (when scanned from 200 to 800 nm). It is a characteristic fingerprint of the nanomaterial that denotes the oscillation of free electrons in the vicinity of the nanoparticles' external surface. It is also related to the morphology and size of nanoparticles. Silver in nanometric form exhibited surface plasmon resonance to be located at several wavelengths such as 408, 430, and 440 nm, owing to the variation of its sizes [81]. Based on the morphology of gold nanoparticles, the absorption band is reported at 500, 550, and 800 nm, whereas the band for the palladium and lead nanoparticles is detected at around 300, 320, and 400 nm.

FTIR spectroscopy examines the presence of functional groups on the surface of nanoparticles and identifies the metabolites that played a key in the reduction and stabilization of nanoparticles. FTIR spectroscopic analysis of nanoparticles revealed that terpenoids are the main bioactive compounds present in the geranium leaves that reduce chloroaurate ions into their stable nano form (triangular, rod-shaped, or flat sheets). Polypeptides/enzymes present in the *Colletotrichum* sp., however, generated spherical gold nanoparticles [24]. The concentration of clove extract (with high eugenol content) is optimized to produce biogenic silver and gold nanoparticles with varying morphology by observing surface plasmon behavior in regular time interval periods. FTIR data analysis establishes the mechanism underlying nanoparticle synthesis. The inductive effect of methoxy and allyl groups of eugenols when two electrons are released from its one molecule plays a key role in the reduction process [82].

There exist several techniques to investigate the size/size distribution of nanostructures, namely DLS, TEM, thermomagnetic measurement, dark-field microscopy, atomic force microscopy (AFM), and acoustic spectrometry measurement. The morphology of the nanoparticles is studied by using SEM and TEM images. The incidence of photons in DLS tells us about the surface charge and hydrodynamic radius of nanoparticles. The hydrodynamic radius of the nanoparticle (R_H) is determined from its diffusion coefficient using the Stokes-Einstein equation:

$$D_f = k_B T / 6 \pi \eta R_H$$

where k_B is the Boltzmann constant, T and η refers to the temperature and the viscosity of the suspension respectively.

Acknowledgment

The authors are thankful to the West Bengal State Government Departmental Fellowship scheme of Jadavpur University for providing the manpower and necessary resources.

References

1. Govindaraju, T. and Ariga, K., *Molecular architectonics and nanoarchitectonics*. Springer Singapore Pte. Limited, 2021.
2. Huston, M., DeBella, M., DiBella, M., Gupta, A., Green synthesis of nanomaterials. *Nanomaterials*, 11, 8, 2130, 2021.
3. Krutyakov, Y.A., Kudrinskiy, A.A., Olenin, A.Y., Lisichkin, G.V., Synthesis and properties of silver nanoparticles: Advances and prospects. *Russ. Chem. Rev.*, 77, 3, 233, 2008.
4. Gao, F. and Gu, Z., Melting temperature of metallic nanoparticles. *Handbook of Nanoparticles*, pp. 661-690, Springer, Cham, 2016.
5. Joshi, D.R. and Adhikari, N., An overview on common organic solvents and their toxicity. *J. Pharm. Res. Int.*, 28, 3, 1–18, 2019.
6. Parashar, M., Shukla, V.K., Singh, R., Metal oxides nanoparticles via sol–gel method: A review on synthesis, characterization and applications. *J. Mater. Sci.: Mater. Electron.*, 31, 5, 3729–3749, 2020.
7. Knauth, P. and Schoonman, J. (eds.), *Nanostructured materials: Selected synthesis methods, properties and applications*, vol. 8, Springer Science & Business Media, Germany, 2006.
8. Walter, J.G., Petersen, S., Stahl, F., Scheper, T., Barcikowski, S., Laser ablation-based one-step generation and bio-functionalization of gold nanoparticles conjugated with aptamers. *J. Nanobiotechnol.*, 8, 1, 1–11, 2010.
9. Hurtado, R.B., Calderon-Ayala, G., Cortez-Valadez, M., Ramírez-Rodríguez, L.P. and Flores-Acosta, M., Green synthesis of metallic and carbon nanostructures, in: *Nanomechanics*, IntechOpen, UK, 2017.
10. Mohanpuria, P., Rana, N.K., Yadav, S.K., Biosynthesis of nanoparticles: Technological concepts and future applications. *J. Nanopart. Res.*, 10, 3, 507–517, 2008.
11. Makarov, V.V. *et al.*, 'Green' nanotechnologies: Synthesis of metal nanoparticles using plants. *Acta Naturae (англоязычная версия)*, 6, 1 (20), 35–44, 2014.
12. Pooja, D., Panyaram, S., Kulhari, H., Reddy, B., Rachamalla, S.S., Sistla, R., Natural polysaccharide functionalized gold nanoparticles as biocompatible drug delivery carrier. *Int. J. Biol. Macromol.*, 80, 48–56, 2015.
13. Parveen, K., Banse, V., Ledwani, L., Green synthesis of nanoparticles: Their advantages and disadvantages. *AIP Conf. Proc.*, 1724, 1, 020048, 2016.
14. Yadav, N., Khatak, S., Sara, U.S., Solid lipid nanoparticles-a review. *Int. J. Appl. Pharm.*, 5, 2, 8–18, 2013.
15. Anastas, P.T. and Warner, J.C., Principles of green chemistry, in: *Green Chemistry: Theory and Practice*, vol. 29, 1998.
16. Pytlik, N. and Brunner, E., Diatoms as potential 'green' nanocomposite and nanoparticle synthesizers: Challenges, prospects, and future materials applications. *MRS Commun.*, 8, 2, 322–331, 2018.

17. Sastry, M., Ahmad, A., Khan, M.I., Kumar, R., Microbial nanoparticle production, in: *Nanobiotechnology: Concepts, Applications and Perspectives*, pp. 126–135, 2004.
18. Mondal, S. *et al.*, Biogenic synthesis of Ag, Au and bimetallic Au/Ag alloy nanoparticles using aqueous extract of mahogany (Swietenia mahogani JACQ) leaves. *Colloids Surf. B: Biointerfaces*, 82, 2, 497–504, 2011.
19. Ahmad, N. *et al.*, Rapid synthesis of silver nanoparticles using dried medicinal plant of basil. *Colloids Surf. B: Biointerfaces*, 81, 1, 81–86, 2010.
20. Kasthuri, J., Veerapandian, S., Rajendiran, N., Biological synthesis of silver and gold nanoparticles using apiin as reducing agent. *Colloids Surf. B: Biointerfaces*, 68, 55–60, 2009.
21. Mandal, S., Selvakannan, P.R., Phadtare, S., Pasricha, R., Sastry, M., Synthesis of a stable gold hydrosol by the reduction of chloroaurate ions by the amino acid, aspartic acid. *J. Chem. Sci.*, 114, 5, 513–520, 2002.
22. Tan, Y.N., Lee, J.Y., Wang, D.I., Uncovering the design rules for peptide synthesis of metal nanoparticles. *J. Am. Chem. Soc.*, 132, 16, 5677–5686, 2010.
23. Shankar, S.S., Rai, A., Ahmad, A., Sastry, M., Rapid synthesis of Au, Ag, and bimetallic Au core–Ag shell nanoparticles using Neem (Azadirachta indica) leaf broth. *J. Colloid Interface Sci.*, 275, 2, 496–502, 2004.
24. Shankar, S.S., Ahmad, A., Pasricha, R., Sastry, M., Bioreduction of chloroaurate ions by geranium leaves and its endophytic fungus yields gold nanoparticles of different shapes. *J. Mater. Chem.*, 13, 7, 1822–1826, 2003.
25. Balasooriya, E.R., Jayasinghe, C.D., Jayawardena, U.A., Ruwanthika, R.W.D., Mendis de Silva, R. and Udagama, P.V., Honey mediated green synthesis of nanoparticles: New era of safe nanotechnology. *J. Nanomater.*, Article ID 5919836, 10 pages, 2017.
26. Rivera-Rangel, R.D., González-Muñoz, M.P., Avila-Rodriguez, M., Razo-Lazcano, T.A., Solans, C., Green synthesis of silver nanoparticles in oil-in-water microemulsion and nano-emulsion using geranium leaf aqueous extract as a reducing agent. *Colloids Surf. A: Physicochem. Eng. Asp.*, 536, 60–67, 2018.
27. Tavakoli, F., Salavati-Niasari, M., Mohandes, F., Green synthesis and characterization of graphene nanosheets. *Mater. Res. Bull.*, 63, 51–57, 2015.
28. Shivaji, S., Madhu, S., Singh, S., Extracellular synthesis of antibacterial silver nanoparticles using psychrophilic bacteria. *Process Biochem.*, 46, 9, 1800–1807, 2011.
29. Ghiuță, I. *et al.*, Characterization and antimicrobial activity of silver nanoparticles, biosynthesized using Bacillus species. *Appl. Surf. Sci.*, 438, 66–73, 2018.
30. Thakkar, K.N., Mhatre, S.S., Parikh, R.Y., Biological synthesis of metallic nanoparticles. *Nanomed.: Nanotechnol. Biol. Med.*, 6, 2, 257–262, 2010.
31. Ahmed, A.-A., Hamzah, H., Maaroof, M., Analyzing formation of silver nanoparticles from the filamentous fungus Fusarium oxysporum and their antimicrobial activity. *Turk. J. Biol.*, 42, 1, 54–62, 2018.

32. Shen, W. et al., Comparison of gold nanoparticles biosynthesized by cell-free extracts of Labrys, Trichosporon montevideense, and Aspergillus. *Environ. Sci. Pollut. Res.*, 25, 14, 13626–13632, 2018.
33. Johnson, P.A., Park, H.J. and Driscoll, A.J., Enzyme nanoparticle fabrication: Magnetic nanoparticle synthesis and enzyme immobilization, in: *Enzyme Stabilization and Immobilization*, pp. 183-191, Humana Press, Totowa, NJ, 2011.
34. Gholami-Shabani, M. et al., Enzymatic synthesis of gold nanoparticles using sulfite reductase purified from Escherichia coli: A green eco-friendly approach. *Process Biochem.*, 50, 7, 1076–1085, 2015.
35. Smuleac, V., Varma, R., Baruwati, B., Sikdar, S., Bhattacharyya, D., Nanostructured membranes for enzyme catalysis and green synthesis of nanoparticles. *ChemSusChem*, 4, 12, 1773–1777, 2011.
36. Lin, Z., Li, M., Lv, S., Zhang, K., Lu, M., Tang, D., In situ synthesis of fluorescent polydopamine nanoparticles coupled with enzyme-controlled dissolution of MnO2 nanoflakes for a sensitive immunoassay of cancer biomarkers. *J. Mater. Chem. B*, 5, 43, 8506–8513, 2017.
37. Nadagouda, M.N. and Varma, R.S., Green and controlled synthesis of gold and platinum nanomaterials using vitamin B2: Density-assisted self-assembly of nanospheres, wires and rods. *Green Chem.*, 8, 6, 516–518, 2006.
38. Shao, Y. et al., Green synthesis of sodium alginate-silver nanoparticles and their antibacterial activity. *Int. J. Biol. Macromol.*, 111, 1281–1292, 2018.
39. Malassis, L., Dreyfus, R., Murphy, R.J., Hough, L.A., Donnio, B., Murray, C.B., One-step green synthesis of gold and silver nanoparticles with ascorbic acid and their versatile surface post-functionalization. *RSC Adv.*, 6, 39, 33092–33100, 2016.
40. Manoj, D., Saravanan, R., Santhanalakshmi, J., Agarwal, S., Gupta, V.K., Boukherroub, R., Towards green synthesis of monodisperse Cu nanoparticles: An efficient and high sensitive electrochemical nitrite sensor. *Sens. Actuators B: Chem.*, 266, 873–882, 2018.
41. Khatami, M., Varma, R.S., Zafarnia, N., Yaghoobi, H., Sarani, M., Kumar, V.G., Applications of green synthesized Ag, ZnO and Ag/ZnO nanoparticles for making clinical antimicrobial wound-healing bandages. *Sustain. Chem. Pharm.*, 10, 9–15, 2018.
42. Rocca, A. et al., Pilot in vivo investigation of cerium oxide nanoparticles as a novel anti-obesity pharmaceutical formulation. *Nanomed.: Nanotechnol. Biol. Med.*, 11, 7, 1725–1734, 2015.
43. Charbgoo, F., Ramezani, M., Darroudi, M., Bio-sensing applications of cerium oxide nanoparticles: Advantages and disadvantages. *Biosens. Bioelectron.*, 96, 33–43, 2017.
44. Abd Elsalam, S.S., Taha, R.H., Tawfeik, A.M., El-Monem, A., Mohamed, O., Mahmoud, H.A., Antimicrobial activity of bio and chemical synthesized cadmium sulfide nanoparticles. *Egypt. J. Hosp. Med.*, 70, 9, 1494–1507, 2018.

45. Bhadwal, A.S., Tripathi, R.M., Gupta, R.K., Kumar, N., Singh, R.P., Shrivastav, A., Biogenic synthesis and photocatalytic activity of CdS nanoparticles. *RSC Adv.*, 4, 19, 9484–9490, 2014.
46. Yoosaf, K., Ipe, B.I., Suresh, C.H., Thomas, K.G., In situ synthesis of metal nanoparticles and selective naked-eye detection of lead ions from aqueous media. *J. Phys. Chem. C*, 111, 34, 12839–12847, 2007.
47. Dwivedi, C., Shah, C.P., Singh, K., Kumar, M. and Bajaj, P.N., An organic acid-induced synthesis and characterization of selenium nanoparticles. *J. Nanotechnol.*, Article ID 651971, 6 pages, 2011.
48. Hosseini-Monfared, H., Parchegani, F., Alavi, S., Carboxylic acid effects on the size and catalytic activity of magnetite nanoparticles. *J. Colloid Interface Sci.*, 437, 1–9, 2015.
49. Amornkitbamrung, L., Pienpinijtham, P., Thammacharoen, C., Ekgasit, S., Palladium nanoparticles synthesized by reducing species generated during a successive acidic/alkaline treatment of sucrose. *Spectrochim. Acta Part A: Mol. Biomol. Spectrosc.*, 122, 186–192, 2014.
50. Li, L. et al., Oxalic acid mediated synthesis of WO3 H2O nanoplates and self-assembled nanoflowers under mild conditions. *J. Solid State Chem.*, 184, 7, 1661–1665, 2011.
51. Kumar, S., Tomar, M.S., Acharya, A., Carboxylic group-induced synthesis and characterization of selenium nanoparticles and its anti-tumor potential on Dalton's lymphoma cells. *Colloids Surf. B: Biointerfaces*, 126, 546–552, 2015.
52. Rajasekharreddy, P., Huang, C., Busi, S., Rajkumari, J., Tai, M.-H., Liu, G., Green synthesized nanomaterials as theranostic platforms for cancer treatment: Principles, challenges and the road ahead. *Curr. Med. Chem.*, 26, 8, 1311–1327, 2019.
53. Mata, R., Nakkala, J.R., Sadras, S.R., Biogenic silver nanoparticles from Abutilon indicum: Their antioxidant, antibacterial and cytotoxic effects *in vitro*. *Colloids Surf. B: Biointerfaces*, 128, 276–286, 2015.
54. Nayak, D., Pradhan, S., Ashe, S., Rauta, P.R., Nayak, B., Biologically synthesised silver nanoparticles from three diverse family of plant extracts and their anticancer activity against epidermoid A431 carcinoma. *J. Colloid Interface Sci.*, 457, 329–338, 2015.
55. Sathishkumar, P. et al., Anti-acne, anti-dandruff and anti-breast cancer efficacy of green synthesised silver nanoparticles using Coriandrum sativum leaf extract. *J. Photochem. Photobiol. B: Biol.*, 163, 69–76, 2016.
56. Piao, M.J. et al., Silver nanoparticles induce oxidative cell damage in human liver cells through inhibition of reduced glutathione and induction of mitochondria-involved apoptosis. *Toxicol. Lett.*, 201, 1, 92–100, 2011.
57. Gao, S. et al., Near-infrared fluorescence imaging of cancer cells and tumors through specific biosynthesis of silver nanoclusters. *Sci. Rep.*, 4, 1, 1–6, 2014.
58. Wang, J. et al., In vivo self-bio-imaging of tumors through *in situ* biosynthesized fluorescent gold nanoclusters. *Sci. Rep.*, 3, 1, 1–6, 2013.

59. Mukherjee, S. *et al.*, Gold-conjugated green tea nanoparticles for enhanced anti-tumor activities and hepatoprotection—Synthesis, characterization and *in vitro* evaluation. *J. Nutr. Biochem.*, 26, 11, 1283–1297, 2015.
60. Vijayakumar, S. *et al.*, Therapeutic effects of gold nanoparticles synthesized using Musa paradisiaca peel extract against multiple antibiotic resistant Enterococcus faecalis biofilms and human lung cancer cells (A549). *Microb. Pathogen.*, 102, 173–183, 2017.
61. Chauhan, A. *et al.*, Fungus-mediated biological synthesis of gold nanoparticles: Potential in detection of liver cancer. *Int. J. Nanomed.*, 6, 2305, 2011.
62. Tan, A., Yildirimer, L., Rajadas, J., De La Peña, H., Pastorin, G., Seifalian, A., Quantum dots and carbon nanotubes in oncology: A review on emerging theranostic applications in nanomedicine. *Nanomedicine*, 6, 6, 1101–1114, 2011.
63. Malavika, J.P., Shobana, C., Sundarraj, S., Ganeshbabu, M., Kumar, P., Selvan, R.K., Green synthesis of multifunctional carbon quantum dots: An approach in cancer theranostics. *Biomater. Adv.*, 136, 212756, May 2022.
64. Shivaji, K. *et al.*, Green-synthesis-derived CdS quantum dots using tea leaf extract: Antimicrobial, bioimaging, and therapeutic applications in lung cancer cells. *ACS Appl. Nano Mater.*, 1, 4, 1683–1693, 2018.
65. Lee, H.U. *et al.*, Photoluminescent carbon nanotags from harmful cyanobacteria for drug delivery and imaging in cancer cells. *Sci. Rep.*, 4, 1, 1–7, 2014.
66. Vahidi, H., Barabadi, H., Saravanan, M., Emerging selenium nanoparticles to combat cancer: A systematic review. *J. Cluster Sci.*, 31, 2, 301–309, 2020.
67. Amiri, H., Hashemy, S.I., Sabouri, Z., Javid, H., Darroudi, M., Green synthesized selenium nanoparticles for ovarian cancer cell apoptosis. *Res. Chem. Intermed.*, 47, 6, 2539–2556, 2021.
68. Ma, X., Zhao, Y., Liang, X.-J., Theranostic nanoparticles engineered for clinic and pharmaceutics. *Acc. Chem. Res.*, 44, 10, 1114–1122, 2011.
69. Alphandéry, E., Bio-synthesized iron oxide nanoparticles for cancer treatment. *Int. J. Pharm.*, 586, 119472, Aug. 2020.
70. Kievit, F.M. and Zhang, M., Surface engineering of iron oxide nanoparticles for targeted cancer therapy. *Acc. Chem. Res.*, 44, 10, 853–862, 2011.
71. Armendariz, V., Herrera, I., Jose-Yacaman, M., Troiani, H., Santiago, P., Gardea-Torresdey, J.L., Size controlled gold nanoparticle formation by Avena sativa biomass: Use of plants in nanobiotechnology. *J. Nanopart. Res.*, 6, 4, 377–382, 2004.
72. Ghodake, G.S., Deshpande, N.G., Lee, Y.P., Jin, E.S., Pear fruit extract-assisted room-temperature biosynthesis of gold nanoplates. *Colloids Surf. B: Biointerfaces*, 75, 2, 584–589, 2010.
73. Kou, J. and Varma, R.S., Beet juice-induced green fabrication of plasmonic AgCl/Ag nanoparticles. *ChemSusChem*, 5, 12, 2435–2441, 2012.

74. Prathna, T.C., Chandrasekaran, N., Raichur, A.M., Mukherjee, A., Kinetic evolution studies of silver nanoparticles in a bio-based green synthesis process. *Colloids Surf. A: Physicochem. Eng. Asp.*, 377, 1–3, 212–216, 2011.
75. Venkatesan, B., Subramanian, V., Tumala, A., Vellaichamy, E., Rapid synthesis of biocompatible silver nanoparticles using aqueous extract of Rosa damascena petals and evaluation of their anticancer activity. *Asian Pac. J. Trop. Med.*, 7, S294–S300, 2014.
76. Vaidyanathan, R., Gopalram, S., Kalishwaralal, K., Deepak, V., Pandian, S.R.K., Gurunathan, S., Enhanced silver nanoparticle synthesis by optimization of nitrate reductase activity. *Colloids Surf. B: Biointerfaces*, 75, 1, 335–341, 2010.
77. Yazdi, M.S. and Khorram, A., Modeling and optimization of milling processby using RSM and ANN methods. *Int. J. Eng. Technol.*, 2, 5, 474, 2010.
78. Bezerra, M.A., Santelli, R.E., Oliveira, E.P., Villar, L.S., Escaleira, L.A., Response surface methodology (RSM) as a tool for optimization in analytical chemistry. *Talanta*, 76, 5, 965–977, 2008.
79. Patil, D.N. et al., Response surface methodology-based optimization of Pancratium parvum Dalzell-mediated synthesis of gold nanoparticles with potential biomedical applications. *Int. Nano Lett.*, 11, 3, 215–232, 2021.
80. Es-Haghi, A. et al., Application of response surface methodology for optimizing the therapeutic activity of ZnO nanoparticles biosynthesized from Aspergillus niger. *Biomimetics*, 6, 2, 34, 2021.
81. Bindhu, M.R. and Umadevi, M., Antibacterial and catalytic activities of green synthesized silver nanoparticles. *Spectrochim. Acta Part A: Mol. Biomol. Spectrosc.*, 135, 373–378, 2015.
82. Singh, A.K., Talat, M., Singh, D.P., Srivastava, O.N., Biosynthesis of gold and silver nanoparticles by natural precursor clove and their functionalization with amine group. *J. Nanopart. Res.*, 12, 5, 1667–1675, 2010.

4

Sustainability: An Emerging Design Criterion in Nanoparticles Synthesis and Applications

Yashtika Raj Singh[1], Abhyavartin Selvam[1], P.E. Lokhande[2] and Sandip Chakrabarti[1]*

[1]*Amity Institute of Nanotechnology, Amity University Uttar Pradesh, Noida, UP, India*
[2]*Department of Mechanical, Manufacturing and Biomedical Engineering, Trinity College Dublin, Dublin, Ireland*

Abstract

The field of nanotechnology has emerged as an exciting field for developing and modifying novel nanostructured materials for a wide array of application possibilities. Nanomaterials are primarily synthesized using physical and chemical methods, using harsh chemicals under hazardous conditions, like chemical precipitation, photochemical, sol-gel, hydrothermal, solvothermal, physical vapor deposition, chemical vapor deposition, and ball milling, rendering them hazardous to human health and the environment. The solution to this problem can be explored in biological methods that incorporate green nanotechnology, which amalgamates green chemistry and engineering principles to devise harmless and eco-friendly nanomaterials. The process of producing stable nanomaterials from enzymes, metabolites, or biomolecules contained in plants and microbes, is known as biosynthesis. The biosynthesis strategy employs biotemplates, namely leaves, fruits, seeds, and flowers, that are rich in alkaloids, flavonoids, saponins, steroids, phenols, and microorganisms. Likewise, biogenic preparation of nanomaterials has been cited as an efficient, low-cost, environmentally friendly, and sustainable technology. Moreover, biosynthesis has gained considerable popularity over the last few decades, with several advantages, notably, easy availability, excellent production rate, and nanoparticles possessing excellent properties with small size, high surface-area-to-volume ratio, non-pathogenic, and safer, which

---

*Corresponding author: schakrabarti@amity.edu

Mousumi Sen and Monalisa Mukherjee (eds.) Bioinspired and Green Synthesis of Nanostructures: A Sustainable Approach, (65–114) © 2023 Scrivener Publishing LLC

make it practically sustainable. Furthermore, synergism between environmental and economic sustainability has been elaborately discussed for the practical implementation of biogenic routes in the synthesis and application of nanomaterials. Moreover, elucidation on strategies to improve synthesis routes by justifying the effects of various parameters and implementation in several applications has been conceptualized in this chapter. This chapter evaluates the emerging nanomaterials possessing copious applications on the accounts of their nature and biological compatibility, high synthesis rate, stability, selectivity, sensitivity, and so on. Similarly, the exploration of the practicality of biogenically developed nanomaterials for many applications has been recently ameliorated. Likewise, inorganic and organic nanomaterials of various classifications are conveniently produced to sustain in different avenues of applications. Biogenic routes play a vital influence in the enhancement of productivity, long-term sustainability, and low cost, enabling cheap products to be produced. Thence, biomedical nanostructures incorporating metal and metal oxide nanoparticles have recently been found to be effective in the biomedical field and diagnostic application. Further, biosynthesized nanoparticles are also strongly focused on the avenue of environmental remediation, using sustainable routes via nanotechnology, focusing on decontaminating soils, sediments, solid wastes, air, water, the textile industry, along with food industries. This chapter recounts sustainable approaches to effectively engineer nanomaterials biogenically to be applied in demanding situations and applications.

Keywords: Biosynthesis, sustainability, plants, microorganisms, applications

4.1 Introduction

A nanotechnology study involves devious engineering at the molecular level across multiple disciplines [1]. In recent years, nanotechnology has ventured away from the confines of the laboratory and has been able to conquer new applications to help us live a better life [2]. These nanoparticles (NPs) possess diverse chemical, optical, mechanical, and magnetic properties as compared to bulk materials because of the increase in the surface area [2]. An influential tenet of nanotechnology is the fabrication of nanoscale materials, as well as their controlled morphology and dimensions. In nanoscale materials, at least one dimension is smaller than 100 nm and consisting a wide array of geometric forms including plates, sheets, tubes, wires, and particles [3, 4]. In addition, they are utilized for storing, producing, and converting energy, ameliorating agricultural productivity, removal of water pollutants, diagnosing diseases, developing drugs, and storing food, they are also used to control pests and mitigate air pollution [5–7]. By converging with various branches of science and impacting all forms of life, this genesis of nanoscience has produced promising results in recent years [8]. Nanostructures can be achieved in two basic ways: top-down and bottom-up approaches [9]. A top-down approach entails reducing bulk material into fine particles by deploying lithographic techniques, for instance, grinding, milling, sputtering, and laser/thermal ablation. The bottom-up method implicates amalgamating smaller entities to form NPs, such as atoms, molecules, and smaller particles, likewise biological and chemical processes are mainly used in bottom-up synthesis. As a result, it gives a greater opportunity to prepare metallic NPs with fewer flaws and greater homogeneous chemical compositions [10, 11]. There are several methods for preparing NPs, including physical, chemical, and biological ones [12]. Physical methods include; lithography, ball milling, physical vapor deposition, pyrolysis, crushing, attrition, and grinding whereas the chemical approach encompasses chemical vapor deposition, microwave-assisted synthesis, sol-gel method, solvothermal method, thermal decomposition, hydrothermal, microwave-assisted synthesis, ultra-sonic assisted reduction through electro-chemical, photocatalysis, and gas-phase inter alia [13]. A wide range of physical and chemical methods are available to synthesize NPs with attractive shapes. Biological syntheses are preferred for the well-built, large-scale synthesis of NPs as they are safe, clean, cheap, and easily scaled up [12]. Physical and chemical processes are primarily used to synthesize nanomaterials, which entail harsh chemicals under perilous conditions that include chemical precipitation, solvothermal methods, electrochemical methods, sol-gel methods, photochemical methods, hydrothermal methods, physical vapor deposition, and sol-gel methods as well as they are exceedingly expensive [13]. During synthesis, these methods are exhibited by even size dispensation, homogeneity, and low energy consumption. The production of these products is, however, energy-intensive, expensive, and yield-deficient [14]. As a result, harmful NPs are produced and environmental pollution occurs due to laborious synthesis procedures and chemical use that results in cytotoxic, genotoxic, and carcinogenic chemicals [15, 16]. Chemical nanomaterials are toxic and unstable, which greatly limits their applications due to their infeasibility and economic viability. In response, green nanotechnology has become a viable and environmentally friendly synthetic strategy. Moreover, biological synthesis eliminates harsh processing conditions by entitling synthesis at physiological pH levels, temperatures, and pressures while simultaneously reducing the cost. Inorganic NP composites are synthesized intracellularly or extracellularly by several microorganisms that aid in alternative methods for generating metal as well as metal oxide NPs including plant extracts, bacteria, fungi, and algae. As a result of developing alternative strategies to reduce bacteria's growth and building their resistance to various antibiotics, nanotechnology has been developed as a new antimicrobial agent [17, 18]. Besides, natural extracts via plants or microbes contribute to reducing, stabilizing, growth terminating, and capping agents as well as swaying the nanomaterials' size and shape [19]. Several methods have been developed recently to enhance the productivity of NPs that vary in size, shape, and stability [18]. Biosynthesis relies on biotemplates leaves, fruits, seeds, and flowers which embrace alkaloids, flavonoids, saponins, steroids, tannins, phenols, and microorganisms [20]. Additionally, biogenic nanomaterials preparation has been touted as a low-cost, environmentally friendly, and sustainable process. Moreover, numerous advantages of biosynthesis have contributed to the popularity of the technology throughout the past few decades, for instance, ease of availability, magnificent yields, and the ability to produce NPs with a range of excellent properties, such as safety, small size, high surface-to-volume ratio, and non-pathogenicity [21]. Moreover, nanomaterials possessing a wide range of applications in this chapter are appraised on the basis of their nature, biological compatibility, high synthesis rate, stability, selectivity, sensitivity, and so forth. In the same vein, recent advancements have been made in exploring the usefulness of biogenically developed nanomaterials for biomedical applications. Likewise, nanomaterials of various classifications, inorganic and organic, can also be readily produced for use in different fields. As a result, biogenic routes improve productivity, long-term sustainability, and low cost, making it possible to manufacture cheap products. In addition to immunotherapy, tissue regeneration, diagnostics, wound healing, dentistry, biosensing platforms, bioimaging, biosensing, tumor-targeting diagnostics, drug delivery, gene delivery, cancer diagnosis, antimicrobial therapy, and stimulus-responsive platforms, biomedical nanostructures with metal oxide nanoparticles have recently proven effective in immunotherapy, tissue regeneration, diagnostics, wound healing, and stimulus-responsive platforms. Nanotechnology is also being used to remediate the environment using biosynthesized nanoparticles via innovative and sustainable routes. This includes tackling soil contamination, sediment, solid wastes, air, water, the textile industry, and food processing. Nanomaterials can be more effectively engineered biogenically to meet the demands of demanding situations and applications in this chapter.

4.2 Biotemplates

Biotemplates have become increasingly enriched over the past few decades, and bionanomaterials have exhibited a variety of functions. Numerous researchers have turned to nature as an inspiration for greener inorganic materials synthesis, utilizing milder conditions to synthesize materials [22–24]. Further, it is possible to produce nanomaterials with minimal heat and solvent use via utilizing templates that exist naturally or are inspired by nature [25, 26]. It is a natural, eco-friendly, inexpensive template, which is capable of binding metal ions, has a high production yield, is stable, and is good for the environment [27]. In Figure 4.1, the green synthesis of NPs is represented schematically.

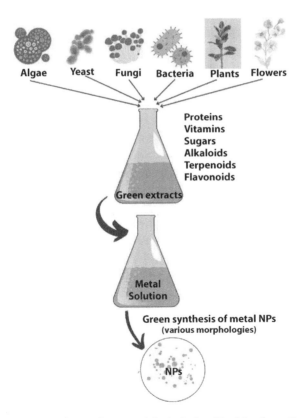

Figure 4.1 The green synthesis of nanoparticles is depicted in this schematic.

4.2.1 Plant-Based Biotemplates

It has recently been discovered that plant-based biotemplate have provided new avenues for the synthesis of nanoparticles, which is an eco-friendly, rapid, stable, and cost-effective way to make nanoparticles. Numerous advantages are associated with using water as a reducing medium, including biocompatibility of plant extracts, sustainability, scalability, and medical applicability [28]. In addition to their reducing capabilities, plant extracts also possess antioxidants that protect nanomaterials from oxidation and act as stabilizers [29]. In light of this, plant-derived nanoparticles can accordingly satisfy the high demand for nanoparticles in biomedical and environmental fields due to readily available plant ingredients and nontoxicity. Figure 4.2 illustrates plant- or extract-based biosynthesis has several applications. Via synthesizing gold and silver nanoparticles from Panax ginseng leaf and root extracts, the medicinal herb is proving to be a valuable source of resources [30–32]. Additionally, metal nanoparticles can be synthesized from a variety of plant parts, such as fruits, leaves, stems, and roots [33]. There is substantial evidence that amino acids, proteins, vitamins, organic acids, and secondary metabolites namely flavonoids, terpenoids, alkaloids, polyphenols, heterocyclic compounds, polyphenols, and polysaccharides reduce metal salts as well as serve as capping and stabilization agents for synthesized nanoparticles. For instance, a study by El-Kassas insinuated that AuNPs can be formed and stabilized by the hydroxyl functional groups in polyphenols and carbonyl functional groups in proteins in Corallina Officinalis extract which are considered excellent anti-cancer as well as anti-tumor agents with great stability [34]. Using Okra plant extract, NiO is synthesized via Sabouri which exhibits antibacterial properties against S. aureus, P. aeruginosa, and E. coli [35]. According to some reports, diverse plant species produce NPs using different mechanisms [36]. For example, certain compounds, namely emodin, a purgative resin with quinone compounds exhibited in xerophytes plants that acculturate to endure in deserts as well as it synthesizes AgNPs; dietchequinone, cyperoquinone, and remirin in mesophytic plants considered to be effective for metal NPs synthesis. Furthermore, AuNPs have been synthesized using eugenol, the main terpenoid derived from Cinnamomum zeylanisum which showed excellent storage stability and thermal stability [37]. In Table 4.1, NPs synthesized by various plants and their parts exhibit a wide range of applications.

Table 4.1 Using plant extracts for the synthesis of biogenic NPs.

Nanoparticle	Plant	Plant part	pH	Time (h)	Temperature (°C)	Applications	Reference
Ag	peganum harmala	Leaf extract	< 4	12	60	Antibacterial, antifungal, pharmaceutical	[38]
Ag	Parsley (*Petroselinum crispum*)	Leaf extract	-	24	Room temperature	Antibacterial	[39]
Ag	*Ficus hispida* Linn. f.	Leaf extract	9	1	90	Antioxidant and antibacterial	[40]
Ag	*Phyllanthus pinnatus*	Stem extract	N/A	20 min	Room temperature	Antibacterial	[41]
Ag	*Salvadora persica*	Root extract	8–10	24	70	Biomedical, water purification, air-filtering	[42]
Ag	*Alpinia nigra*	Fruit extract	N/A	2	Room temperature	Antibacterial, and catalytic agent	[43]

(*Continued*)

Table 4.1 Using plant extracts for the synthesis of biogenic NPs. (*Continued*)

Nanoparticle	Plant	Plant part	pH	Time (h)	Temperature (°C)	Applications	Reference
Ag	*Citrullus lanatus*	Fruit rind extract	10	10 min	80	Nanomedicine, sensing	[44]
Ag	*Catharanthus roseus*	Bark extract	4–5	N/A	N/A	Water remediation	[45]
Ag	*Cannabis sativa*	Stem fiber	4–10	3	550	Medical applications	[46]
Ag	*Phaseolus vulgaris*	Seed	10	30 min	Room temperature in a dark room	Antibacterial	[47]
Au	*Gnidia glauca*	Flower extract	7	20 min	40	Biomedical diagnostics	[48]
Au	*Plumeria alba*	Flower extract	N/A	2 min (change in color)	Room temperature	Biomedical, pharmaceutical, catalysis, cosmetics and water treatment	[49]

(*Continued*)

Table 4.1 Using plant extracts for the synthesis of biogenic NPs. (Continued)

Nanoparticle	Plant	Plant part	pH	Time (h)	Temperature (°C)	Applications	Reference
Au	Moringa oleifera	Petal extract	7	N/A	Room temperature	Cancer and catalytic	[50]
Au	Garcinia mangostana	Fruit peel extract	N/A	N/A 30min (For extract)	40 60 (for extract)	Biomedical	[51]
Au	Elettaria cardamomum	Seed extract	N/A	N/A	Room temperature	Biomedical, biochemical sensors, catalysis	[52]
Au	Lantana camara	Berry extract	N/A	72	Room temperature	Biomedical	[53]
Au	Dracocephalum kotschyi	Leaf extract	N/A	10min	Room temperature	Antibacterial, antioxidant, catalytic and cytotoxic	[54]
ZnO	Trianthema portulacastrum	Plant extract	N/A	10 min	Room temperature	Antibacterial and catalytic	[55]

(Continued)

Table 4.1 Using plant extracts for the synthesis of biogenic NPs. (*Continued*)

Nanoparticle	Plant	Plant part	pH	Time (h)	Temperature (°C)	Applications	Reference
Au	*Artemisia vulgaris* L.	Leaf extract	N/A	24	Room temperature	Against dengue fever	[56]
Au	*Panax ginseng*	Leaf extract	N/A	45min	80	Anticancer	[57]
ZnO	*Punica granatum*	Peel extract	N/A	1	25	Anticancerr and antibacterial	[58]
Au	*Annona muricata*	Fruit pulp extract	6.81	N/A	100	Agriculture	[59]

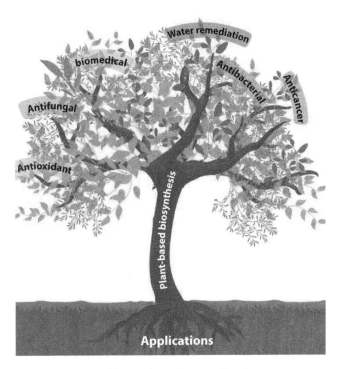

Figure 4.2 Plant or extract-based biosynthesis has several applications.

4.2.2 Microorganism-Based Biotemplates

4.2.2.1 Bacteria

Bacteria have been extensively researched for the synthesis of NPs due to their rapid growth, ease of purification, optimum conditions, and relatively easier genetic manipulation [60]. Due to their ability to produce a wide range of inorganic substances either intracellularly or extracellularly, bacteria are efficient biofactories for the synthesis of metallic nanoparticles, such as silver and gold. Biocidal properties make silver prominent. However, few bacteria are resistant to it and can accrue silver on their cell walls of approx. 25% of their dry weight biomass intimated the possibility that they could employ silver for recovery in the industrial sector [61]. The first instance of bacteria synthesizing silver nanoparticles was manifested using Pseudomonas stutzeri AG259 from a silver mine of size ranging from 35 to 46 nm [62]. The presence of higher concentrations of metal ions can induce toxicity in these organisms despite their ability to grow in lower concentrations [63]. Recent

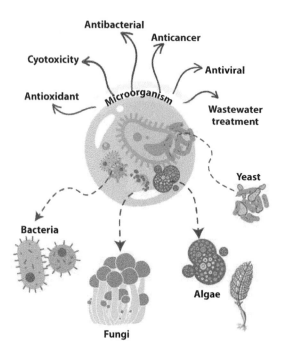

Figure 4.3 Microorganism-based biosynthesis has several applications.

research has used Bacillus thuringiensis to synthesize Ag-NPs with size ranges from 43.52 to 142.97 nm. Moreover, Ag-NPs were synthesized using *Bacillus licheniformis, Morganella psychrotolerans,* and *Klebsiella pneumonia* bacteria showed high sustainability as well as high stability at room temperature [64]. In contrast, *Bacillus subtilis* and Lactobacillus species synthesized Ti_2O that exhibit to stabilize as well as inhibit the accumulation of the particles [65]. Moreover, in order to syntto nanoparticles, *Rhodopseudomonas capsulata, Pseudomonas aeruginosa, Escherichia coli* DH5α, Bacillus subtilis, and Bacillus licheniformis were employed [66], whereas in the past, cadmium NPs were synthesized using *Clostridium thermoaceticum, Rhodopseudomonas palustris,* and *E. coli* thus, both show sustainability when applied in bioremediation applications [67]. In addition, there is the possibility of using bacteria as biocatalysts to synthesize inorganic materials via serving as a bioscaffold for mineralization or participating in the synthesis of NPs. Furthermore, a major role played by the PH in controlling the morphology of bacteria-causing nanoparticles and their location was to control where they

Table 4.2 Biogenic nanoparticles synthesized from bacteria.

Nanoparticle	Synthesized by	Size (nm)	Locations	Applications	Reference
Ag	*Pseudomonas* sp.	20–70	Intracellular	Antibacterial activity	[64]
Ag	*Bacillus thuringiensis*	43.5–142.9	Extracellular	Larvicidal activity against *Culex quinquefasciatus* and *Aedes aegypti*	[60]
Ag	*Bacillus licheniformis*	40	Extracellular	-	[70]
Ag	*Ochrobactrum anhtropi*	38–85	Intracellular	Antibacterial activity	[71]
Ag	*Bacillus* spp.	77–92	Extracellular	Antimicrobial activity, antiviral activity	[72]
Ag	*Pantoea ananatis*	8.06–91.31	Extracellular	Antibacterial against multidrug-resistant	[73]
Ag	*Bacillus mojavensis* BTCB15	105	Extracellular	Antibacterial activity against multidrug resistant	[74]
Cu	*Shewanella loihica*	10–16	Extracellular	Antibacterial	[75]
Au	*Micrococcus yunnanensis*	53	-	Antibacterial and Anticancer	[76]

(*Continued*)

Table 4.2 Biogenic nanoparticles synthesized from bacteria. (*Continued*)

Nanoparticle	Synthesized by	Size (nm)	Locations	Applications	Reference
Au	*Mycobacterium sp.*	5-55	Extracellular	Anticancer	[77]
TiO$_2$	*Aeromonas hydrophila*	28-54	Extracellular	Antibacterial	[78]
ZnO	*Halomonas elongata* IBRC-M 10,214	18.1	Extracellular	Antibacterial	[79]
ZnO	*Sphingobacterium thalpophilum*	40	-	Antibacterial	[80]

accumulated [68]. Nanomaterials can be synthesized either intracellularly or extracellularly by bacteria after an incubation period in broth media. As a consequence, the biosynthesis of NPs using bacteria is a flexible, reasonable, and suitable method for producing NPs on a large scale [69]. NPs synthesized by various bacterial species have a wide range of applications as shown in Table 4.2.

4.2.2.2 Fungi

It has been found that fungi have a higher tolerance for environmental conditions, a greater bioaccumulation property, self-sustaining, ease of cultivation, and a higher metabolite efficiency [81, 82]. In addition, fungi are capable of secreting proteins or extracellular enzymes [83]. Since fungal metabolites can produce different NPs at high efficiency, they have been widely used for NPs biosynthesis [84–86]. Among the microorganisms used in NP fabrications, fungi are an excellent current addition [87]. There have been several fungi used to produce nanoparticles, for instance, *Fusarium oxysporum* can synthesize AgNPs with diameters ranging from 5 to 15 nm, and the particles have been coated with mycological proteins which make them sustainable to use in antibacterial applications. In studies, cadmium, molybdenum, zinc, and cadmium sulphide NPs were synthesized in the cytoplasm and intracellular production of these particles has been reported [88]. A variety of intracellular enzymes make them superior abiotic factors for generating metal NPs and metal oxide NPs. Competent fungi may produce more nanoparticles than bacteria [89]. The most likely technique for creating metal nanoparticles is enzymatic reduction (reductase) inside infectious cells or in cell walls. Additionally, infectious enzymes enhance the production of stable NP and accelerate the reductive process [90]. As Pt NPs synthesized extracellularly at room temperature with *Fusarium oxysporum* extract, extracellularly synthesized nanoparticles are often scrutinized less or non-toxic as well as help to stabilize Pt [91]. Several fungal species synthesize different types of nanoparticles as shown in Table 4.3.

4.2.2.3 Yeast

Akin to fungi, yeast is proficient in handling high levels of metal ions, allowing high levels of metal nanoparticle deposits [103]. In addition, yeasts have been becoming increasingly considerable in nanotechnology by their excellent properties for producing NPs, due to the ease of governing under laboratory conditions, growth at unusual temperatures, pHs,

Table 4.3 Biogenic nanoparticles synthesized from fungi.

Nanoparticles	Organism	Locations	Applications	Reference
Ag	Candida glabrata	Extracellular	Antibacterial	[92]
Ag	Trichoderma longibrachiatum	Extracellular	Antibacterial	[93]
Ag	Fusarium oxysporum	Extracellular	Antibacterial	[94]
Ag	Ganoderma sessiliforme	Extracellular	Antibacterial, antioxidant, and anticancer	[95]
Ag	Aspergillus sp.	Extracellular	Antibacterial and cytotoxicity	[96]
Au	Rhizopus oryzae	Cell surface	Antibacterial and Wastewater treatment	[97]
Au	Aureobasidium pullulans	Intracellular	Antibacterial	[98]
Au	Cladosporium cladosporioides	Extracellular	Antibacterial and Antioxidant	[99]
ZnO	Aspergillus terreus	Extracellular	Antibacterial and cytotoxic	[100]
ZnO	Aspergillus fumigatus	Extracellular	Antibacterial	[101]
Al_2O_3	Colletotrichum sp	Extracellular	Antibacterial	[102]

Table 4.4 Biogenic nanoparticles synthesized from yeast.

Nanoparticles	Yeast	Location	Application	Reference
Ag	*Yarrowia lipolytica*	Intracellular	Antibacterial	[107]
Sb_2O_3	*Saccharomyces cerevisiae*	N/A	Antibacterial	[108]
ZnO	*Pichia kudriavzevii*	Extracellular	Antioxidant and Antibacterial	[109]
Se	*Nematospora coryli*	Intracellular	Antioxidant, and anti-candida	[110]
Au	Baker's yeast	Extracellular	Anticancer	[111]
Au	*Candida albicans*	Cell-free extract	Antibacterial	[112]
Fe_3O_4	Yeast cell	Extracellular	Biomedical	[113]
Ni/NiO	*Rhodotorula mucilaginosa*	Extracellular	Biomedical and magnetic resonance imaging	[114]

and nutrients, rapid growth, enzyme production, easy scaling up, cost-effectiveness, easy handling, and easy processing [104–106]. Research on yeast-based biosynthesis of metal nanoparticles has been conducted in several studies (Table 4.4). Scientists have demonstrated that yeast can synthesize Ag nanoparticles using melanin, a brown pigment found in yeast *Yarrowia lipolytica* that exhibits antibacterial properties against pathogenic bacteria while being resistant to radiation, high temperatures, free radical attacks, and metal poisoning [107]. In this study, *Saccharomyces cerevisiae* was used to synthesize Sb_2O_3 NPs are considered to be eco-friendly and show a reproducible approach [108]. A novel method is described for the synthesis of ZnO-NPs extracellularly through the use of *Pichia kudriavzevii* that showed a variety of antioxidant and antibacterial properties exhibited via the prepared NPs [109]. Additionally, *Nematospora coryli* was considered for its ability to produce SeNPs which convert the toxic oxyanion of Se into vital SeNPs for human health. Moreover, the synthesized SeNPs exhibited anti-candida and anti-oxidant activities [110]. Magnetic NPs of Ni/NiO synthesized via *Rhodotorula mucilaginous* yeast are taken up by the dead organic matrix. A natural strategy to synthesize nano-metals from yeast dead organic matrix can result in magnetic nanostructured Ni/NiO nanoparticles in films that may be very promising for the sustainable synthesis of nano-metals in the future [114].

4.2.2.4 Algae

Algae can be unicellular or multicellular organisms, possessing a nuclear envelope, and membranous organelles can inhibit in all types of environments where light and moisture are available, and can be classified as microalgae or macroalgae. As photosynthetic organisms, they live in freshwater, marine water, and even on moist rocks as well as on snow also [115, 116]. Several advantages can be attributed to algae, such as low-temperature NP synthesis, low toxicity, and ease to handle, besides that they are capable of accumulating metal and reducing metal ions [117]. Algae can synthesize metallic nanoparticles through the integration of functional groups and enzymes found in their cell walls; even edible algae can be used. Using a highly efficient method to synthesize AuNPs from Cystoseria baccatta extract (brown-algae), which exhibits excellent biocompatibility, strong cytotoxic activity, and great stability of AuNPs [118]. Researchers found that brown algae extracts (Turbinaria conoides and Sargassum tenerrimum) are capable of reducing Au^{+3} ions to AuNPs and stabilizing them. The catalytic potential of Au synthesized from *Turbinaria conoides* was significantly higher than

Table 4.5 Biogenic nanoparticles synthesized from algae.

NPs	Algae	Location	Synthesis condition	Application	Reference
Au	*Cystoseira baccata*	Extracellular	24 h stirring at room temperature	Biomedical	[118]
Au	*Turbinaria conoides*	Extracellular	300s at room temperature, pH 7	Biomedical	[119]
Au	*Sargassum tenerrimum*	Extracellular	300s at room temperature, pH 7	Biomedical	[119]
Ag	*Cystophora moniliformis*	Extracellular	30 min incubation at 65 °C	Water purification	[120]
Ag and ZnO	*Gracilaria edulis*	Extracellular	Centrifugation at 10,000 rpm	Anticancer	[121]
Ag	*Gelidium amansii*	Extracellular	48 h incubation at room temperature and centrifugation at 13,000 rpm	Antibacterial agent, and microfouling bacteria	[122]

that of *Sargassum tenerrimum* [119]. Scientists synthesized AgNPs using extracts of the Australasian brown alga *Cystophora moniliformis*, which showed great sustainability in water purification [120]. AgNPs synthesized with red marine algae *Gelidium amansii* manifest antibacterial effects—especially against microfouling bacteria [122]. Various algae species synthesize different types of NPs as shown in Table 4.5. Figure 4.3 illustrates how microorganism-based biosynthesis has multiple applications in different fields.

4.3 Synthesis Routes

As NPs are biosynthesized, characterized, and morphologically shaped, several parameters are involved like effects of pH, temperature, reaction time, concentration, biomolecules and so on [123]. A key challenge in the biosynthesis of NPs is controlling particle shape as well as size, and achieving monodispersed in solution. Researchers have shown that control of the growth environment, as well as changing the function of the molecules, can be used to control the morphology and size of biologically synthesized nanoparticles [124].

4.3.1 Effect of pH

Earlier studies have demonstrated that a significant influence of pH is observed amid the green synthesis procedure of preparing MNPs. Several studies have manifested that varying the pH of a solution medium enables NPs to be synthesized in various shapes and sizes. As an example, at lower acidic pH, larger particles can be produced than at higher acidic/basic pH. This was proved by the researcher, in order to synthesize Au nano-rods, researchers used biomass from *Avena sativa* (Oat). Consequently, nano-rods with sizes between 25 and 85 nm at a pH of 2, and those with sizes between 5 and 20 nm at a pH of 3 to 4. This is because at pH 2, there were few functional groups, which result in larger Au particles aggregating than in pH 3–4 where more functional groups are available for smaller Au NPs to form [123]. In contrast, Pd NPs synthesized from *Cinnamon zeylanicum* bark extract have the opposite effect, pH less than 5, Pd NPs were obtained at 15 to 20 nm, whereas at pH greater than 5, the Pd NPs were obtained at 20 to 25 nm [124].

Further, researchers demonstrated that AgNPs at stronger acidic pH, show lower zeta potential (-26 mV) in comparison with alkaline pH which exhibits greater stability and smaller size at basic pH. According to the results of their study, the absorbance peaks in SPR spectra at different pH levels corroborate this conclusion [125]. Moreover, Andreescu also observed a negative Zeta potential of silver nanoparticles synthesized at different pH levels after rapid and entire reduction of silver at elevated pH. In this study, the researchers found that higher pH results in higher negative zeta potential values, which result in highly dispersed NPs [126]. On the contrary, as evidenced by Dwivedi's study using Chenopodium album, they observed that NPs of silver and gold are stable over a wide pH range [127]. Moreover, the formation of hexagonal and triangular gold nanoplates at alkaline pH values has been demonstrated using extracts from pears albeit NPs are not formed at acidic pH [128].

Furthermore, the fungi *Fusarium oxysporum* and *Chrysosporium tropicum* were used to synthesize Ag, resulting in larger particles and a decrease in pH in conjunction with increased absorption. The equation represents a relationship between pH and absorbance (A) [129].

$$A \alpha\ 1/pH$$

4.3.2 Effect of Temperature

Temperature is another considerable parameter that influences the synthesis of NPs vis-a-vis physical, chemical, and biological methods. For the physical synthesis of NPs, the temperature should be higher than 350°C, while for the chemical synthesis, the temperature should be lower than 350°C, and for the biological synthesis, the temperature should not exceed 100°C or ambient temperature [129, 130]. The nature of the NP formed depends on the temperature of the reaction medium [129]. Cruz *et al.* found that AgNPs were synthesized from lemon verbena extracts (*Aloysia citrodora*) at higher temperatures, demonstrating that silver ion reduction efficiency increases with reaction temperature [131]. Moreover, a higher temperature is also conducive to the formation of crystal particles more frequently than at room temperature. The nucleation rate is believed to increase with increasing temperature [132]. It was found that temperature can influence the structure of the synthesized nanoparticles from *Cassia fistula* (golden shower tree) extracts; at room temperature, silver nanoribbons predominate, while spherical nanoparticles are most common at temperatures over 60°C. Consequently, the higher temperature reduces the ability of nearby NPs to be incorporated into nanoribbon structures as a result of phytochemical interactions with the NP surfaces [133]. Gericke *et al.* demonstrated that in the biosynthesis of Au NPs by fungal cultures, at lower temperatures, spherical morphology was observed, while at a higher temperature, it formed as stick and plate shapes [134].

4.3.3 Effect of Biomolecules

A wide variety of biological resources have been studied for NP synthesis in intracellular and extracellular environments, including plant extracts and microorganism-based biomolecules. There is ample evidence that biosynthesis can be cheaper and more environmentally friendly than the conventional physical or chemical synthesis of NPs. Sustainably nourished microorganisms and plants (or their extracts) are employed as bio-reductants for capping and stabilizing agents, which eliminates the need for additional agents [135].

4.3.3.1 Plant-Based

The significant plant biomolecules that serve as reducing as well as stabilizing agents and hinder the conglomeration and agglomeration of the new metallic NPs by unperilous means are secondary metabolites, which are found in plant extract, including terpenoids, flavonoids, protein, polysaccharides, tannins, alkaloids, steroids, phenolic acids, saponins, and other plant parts originated nutritional content like bark, leaves, fruits, stems, roots, seeds [136–138]. In terms of both reductive and protective effects, polyphenols have the greatest effect [139] due to their ability to protect plants from reactive oxygen species (ROS) produced amid photosynthesis and impurities derived from human activities [140]. A significant factor in the preparation of NPs, especially AuNPs, is the concentration of polyphenols with the carbonyl groups and phenolic hydroxyl competing for oxygen, causing polyphenol to oxidize, thereby affecting NP size and morphology [139]. Phytochemicals and plant polysaccharides used for reducing and stabilizing AuNPs and AgNPs in the green synthesis have been discovered admirably by Park *et al*. In addition, he demonstrated the oxidative reduction of metal salts into NPs is mediated by the polysaccharide hydroxyl group in the carbonyl group [140]. Likewise, saponins contained in Memecylon edule leaf extract were used in the reduction of AgNPs and AuNPs [141]. Similarly, Philip and Armol demonstrated the ability of fenugreek seed extract to execute both the function of reducing and stabilization of AuNPs. In addition, it was also discovered that flavonoids present in seed extracts may contribute to $HAuCl_4$ reduction by acting as strong reducing agents. On the other hand, proteins can serve as surfactants by electrostatically stabilizing AuNPs [142]. Song *et al*. demonstrated the synthesis of AuNPs from terpenoids present in Magnolia Kobus was responsible for the stabilization of AuNPs [143].

4.3.3.2 Microorganism-Based

Biological interactions between inorganic molecules are increasingly attracting scientists' attention. Researchers have discovered that intracellular or extracellular production of inorganic NPs has been demonstrated by various microbes [144]. In biosynthesis, NPs are formed by using enzymes generated via cellular activity, microorganisms synthesize NPs by grabbing target ions from their environment [145]. Various prokaryotic and eukaryotic microorganisms are used to synthesize metallic NPs [146]. Microorganisms encompassing bacteria, yeast, fungi, and algae are capable of synthesizing NPs either extracellularly or intracellularly depending on the location where they are formed [145]. In intracellular methodologies, NP recovery requires multiple centrifugation and washing steps to remove the cell wall, and sonication for breaking down the wall for NP removal. In contrast, extracellular biosynthesis of NPs is an extremely efficient method, which eliminates downstream processing steps amid NP purification [147].

When extracellular synthesis is performed, microorganisms that have been cultivated in shakers under optimal growth conditions, centrifugation is used to harvest biomass. As a result of centrifugal separation, the supernatant is incubated with a metal salt solution to synthesize interested metallic NPs. The first qualitative evidence of NP synthesis is a color change in the reaction mixture. For instance: Au NPs color changes from ruby red to deep purple [147]. Detecting nanoparticles in solution uses the Tyndall effect, an effect that colloidal particles exhibit in solution [148]. An ethanol/methanol solution is used to wash the NPs and bottom pellets are collected following the reaction [147].

When intercellular synthesis is performed, a harvest of microorganism biomass occurs after optimal conditions for growth have been achieved. Incubation with metal ion solutions occurs after thorough washing of the biomass via sterile water. As nanoparticles are synthesized, their color changes. In the following steps, ultrasonication, washing, and centrifugation are applied to the biomass, amid which the cell wall is broken down and releases NPs. A final step involves washing, centrifuging, and collecting the NPs [147].

Using an intracellular method, enzymes are used to form NPs from ions in the microbe's cell whereas in extracellular metal ions are trapped on the surface of the cells in the presence of enzyme and reduced into NPs [149].

4.4 Applications

In various sectors, biosynthesized NPs are proving to be very beneficial. A wide variety of NP-based products are manufactured these days, including desalination, inks, sun filters, and blemish clothing, agribusiness and pharmaceuticals, textile, as well as in wound treatments [150]. The characteristics of nanoparticles have attracted a lot of interest in biomedical research. Moreover, there are some biological applications of nanoparticles, including cancer treatment, drug delivery, antibacterial therapy, theranostic therapy, implants, wound healing, and many more [151]. Moreover, it showed interest in many other fields, namely agriculture, wastewater treatment, environmental remediation, and many more.

4.4.1 Biomedical Application

In the biomedical sector, metallic NPs are used widely, enabling constant development. Nanomedicine is one of the most rapidly growing fields of study due to its numerous opportunities to improve the diagnosis as well as treatment of human illnesses [152]. The use of NPs in biomedicine has been extensively reviewed and reported in numerous journals [153, 154]. Many biomedical applications are possible with NPs, which exhibit a number of pharmacological activities [155]. Antibacterial, antifungal, antioxidative, and cytotoxic activities are most commonly observed in Ag NPs and Au NPs [156]. These biological activities have also been observed with other NPs, but in a much lower percentage [156]. It has been found that biosynthesized NPs can be applied to a wide variety of applications, including biosensors, bioimaging, anti-diabetics, anticancer, and many others. In the following, paragraphs detailed descriptions of all these biological activities will be mentioned.

4.4.1.1 Antimicrobial Activity

New drug-resistant microbes pose the greatest challenge to medical practitioners. For the treatment of different diseases, it is essential to develop new drugs [150]. Public health concerns are growing due to the emergence of multidrug-resistant bacteria [151]. Bacteria are becoming increasingly resistant to advanced antimicrobial drugs, and they are a bit expensive, making it desirable for scientists to find more cost-effective, efficient, sustainable, and widely applicable antimicrobial drugs [157]. In order to combat resistant pathogens, new antimicrobial compounds or modifications of existing compounds are promptly requisite [158]. Ag NPs have the capability of overcoming antibiotic resistance; they have been extensively used as antimicrobials as well as have the capability to perform expeditiously against gram-negative and gram-positive bacteria. Several reports have shown that Ag NP binds and penetrates the bacterial cell wall, causing serious damage to cell functions and resulting in cell death [159–161]. Ag NPs have antimicrobial properties influenced via various parameters, including size, colloidal state, shape, concentration, and surface charge [162]. A study by Sondi et al. demonstrated that Ag NP attaches to *Escherichia coli*'s cell wall and creates holes in its membrane, eventually resulting in cell death [163]. Using leaf extract of Parkia speciosa to reduce $AgNO_3$, to obtain spherical Ag NPs of an average size of 31nm as well as it shows antibacterial activity that is highly effective against *S. aureus*, *P. aeruginosa*, *E. coli*, and *B. subtilis* [164]. Likewise, spherical and highly stable Ag NPs were also synthesized via latex extracted from immature Papaya carica. Consequently, compared to Gram-negative bacteria, namely *V. cholerae*, *P. mirabilis*, *E. coli*, and *K pneumonia*, Gram-positive bacteria *E. faecalis* and *B. subtilis* showed a smaller reduction [165]. The antibacterial properties of NPs are complex and encompass a variety of mechanisms. Due to this lack of knowledge, the mechanism underlying antibacterial activity remains obscure. It has been demonstrated that NPs are antibacterial at low concentrations. Researchers have found that ZnO nanopowders are less effective as antimicrobials and have a lower inhibitive and microbial concentration than zinc acetate [166, 167]. A study evaluating the antimicrobial efficacy of Cu, Zn, and Fe_2O_3 nanopowders demonstrated that ZnO nanopowders are incompatible with Gram-positive and Gram-negative microbes [167]. In agriculture and the food industry, NPs are found to interact negatively with bacteria, subsequently resulting in antimicrobial applications [168]. As an alternative to some traditional antibiotics, bactericidal nanoparticles can address the issue of antibiotic resistance strains, that emanated from the transmission of antibiotic resistance genes across bacteria [169, 170]. In comparison to chemical synthesis, green synthesis of ZnO nanoparticles promises to be a more sustainable strategy. Consequently, the use of Trifolium pratense flower water extract demonstrates a green synthesis of ZnO NPs. This results in a stronger inhibitory effect and was found to be effective against *P. aeruginosa* [171]. Moreover, orange juice was used to synthesize ZnO NPs, which prevented the spread of pathogenic microbes at low concentrations [172]. It is highly favorable to employ NPs as antimicrobial agents since they have multiple mechanisms of action, withal most of their major lethal pathways occurring simultaneously. The preponderance of metal-oxide NPs manifests bactericidal properties by producing ROS or emancipating ions. Similarly, photocatalytic toxicity was shown to cause lipid peroxidation under UV-lamp, leading to respiratory dysfunction and death in *E. coli* [173] as well as synergistic antibacterial effects, when amalgamated with Ag nanoparticles [174]. Moreover, NPs alter the potential and integrity of the cell membrane [175], as well as amend the uptake of metal ions into the cell, which leads to an increase in adenosine triphosphate (ATP) production and a decrease in DNA replication [176].

4.4.1.2 Biomedication

It has been found that biosynthesized Ag NPs are interested in several health areas, namely bio-labeling, antimicrobial agents, and the pharmaceutical industry [177]. The advent of antibiotic resistance has led to the development of new therapeutic options, which include nanoparticles added to composite fibers to enhance anti-inflammatory properties. Further, Ag NPs exhibit a wide spectrum of activity, not just effective against infections, but also in treating tumours, particularly those with multidrug resistance [178]. It is interesting to note that drug delivery is one of the most promising sectors of biomedical applications of nanoparticles, which endeavors to deliver drugs safely to specific locations. Additionally, NPs have shown extensive use in medical imaging, for instance, Ag NPs synthesized by the fungus *Trichoderma viride*, demonstrated photoluminescence measurements when subjected to laser excitation, enabling their use as labels and imaging agents [179]. As a result, Ag NPs are becoming progressively popular in modern medicine for dressing various kinds of wounds [180].

4.4.1.3 Vaccines

It is possible for inactivated pathogen vaccine frequently results in a weak immune response in contrast to conventional live-attenuated vaccines, which are prone to reversion to pathogenic virulence. Further, a new approach to the vaccine based on NPs has shown great promise for overcoming conventional vaccine limitations. By virtue of advances in chemical and biological engineering, nanoparticles can now be more precisely controlled in size, shape, functionality, and surface properties, resulting in improved antigen presentation and great immunogenicity. The synthesis of eco-friendly silver nitrate nanoparticles (SNM) from Streptomyces sp Al-Dhabi-91 is applicable in nanomedicine for the killing of disease-causing bacteria. The marine actinobacterium Streptomyces strain Al-Dhabi-91 can synthesize SNM in an eco-friendly way, making it a promising source for the development of new pharmaceuticals [181].

4.4.1.4 Antidiabetic

A metabolic disorder, diabetes (a group of diseases) results in elevated blood sugar levels due to insufficient insulin production, and the cells are unable to respond to insulin. So, the severity of diabetic complications can vary depending on whether it is insulin-dependent or not [182]. It has now become widely acknowledged that green-synthesized silver nanoparticles are more effective than conventional insulin for the treatment of diabetes since they suppress the reaction of enzymes at the secretary level. Various complex diseases are caused by several enzymes, comprising α-glucosidase and α-amylase. The following examples illustrate this point [182]. Researchers synthesized AgNPs from the aqueous leaf extract of Pouteria sapota using the green route. A substantial reduction in blood sugar level was shown in rats when they were treated with leaf extract (100 mg/kg) of AgNPs (10 mg/kg) [183]. Likewise, with *Momordica charantia* extract, AgNPs were synthesized via green synthetic methods. Further, synthesized AgNPs are considered effective against several enzymes for Diabetes that were inhibited by it, including α-amylase (an enzyme associated with diabetes). According to the current study, researchers demonstrated that Ag NPs can inhibit the diabetic enzyme amylase using a green method of synthesis [184]. Furthermore, researchers pointed out that the anti-diabetic properties of the green-synthesized AgNPs could have applications in diabetes treatment [183].

4.4.1.5 Diagnostic Applications

A novel technology called advanced bioimaging allows researchers to visualize cellular compartments and manipulate functional changes within cells in order to diagnose diseases with greater accuracy. In order to obtain more accurate images of biological sy stems, nanotechnology plays a crucial role. In the first instance, scientists showed that biosynthesized silver nanoparticles containing Olax scandens leaf extract (b-AgNPs) can be used in bioimaging applications [185]. Further, when the AgNPs are incubated with B16F10 cells, they manifested a red color fluorescence that can be utilized in bioimaging processes. Intriguingly, these AgNPs internalize into cancer cells and emit a red fluorescent color; normal cells do not emit the same color of fluorescence. Consequently, it has been suggested that plant-based fluorescent molecules could be used in biomedical applications owing to the bioimaging properties of b-AgNPs [186].

Biosensors are self-reliant analytical devices with the capability of converting biological signals into analytical signals that can be measured and processed [187]. Sensors based on biological molecules are composed of enzymes, nucleic acids, cells, or tissues that interact with analytes, while a physical probe converts the signals from the analyte-based receptors into output signals and then is displayed by the device [188]. Consequently, they are categorized according to their transducers or biorecognition elements [189]. Due to their ease, speed, affordability, sensitivity, and high selectivity, they are commonly used in biomedical diagnostics [190]. Due to the remarkable properties of nanomaterials, it serves as transducers in biosensor development, resulting in improved analytical signals and increased sensitivities and selectivity [191]. Sustainability and eco-friendliness of biosensors can be achieved by using green synthesized nanomaterials. Nanomaterials, for instance, AuNPs and AgNPs have been found to be antagonists to a wide range of biomolecules present in physiological fluids like blood, saliva, and urine, including proteins, DNA, enzymes, amino acids, lipids, and ions [189]. A viable solution to this problem depends on the capping agents, layer thickness, and surface chemistry used during the fabrication of nano-biosensors [189]. In nanobiosensors, as an analyte interacts with the capping agents and surface molecules, nanobiosensors modulate their signals by their interactions across the capping layer based on the distance. Consequently, nanomaterials' low toxicity, biocompatibility, and exceptional properties have made them perfect for biosensing [192]. For instance, Bollella *et al.* synthesized Ag NPs via quercetin as a reducing agent to amend graphite electrodes, which were then used to fabricate a lactose biosensor that demonstrated good linearity, enhanced sensitivity based on low detection limits, excellent stability, and quick response times. Similarly, the detection of toxic mercury in liquids has been produced using Ag NPs synthesized from onion peel [193]. Zamarchi and Vieira *et al.* demonstrated that an electrochemical sensor for paracetamol determination was made via Ag NPs synthesized using pine nut extract (*Araucaria angustifolia*) as a reducing and stabilizing agent, and its performance was shown by a good linear range and a detection limit of 8.50×10^{-8} MM, as well as great stability [194].

4.4.2 Environmental Application

Environmental cleanup using NPs has shown significant promise in the treatment of soils, sediments, surface waters, wastewater, and groundwater contaminated having heavy metals, organic and inorganic solutes [195]. Research is focused on improving water quality; nanotechnology plays a key role in water purification [196]. Due to their high durability and xenobiotic nature, dyes like azo dyes, acid dyes, cationic dyes, and others, have received increased attention in bioprocessing [197]. As a result of these pollutants being disposed of in water bodies, this will lead to a greater level of pollution in rivers, lakes, and other water bodies, which affect aquatic life such as rivers, lakes, and streams, aquatic life becomes contaminated and altered [197]. These pollutants can be treated greenly with NPs [198]. There have been several reports evaluating the catalytic qualities of some nanomaterials to reduce the dangers of environmental materials through biotreatment techniques in combination with nanotechnology [198, 199]. Moreover, in an effort to minimize the effect of organophosphorus pesticides, AuNPs are synthesized via the *Rhizopus oryzae* fungus, which adsorbs these pesticides [200].

4.4.2.1 Environmental Remediation

It has been demonstrated environmental contamination can be effectively remedied by nanoparticles that include heavy metal contamination in soil, organic and inorganic solute contamination in surface water, and contaminated soil contamination in wastewater [201, 202]. The quest for high water quality is a foremost concern for researchers; the biosynthesis of NPs and their sustainability has enhanced purification processes through biosynthesized NPs [203]. A promising application of nanotechnology in bioremediation has been demonstrated; by using fungi dead biomass to clean waste while generating NPs simultaneously [204]. The biosynthesis of metal NPs has primarily excellent environmental repercussions, like the removal of pollutants like metal ions, dyes, and biological substances [205]. As a major environmental concern, pesticide contamination is becoming increasingly common, before pesticides may pose health risks and enter human as well as animal food chains, there is serious concern over the extent of pesticide contamination in water [206], and biosynthesized NPs are being used to minimize the impact of pesticides on aquatic and terrestrial ecosystems. Consequently, the use of AuNPs produced by Rhizopus oryzae is capable of adsorbing organophosphorus pesticides and is more sustainable [207].

4.4.2.2 Catalytic Removal of Textile Dyes

This topic will focus on recent advances in the biosynthesis of NPs with applications that are used to degrade textile pollutants, and dyes that can be harmful to human health and the environment [208]. Ni was synthesized using leaf extracts from the *Ocimum sanctum* by Pandian *et al.* The green method was used to characterize the physicochemical properties of Ni NPs, which were then implemented in the removal of industrial dyes, including eosin Y, crystal violet, sulfate, nitrate anionic pollutants, and orange II azo [209]. The jackfruit leaf extract (Artocarpus heterophyllus) was used by Vidya et al. to synthesize ZnO NPs. A rich source of steroids, flavonoids, terpenoids, phenols, and carbohydrates is present in jackfruit leaves, which are used in the oxidative combustion of $Zn(NO_3)_2 \cdot 6H_2O$. Rose Bengal dye was photocatalytic degraded using biosynthesized NPs, which discolored 85% in 100 minutes of reaction [210]. Salvia hydrangea extract was used to synthesize Pd NPs was researched by Khodadadi et al. At room temperature, four textile dyes were applied to the fabrics, Congo Red (CR), Methylene Blue (MB), Methyl Orange (MO), and Rhodamine B (RB). As a result, it appears that Pd NPs/apricot kernel peel requires more reduction and it took about 16 seconds for the dyes investigated to discolor completely [211].

4.4.2.3 Wastewater Treatment

Recently, contaminated water sources have been a major concern [212]. Wastewater treatment can be part of the solution to this ardent issue [213]. Due to limitations in time and toxic contaminants, traditional methods are unable to remove pollutants completely from water [213]. Polluted water can be treated effectively using NPs [214]. For the degradation of dyes, Fe and Fe/Pd bimetallic NPs were produced using a green method. Researchers studied the effects of grape leaf extract on the reduction of orange II dye [215], and green tea extract on the reduction of malachite green dye [216]. Using the extract of eucalyptus leaf to prepare Fe NPs which showed promise in treating swine wastewater. As a result, these particles effectively removed ammonia nitrogen, chemical oxygen demand, and total phosphates from wastewater [217]. Das et al. synthesized CuO NPs by using extract of the *Madhuca longifolia* plant which possesses splendid photocatalytic properties that degrades methylene blue dye and is demonstrated as an excellent photocatalyst for the treatment of wastewater [214]. The potential for NPs to treat wastewater alternatively and environmentally friendly is, therefore, astonishing [218].

4.4.2.4 Agriculture

It has been reported that NPs can control the uptake of nutrients, pest infestations, and herbicides, resulting in plant growth, productivity, and development [219]. There have been numerous harmful effects on ecosystems and human health caused by the uncontrolled use of pesticides [220]. The application of NPs in agriculture is depicted in Figure 4.4. Moreover, the

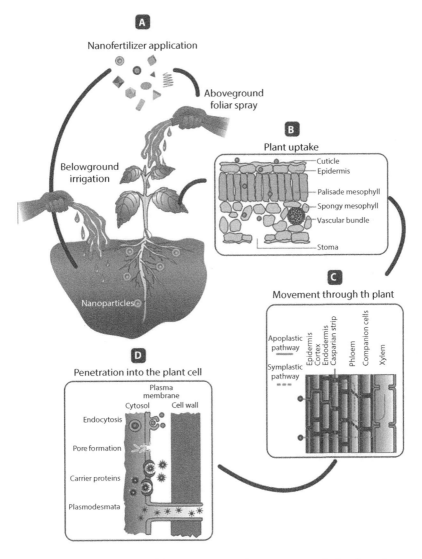

Figure 4.4 Mechanisms involved during nanoparticle interaction with the plant. A, nanofertilizer application aboveground or belowground. B, nanoparticle uptake by the plant. C, nanoparticle transportation pathways in the plant. D, internalization of nanoparticles within the plant cell (modified from Pérez-de-Luque 2017)

use of NPs in sustainable agriculture has been applied to monitoring water quality, pesticide use, as well as herbicide use [221]. In this study, Cornus mas fruit extract is used to fabricate Fe_2O_3 NPs in light of green nanotechnology, environmental pollution, and sustainable agriculture [222]. It was also examined whether the application of bulk form Fe2O3 altered the growth of barley (*Hordeum vulgare*). When bulk Fe_2O_3 was applied to shoots, the shoot biomass increased by 32.6%, while the synthesized nano-type increased it by 68%, suggesting that nano-Fe_2O_3 is more efficient at improving plant growth than bulk Fe_2O_3 [222]. Using aqueous root extract from *Sphagneticola trilobata* Lin et al., ZnO NPs were synthesized to remove Cr(VI) in an aqueous solution, and the NPs were found to have a higher removal efficiency of Cr(VI) metal was found to exceed 80% within 8 hours of interaction [223]. In addition, plant growth and seed germination were greatly enhanced by the synthesized ZnO NPs [224].

4.5 Conclusion and Outlook

A comprehensive study of current developments in the synthesis of NPs and their applications has been presented in this chapter. NPs synthesized by plants and microorganisms are more prominently featured in this chapter. NPs are synthesized using physical and chemical methods, those methods require toxic chemicals, which are harmful to the environment, while biosynthesis produces NPs in a defined size and shape. In the biosynthesis of NPs, controlling particle shape, as well as size and ensuring monodispersed in solution are key challenges.

A variety of methods have been explored for controlling the morphology and size of biologically synthesized NPs by interfering with the growth environment and changing the function of molecules. Controlling biomolecules, pH, and temperature can help overcome these challenges. Moreover, biosynthesis routes offer other benefits to humans, such as, reducing the production and consumption of hazardous wastes and materials, encouraging the use of organic solvent-free media, as well as a one-step, eco-friendly, and low-cost method for conducting synthesis at ambient pressure and temperature. There is substantial evidence that secondary metabolites namely amino acids, proteins, vitamins, organic acids, and secondary metabolites namely flavonoids, terpenoids, alkaloids, polyphenols, heterocyclic compounds, polyphenols, and polysaccharides reduce metal salts as well as serve as capping and stabilization agents for synthesized NPs. For NP synthesis, microbes offer the advantages of reducing downstream processing and high metal tolerance, and easy nanomaterials harvesting, unlike intracellular synthesis. Laboratory conditions also make it smooth to control microbial growth. Despite this, it requires monotonous maintenance via culturing, microbial isolation, and incubation to carry out a microorganism-based synthesis.

A variety of applications demonstrate the biocompatibility, sustainability, and renewability of biosynthesized NPs that have made significant progress. The production of metal and metal oxide NPs using green technology can be applied to a variety of applications, like anticancer, antimicrobial, environmental, wastewater treatment, biomedical, diagnostic, and many more.

A focus of future research must be on identifying effective methods for eliminating these NP residues without compromising the delivery of adequate doses of medication. It is estimated that the research and use of nanomaterials will continue to grow presently. Extensive research needs to be conducted, especially to find eco-friendly, efficient synthesis methods. However, certain areas require further research and development, for biogenic NPs to fully maximize their benefits:

1. NMs are still relatively new in fields of research, so the potentiality of biopolymers, microbes, or waste materials to synthesize biogenic nanomaterials has not been investigated extensively. Since chemical and physical synthesis cannot replace the benefits of green synthesis, it must be expanded.
2. Despite improvements in the manipulation of nanomaterials morphologies, it remains difficult to manipulate nanomaterials dimensions, particularly size. Over a certain range, the advantages of nanomaterials are limited. Due to their new properties, NMs require more flexibility in controlling their morphologies.
3. A focus on factors affecting the growth of microbes and the synthesis of natural products (NMs) is necessary to overcome longer incubation times and high maintenance costs associated with microbial synthesis.
4. In order to ensure that raw materials are disposed of appropriately, the entire life cycle should be rigorously examined. Additionally, risk management at various stages of production, storing, handling, and disposing needs more attention.

It is vital that future research focuses on filling these gaps in order to commercialize green synthesis of nanomaterials and take full advantage of their benefits. Moreover, nanomaterials pose adverse effects on human health and ecosystems, so better management and planning is imperative to mitigating those impacts.

References

1. Sepeur, S., *Nanotechnology: Technical basics and applications*, Vincentz Network GmbH & Co KG, Hannover, 2008.
2. El Shafey, A.M., Green synthesis of metal and metal oxide nanoparticles from plant leaf extracts and their applications: A review. *Green Process. Synth.*, 9, 1, 304–339, 2020.
3. Goodsell, D.S., *Bionanotechnology: Lessons from nature*, John Wiley & Sons, USA, 2004.
4. Bogunia-Kubik, K. and Sugisaka, M., From molecular biology to nanotechnology and nanomedicine. *Biosystems*, 65, 2-3, 123–138, 2002.
5. Rzayev, F.H., Gasimov, E.K., Agayeva, N.J. et al., Microscopic characterization of bioaccumulated aluminium nanoparticles in simplified food chain of aquatic ecosystem. *J. King Saud Univ.-Sci.*, 34, 1, 101666, 2022.
6. Ramalingam, M., Kokulnathan, T., Tsai, P.C. et al., Ultrasonication-assisted synthesis of gold nanoparticles decorated ultrathin graphitic carbon nitride nanosheets as a highly efficient electrocatalyst for sensitive analysis of caffeic acid in food samples. *Appl. Nanosci.*, 2021. https://doi.org/10.1007/s13204-021-01895-4.
7. Murthy, H.C., Ghotekar, S., Vinay Kumar, B. et al., Graphene: A multifunctional nanomaterial with versatile applications. *Adv. Mater. Sci. Eng.*, 2021, 2418149, 2021.
8. Murty, B.S., Shankar, P., Raj, B. et al., Applications of nanomaterials, in: *Textbook of Nanoscience and Nanotechnology*, Springer, Heidelberg, Berlin, 2013.
9. Sepeur, S., *Nanotechnology: Technical basics and applications*, Vincentz Network GmbH & Co KG, Hannover, 2008.
10. Mukherjee, P., Ahmad, A., Mandal, D. et al., Fungus-mediated synthesis of silver nanoparticles and their immobilization in the mycelial matrix: A novel biological approach to nanoparticle synthesis. *Nano Lett.*, 1, 10, 515–519, 2001.
11. Thakkar, K.N., Mhatre, S.S., Parikh, R.Y., Biological synthesis of metallic nanoparticles. *Nanomed.: Nanotechnol. Biol. Med.*, 6, 2, 257–262, 2010.
12. Chen, H., Roco, M.C., Li, X. et al., Trends in nanotechnology patents. *Nat. Nanotechnol.*, 3, 3, 123–125, 2008.
13. Khan, F.A., Synthesis of nanomaterials: Methods & technology, in: *Applications of Nanomaterials in Human Health*, pp. 15–21, Springer, Singapore, 2020.
14. Shedbalkar, U., Singh, R., Wadhwani, S. et al., Microbial synthesis of gold nanoparticles: Current status and future prospects. *Adv. Colloid Interface Sci.*, 209, 40–48, 2014.
15. Arshad, A., Iqbal, J., Mansoor, Q. et al., Graphene/SiO2 nanocomposites: The enhancement of photocatalytic and biomedical activity of SiO2 nanoparticles by graphene. *J. Appl. Phys.*, 121, 24, 244901, 2017.

16. Kharisov, B.I., Kharissova, O.V., Ortiz Mendez, U. et al., Decoration of carbon nanotubes with metal nanoparticles: Recent trends. *Synth. React. Inorg. Met.-Org. Nano-Metal Chem.*, 46, 1, 55–76, 2016.
17. Sharma, V.K., Yngard, R.A., Lin, Y., Silver nanoparticles: Green synthesis and their antimicrobial activities. *Adv. Colloid Interface Sci.*, 145, 1-2, 83–96, 2009.
18. Pandit, C., Roy, A., Ghotekar, S. et al., Biological agents for synthesis of nanoparticles and their applications. *J. King Saud Univ.-Sci.*, 34, 3, 101869, 2022.
19. Hussain, I., Singh, N.B., Singh, A. et al., Green synthesis of nanoparticles and its potential application. *Biotechnol. Lett.*, 38, 4, 545–560, 2016.
20. Chung, I.M., Park, I., Seung-Hyun, K. et al., Plant-mediated synthesis of silver nanoparticles: Their characteristic properties and therapeutic applications. *Nanoscale Res. Lett.*, 11, 1, 1–14, 2016.
21. Ankamwar, B., Kirtiwar, S., Shukla, A.C., Plant-mediated green synthesis of nanoparticles, in: *Advances in Pharmaceutical Biotechnology*, J.K. Patra, A.C. Shukla, G. Das, (Eds.), pp. 221–234, Springer, Singapore, 2020.
22. Dahl, J.A., Maddux, B.L., Hutchison, J.E., Toward greener nanosynthesis. *Chem. Rev.*, 107, 6, 2228–2269, 2007.
23. Gasparotto, A., Barreca, D., Maccato, C. et al., Manufacturing of inorganic nanomaterials: Concepts and perspectives. *Nanoscale*, 4, 9, 2813–2825, 2012.
24. Pileni, M.P., The role of soft colloidal templates in controlling the size and shape of inorganic nanocrystals. *Nat. Mater.*, 2, 3, 145–150, 2003.
25. Yi, Z., Li, X., Xu, X. et al., Green, effective chemical route for the synthesis of silver nanoplates in tannic acid aqueous solution. *Colloids Surf. A: Physicochem. Eng. Asp.*, 392, 1, 131–136, 2011.
26. Das, S.K. and Marsili, E., A green chemical approach for the synthesis of gold nanoparticles: Characterization and mechanistic aspect. *Rev. Environ. Sci. Bio/Technol.*, 9, 3, 199–204, 2010.
27. Landis, W.J., The strength of a calcified tissue depends in part on the molecular structure and organization of its constituent mineral crystals in their organic matrix. *Bone*, 16, 5, 533–544, 1995.
28. Noruzi, M., Biosynthesis of gold nanoparticles using plant extracts. *Bioprocess Biosyst. Eng.*, 38, 1, 1–14, 2015.
29. Veisi, H., Kavian, M., Hekmati, M. et al., Biosynthesis of the silver nanoparticles on the graphene oxide's surface using Pistacia atlantica leaves extract and its antibacterial activity against some human pathogens. *Polyhedron*, 161, 338–345, 2019.
30. Singh, P., Kim, Y.J., Yang, D.C., A strategic approach for rapid synthesis of gold and silver nanoparticles by Panax ginseng leaves. *Artif. Cells Nanomed. Biotechnol.*, 44, 8, 1949–1957, 2016.
31. Singh, P., Kim, Y.J., Wang, C., Mathiyalagan, R., Yang, D.C., The development of a green approach for the biosynthesis of silver and gold nanoparticles by using Panax ginseng root extract, and their biological applications. *Artif. Cells Nanomed. Biotechnol.*, 44, 4, 1150–1157, 2016.

32. Singh, P., Kim, Y.J., Wang, C. et al., Biogenic silver and gold nanoparticles synthesized using red ginseng root extract, and their applications. *Artif. Cells Nanomed. Biotechnol.*, 44, 3, 811–816, 2016.
33. Duan, H., Wang, D., Li, Y., Green chemistry for nanoparticle synthesis. *Chem. Soc. Rev.*, 44, 16, 5778–5792, 2015.
34. El-Kassas, H.Y. and El-Sheekh, M.M., Cytotoxic activity of biosynthesized gold nanoparticles with an extract of the red seaweed Corallina officinalis on the MCF-7 human breast cancer cell line. *Asian Pac. J. Cancer Prev.*, 15, 10, 4311–4317, 2014.
35. Sabouri, Z., Akbari, A., Hosseini, H.A. et al., Eco-friendly biosynthesis of nickel oxide nanoparticles mediated by okra plant extract and investigation of their photocatalytic, magnetic, cytotoxicity, and antibacterial properties. *J. Clust. Sci.*, 30, 6, 1425–1434, 2019.
36. Baker, S., Rakshith, D., Kavitha, K.S. et al., Plants: Emerging as nanofactories towards facile route in synthesis of nanoparticles. *BioImpacts: BI*, 3, 3, 111, 2013.
37. Makarov, V.V., Love, A.J., Sinitsyna, O.V. et al., "Green" nanotechnologies: Synthesis of metal nanoparticles using plants. *Acta Naturae (англоязычная версия)*, 6, 1 (20), 35–44, 2014.
38. Alomar, T.S., AlMasoud, N., Awad, M.A. et al., An eco-friendly plant-mediated synthesis of silver nanoparticles: Characterization, pharmaceutical and biomedical applications. *Mater. Chem. Phys.*, 249, 123007, 2020.
39. Roy, K., Sarkar, C.K., Ghosh, C.K., Plant-mediated synthesis of silver nanoparticles using parsley (*Petroselinum crispum*) leaf extract: Spectral analysis of the particles and antibacterial study. *Appl. Nanosci.*, 5, 8, 945–951, 2015.
40. Ramesh, A.V., Devi, D.R., Battu, G. et al., A Facile plant mediated synthesis of silver nanoparticles using an aqueous leaf extract of Ficus hispida Linn. f. for catalytic, antioxidant and antibacterial applications. *S. Afr. J. Chem. Eng.*, 26, 25–34, 2018.
41. Balachandar, R., Gurumoorthy, P., Karmegam, N. et al., Plant-mediated synthesis, characterization and bactericidal potential of emerging silver nanoparticles using stem extract of Phyllanthus pinnatus: A recent advance in phytonanotechnology. *J. Cluster Sci.*, 30, 6, 1481–1488, 2019.
42. Arshad, H., Sami, M.A., Sadaf, S. et al., Salvadora persica mediated synthesis of silver nanoparticles and their antimicrobial efficacy. *Sci. Rep.*, 11, 1, 1–11, 2021.
43. Baruah, D., Yadav, R.N.S., Yadav, A. et al., Alpinia nigra fruits mediated synthesis of silver nanoparticles and their antimicrobial and photocatalytic activities. *J. Photochem. Photobiol. B: Biol.*, 201, 111649, 2019.
44. Ndikau, M., Noah, N.M., Andala, D.M. et al., Green synthesis and characterization of silver nanoparticles using Citrullus lanatus fruit rind extract. *Int. J. Anal. Chem.*, 2017, 8108504, 2017.

45. Rohaizad, A., Shahabuddin, S., Shahid, M.M. et al., Green synthesis of silver nanoparticles from Catharanthus roseus dried bark extract deposited on graphene oxide for effective adsorption of methylene blue dye. *J. Environ. Chem. Eng.*, 8, 4, 103955, 2020.
46. Singh, P., Pandit, S., Garnæs, J. et al., Green synthesis of gold and silver nanoparticles from Cannabis sativa (industrial hemp) and their capacity for biofilm inhibition. *Int. J. Nanomed.*, 13, 3571, 2018.
47. Rani, P., Kumar, V., Singh, P.P. et al., Highly stable AgNPs prepared via a novel green approach for catalytic and photocatalytic removal of biological and non-biological pollutants. *Environ. Int.*, 143, 105924, 2020.
48. Ghosh, S., Patil, S., Ahire, M. et al., Gnidia glauca flower extract mediated synthesis of gold nanoparticles and evaluation of its chemocatalytic potential. *J. Nanobiotechnol.*, 10, 1, 1–9, 2012.
49. Mata, R., Bhaskaran, A., Sadras, S.R., Green-synthesized gold nanoparticles from Plumeria alba flower extract to augment catalytic degradation of organic dyes and inhibit bacterial growth. *Particuology*, 24, 78–86, 2016.
50. Anand, K., Gengan, R.M., Phulukdaree, A. et al., Agroforestry waste Moringa oleifera petals mediated green synthesis of gold nanoparticles and their anti-cancer and catalytic activity. *J. Ind. Eng. Chem.*, 21, 1105–1111, 2015.
51. Lee, K.X., Shameli, K., Miyake, M. et al., Gold nanoparticles biosynthesis: A simple route for control size using waste peel extract. *IEEE Trans. Nanotechnol.*, 16, 6, 954–957, 2017.
52. Rajan, A., Rajan, A.R., Philip, D., Elettaria cardamomum seed mediated rapid synthesis of gold nanoparticles and its biological activities. *OpenNano*, 2, 1–8, 2017.
53. Kumar, B., Smita, K., Cumbal, L. et al., Extracellular biofabrication of gold nanoparticles by using Lantana camara berry extract. *Inorg. Nano-Metal Chem.*, 47, 1, 138–142, 2017.
54. Chahardoli, A., Karimi, N., Fattahi, A. et al., Biological applications of phyto-synthesized gold nanoparticles using leaf extract of Dracocephalum kotschyi. *J. Biomed. Mater. Res. Part A*, 107, 3, 621–630, 2019.
55. Khan, Z.U.H., Sadiq, H.M., Shah, N.S. et al., Greener synthesis of zinc oxide nanoparticles using Trianthema portulacastrum extract and evaluation of its photocatalytic and biological applications. *J. Photochem. Photobiol. B: Biol.*, 192, 147–157, 2019.
56. Sundararajan, B. and Kumari, B.R., Novel synthesis of gold nanoparticles using Artemisia vulgaris L. leaf extract and their efficacy of larvicidal activity against dengue fever vector Aedes aegypti L. *J. Trace Elem. Med. Biol.*, 43, 187–196, 2017.
57. Singh, P., Singh, H., Ahn, S. et al., Pharmacological importance, characterization and applications of gold and silver nanoparticles synthesized by Panax ginseng fresh leaves. *Artif. Cells Nanomed. Biotechnol.*, 45, 7, 1415–1424, 2017.

58. Sukri, S.N.A.M., Shameli, K., Wong, M.M.T. et al., Cytotoxicity and antibacterial activities of plant-mediated synthesized zinc oxide (ZnO) nanoparticles using Punica granatum (pomegranate) fruit peels extract. *J. Mol. Struct.*, 1189, 57–65, 2019.
59. Singh, K., Kumar, V., Kukkar, B. et al., Facile and efficient colorimetric detection of cadmium ions in aqueous systems using green-synthesized gold nanoparticles. *Int. J. Environ. Sci. Technol.*, 19, 6, 4673–4690, 2022.
60. Banu, A.N. and Balasubramanian, C., Optimization and synthesis of silver nanoparticles using Isaria fumosorosea against human vector mosquitoes. *Parasitol. Res.*, 113, 10, 3843–3851, 2014.
61. Paulkumar, K., Rajeshkumar, S., Gnanajobitha, G., et al., Biosynthesis of silver chloride nanoparticles using Bacillus subtilis MTCC 3053 and assessment of its antifungal activity. *Int. Sch. Res. Notices*, 2013, 317963, 2013.
62. Lovley, D.R. and Woodward, J.C., Mechanisms for chelator stimulation of microbial Fe (III)-oxide reduction. *Chem. Geol.*, 132, 1-4, 19–24, 1996.
63. Chen, X. and Schluesener, H.J., Nanosilver: A nanoproduct in medical application. *Toxicol. Lett.*, 176, 1, 1–12, 2008.
64. John, M.S., Nagoth, J.A., Ramasamy, K.P. et al., Synthesis of bioactive silver nanoparticles by a Pseudomonas strain associated with the antarctic psychrophilic protozoon Euplotes focardii. *Mar. Drugs*, 18, 1, 38, 2020.
65. Khan, R. and Fulekar, M.H., Biosynthesis of titanium dioxide nanoparticles using Bacillus amyloliquefaciens culture and enhancement of its photocatalytic activity for the degradation of a sulfonated textile dye Reactive Red 31. *J. Colloid Interface Sci.*, 475, 184–191, 2016.
66. Srinath, B.S., Namratha, K., Byrappa, K., Eco-friendly synthesis of gold nanoparticles by Bacillus subtilis and their environmental applications. *Adv. Sci. Lett.*, 24, 8, 5942–5946, 2018.
67. Sweeney, R.Y., Mao, C., Gao, X. et al., Bacterial biosynthesis of cadmium sulfide nanocrystals. *Chem. Biol.*, 11, 11, 1553–1559, 2004.
68. Huang, J., Li, Q., Sun, D. et al., Biosynthesis of silver and gold nanoparticles by novel sundried Cinnamomum camphora leaf. *Nanotechnology*, 18, 10, 105104, 2007.
69. Salem, S.S. and Fouda, A., Green synthesis of metallic nanoparticles and their prospective biotechnological applications: An overview. *Biol. Trace Elem. Res.*, 199, 1, 344–370, 2021.
70. Kalishwaralal, K., Deepak, V., Ramkumarpandian, S. et al., Extracellular biosynthesis of silver nanoparticles by the culture supernatant of Bacillus licheniformis. *Mater. Lett.*, 62, 29, 4411–4413, 2008.
71. Thomas, R., Janardhanan, A., Varghese, R.T. et al., Antibacterial properties of silver nanoparticles synthesized by marine Ochrobactrum sp. *Braz. J. Microbiol.*, 45, 1221–1227, 2014.
72. Elbeshehy, E.K., Elazzazy, A.M., Aggelis, G., Silver nanoparticles synthesis mediated by new isolates of Bacillus spp., nanoparticle characterization and

their activity against Bean Yellow Mosaic Virus and human pathogens. *Front. Microbiol.*, 6, 453, 2015.
73. Monowar, T., Rahman, M.S., Bhore, S.J. *et al.*, Silver nanoparticles synthesized by using the endophytic bacterium Pantoea ananatis are promising antimicrobial agents against multidrug resistant bacteria. *Molecules*, 23, 12, 3220, 2018.
74. Iqtedar, M., Aslam, M., Akhyar, M. *et al.*, Extracellular biosynthesis, characterization, optimization of silver nanoparticles (AgNPs) using Bacillus mojavensis BTCB15 and its antimicrobial activity against multidrug resistant pathogens. *Prep. Biochem. Biotechnol.*, 49, 2, 136–142, 2019.
75. Lv, Q., Zhang, B., Xing, X. *et al.*, Biosynthesis of copper nanoparticles using Shewanella loihica PV-4 with antibacterial activity: Novel approach and mechanisms investigation. *J. Hazard. Mater.*, 347, 141–149, 2018.
76. Jafari, M., Rokhbakhsh-Zamin, F., Shakibaie, M. *et al.*, Cytotoxic and antibacterial activities of biologically synthesized gold nanoparticles assisted by Micrococcus yunnanensis strain J2. *Biocatal. Agric. Biotechnol.*, 15, 245–253, 2018.
77. Camas, M., Sazak Camas, A., Kyeremeh, K., Extracellular synthesis and characterization of gold nanoparticles using Mycobacterium sp. BRS2A-AR2 isolated from the aerial roots of the Ghanaian mangrove plant, Rhizophora racemosa. *Indian J. Microbiol.*, 58, 2, 214–221, 2018.
78. Jayaseelan, C., Rahuman, A.A., Roopan, S.M. *et al.*, Biological approach to synthesize TiO2 nanoparticles using Aeromonas hydrophila and its antibacterial activity. *Spectrochim. Acta Part A: Mol. Biomol. Spectrosc.*, 107, 82–89, 2013.
79. Taran, M., Rad, M., Alavi, M., Biosynthesis of TiO2 and ZnO nanoparticles by Halomonas elongata IBRC-M 10214 in different conditions of medium. *BioImpacts: BI*, 8, 2, 81, 2018.
80. Rajabairavi, N., Raju, C.S., Karthikeyan, C. *et al.*, Biosynthesis of novel zinc oxide nanoparticles (ZnO NPs) using endophytic bacteria Sphingobacterium thalpophilum, in: *Recent Trends in Materials Science and Applications*, pp. 245–254, Springer, Cham, 2017.
81. Ahmad, S., Munir, S., Zeb, N. *et al.*, Green nanotechnology: A review on green synthesis of silver nanoparticles—An ecofriendly approach. *Int. J. Nanomed.*, 14, 5087, 2019.
82. Dhankhar, R. and Hooda, A., Fungal biosorption–an alternative to meet the challenges of heavy metal pollution in aqueous solutions. *Environ. Technol.*, 32, 5, 467–491, 2011.
83. Salem, S.S. and Fouda, A., Green synthesis of metallic nanoparticles and their prospective biotechnological applications: An overview. *Biol. Trace Elem. Res.*, 199, 1, 344–370, 2021.
84. Mohamed, A.A., Fouda, A., Abdel-Rahman, M.A. *et al.*, Fungal strain impacts the shape, bioactivity and multifunctional properties of green synthesized zinc oxide nanoparticles. *Biocatal. Agric. Biotechnol.*, 19, 101103, 2019.

85. Fouda, A., Saad, E.L., Salem, S.S. et al., In-vitro cytotoxicity, antibacterial, and UV protection properties of the biosynthesized Zinc oxide nanoparticles for medical textile applications. *Microb. Pathogen.*, 125, 252–261, 2018.
86. Dhillon, G.S., Brar, S.K., Kaur, S. et al., Green approach for nanoparticle biosynthesis by fungi: Current trends and applications. *Crit. Rev. Biotechnol.*, 32, 1, 49–73, 2012.
87. Spagnoletti, F.N., Spedalieri, C., Kronberg, F. et al., Extracellular biosynthesis of bactericidal Ag/AgCl nanoparticles for crop protection using the fungus Macrophomina phaseolina. *J. Environ. Manage.*, 231, 457–466, 2019.
88. Al-Mubaddel, F.S., Haider, S., Al-Masry, W.A. et al., Engineered nanostructures: A review of their synthesis, characterization and toxic hazard considerations. *Arabian J. Chem.*, 10, S376–S388, 2017.
89. Gudikandula, K., Vadapally, P., Charya, M.S., Biogenic synthesis of silver nanoparticles from white rot fungi: Their characterization and antibacterial studies. *OpenNano*, 2, 64–78, 2017.
90. Mohanpuria, P., Rana, N.K., Yadav, S.K., Biosynthesis of nanoparticles: Technological concepts and future applications. *J. Nanopart. Res.*, 10, 3, 507–517, 2008.
91. Castro, L., Blázquez, M.L., González, F.G. et al., Mechanism and applications of metal nanoparticles prepared by bio-mediated process. *Rev. Adv. Sci. Eng.*, 3, 3, 199–216, 2014.
92. Jalal, M., Ansari, M.A., Alzohairy, M.A. et al., Biosynthesis of silver nanoparticles from oropharyngeal Candida glabrata isolates and their antimicrobial activity against clinical strains of bacteria and fungi. *Nanomaterials*, 8, 8, 586, 2018.
93. Elamawi, R.M., Al-Harbi, R.E., Hendi, A.A., Biosynthesis and characterization of silver nanoparticles using Trichoderma longibrachiatum and their effect on phytopathogenic fungi. *Egypt. J. Biol. Pest Control*, 28, 1, 1–11, 2018.
94. Ahmed, A.A., Hamzah, H., Maaroof, M., Analyzing formation of silver nanoparticles from the filamentous fungus Fusarium oxysporum and their antimicrobial activity. *Turk. J. Biol.*, 42, 1, 54–62, 2018.
95. Mohanta, Y.K., Nayak, D., Biswas, K. et al., Silver nanoparticles synthesized using wild mushroom show potential antimicrobial activities against food borne pathogens. *Molecules*, 23, 3, 655, 2018.
96. Mohmed, A.A., Saad, E., Fouda, A. et al., Extracellular biosynthesis of silver nanoparticles using Aspergillus sp. and evaluation of their antibacterial and cytotoxicity. *J. Appl. Life Sci. Int.*, 11, 2, 1–12, 2017.
97. Das, S.K., Das, A.R., Guha, A.K., Gold nanoparticles: Microbial synthesis and application in water hygiene management. *Langmuir*, 25, 14, 8192–8199, 2009.
98. Zhang, Y.X., Zheng, J., Gao, G. et al., Biosynthesis of gold nanoparticles using chloroplasts. *Int. J. Nanomed.*, 6, 2899, 2011.
99. Joshi, C.G., Danagoudar, A., Poyya, J. et al., Biogenic synthesis of gold nanoparticles by marine endophytic fungus-Cladosporium cladosporioides

isolated from seaweed and evaluation of their antioxidant and antimicrobial properties. *Process Biochem.*, 63, 137–144, 2017.
100. Fouda, A., Saad, E.L., Salem, S.S. *et al.*, In-vitro cytotoxicity, antibacterial, and UV protection properties of the biosynthesized Zinc oxide nanoparticles for medical textile applications. *Microb. Pathogen.*, 125, 252–261, 2018.
101. Raliya, R. and Tarafdar, J.C., ZnO nanoparticle biosynthesis and its effect on phosphorous-mobilizing enzyme secretion and gum contents in Clusterbean (Cyamopsis tetragonoloba L.). *Agric. Res.*, 2, 1, 48–57, 2013.
102. Suryavanshi, P., Pandit, R., Gade, A. *et al.*, Colletotrichum sp.-mediated synthesis of sulphur and aluminium oxide nanoparticles and its *in vitro* activity against selected food-borne pathogens. *LWT-Food Sci. Technol.*, 81, 188–194, 2017.
103. Shah, M., Fawcett, D., Sharma, S. *et al.*, Green synthesis of metallic nanoparticles via biological entities. *Materials*, 8, 11, 7278–7308, 2015.
104. Anand, P., Isar, J., Saran, S., Saxena, R.K., Bioaccumulation of copper by Trichoderma viride. *Bioresour. Technol.*, 97, 8, 1018–1025, 2006.
105. Kumar, D., Karthik, L., Kumar, G. *et al.*, Biosynthesis of silver nanoparticles from marine yeast and their antimicrobial activity against multidrug resistant pathogens. *Pharmacologyonline*, 3, 1100–1111, 2011.
106. Varshney, R., Bhadauria, S. and Gaur, M.S., A review: Biological synthesis of silver and copper nanoparticles. *Nano Biomed. Eng.*, 4, 2, 99-106, 2012.
107. Apte, M., Sambre, D., Gaikawad, S. *et al.*, Psychrotrophic yeast Yarrowia lipolytica NCYC 789 mediates the synthesis of antimicrobial silver nanoparticles via cell-associated melanin. *Amb Express*, 3, 1, 1–8, 2013.
108. Kaur, K. and Thombre, R., Nanobiotechnology: Methods, applications, and future prospects, in: *Nanobiotechnology*, pp. 1–20, Elsevier, Netherlands, 2021.
109. Vithiya, K. and Sen, S., Biosynthesis of nanoparticles. *Int. J. Pharm. Sci. Res.*, 2, 11, 2781, 2011.
110. Rasouli, M., Biosynthesis of selenium nanoparticles using yeast Nematospora coryli and examination of their anti-candida and anti-oxidant activities. *IET Nanobiotechnol.*, 13, 2, 214–218, 2019.
111. Attia, Y.A., Farag, Y.E., Mohamed, Y.M. *et al.*, Photo-extracellular synthesis of gold nanoparticles using Baker's yeast and their anticancer evaluation against Ehrlich ascites carcinoma cells. *New J. Chem.*, 40, 11, 9395–9402, 2016.
112. Ahmad, T., Wani, I.A., Manzoor, N. *et al.*, Biosynthesis, structural characterization and antimicrobial activity of gold and silver nanoparticles. *Colloids Surf. B: Biointerfaces*, 107, 227–234, 2013.
113. Zhou, W., He, W., Zhong, S. *et al.*, Biosynthesis and magnetic properties of mesoporous Fe_3O_4 composites. *J. Magn. Magn. Mater.*, 321, 8, 1025–1028, 2009.
114. Salvadori, M.R., Ando, R.A., Muraca, D. *et al.*, Magnetic nanoparticles of Ni/NiO nanostructured in film form synthesized by dead organic matrix of yeast. *RSC Adv.*, 6, 65, 60683–60692, 2016.

115. Thajuddin, N. and Subramanian, G., Cyanobacterial biodiversity and potential applications in biotechnology. *Curr. Sci.*, 89, 47–57, 2005.
116. Oscar, F.L., Bakkiyaraj, D., Nithya, C. *et al.*, Deciphering the diversity of microalgal bloom in wastewater-an attempt to construct potential consortia for bioremediation. *J. Curr. Perspect. Appl. Microbiol.*, 2278, 92, 2014.
117. LewisOscar, F., Vismaya, S., Arunkumar, M. *et al.*, Algal nanoparticles: Synthesis and biotechnological potentials, in: *Algae-Organisms for Imminent Biotechnology*, vol. 7, pp. 157–182, 2016.
118. González-Ballesteros, N., Prado-López, S., Rodríguez-González, J.B. *et al.*, Green synthesis of gold nanoparticles using brown algae Cystoseira baccata: Its activity in colon cancer cells. *Colloids Surf. B: Biointerfaces*, 153, 190–198, 2017.
119. Ramakrishna, M., Rajesh Babu, D., Gengan, R.M. *et al.*, Green synthesis of gold nanoparticles using marine algae and evaluation of their catalytic activity. *J. Nanostructure Chem.*, 6, 1, 1–13, 2016.
120. Prasad, T.N., Kambala, V.S.R., Naidu, R., Phyconanotechnology: Synthesis of silver nanoparticles using brown marine algae Cystophora moniliformis and their characterisation. *J. Appl. Phycol.*, 25, 1, 177–182, 2013.
121. Priyadharshini, R.I., Prasannaraj, G., Geetha, N. *et al.*, Microwave-mediated extracellular synthesis of metallic silver and zinc oxide nanoparticles using macro-algae (Gracilaria edulis) extracts and its anticancer activity against human PC3 cell lines. *Appl. Biochem. Biotechnol.*, 174, 8, 2777–2790, 2014.
122. Pugazhendhi, A., Prabakar, D., Jacob, J.M. *et al.*, Synthesis and characterization of silver nanoparticles using Gelidium amansii and its antimicrobial property against various pathogenic bacteria. *Microb. Pathogen.*, 114, 41–45, 2018.
123. Armendariz, V., Herrera, I., Jose-Yacaman, M. *et al.*, Size controlled gold nanoparticle formation by Avena sativa biomass: Use of plants in nanobiotechnology. *J. Nanopart. Res.*, 6, 4, 377–382, 2004.
124. Munir, H., Bilal, M., Mulla, S.I. *et al.*, Plant-mediated green synthesis of nanoparticles, in: *Advances in Green Synthesis*, Springer, Cham, 2021.
125. Kathiresan, K., Manivannan, S., Nabeel, M.A. *et al.*, Studies on silver nanoparticles synthesized by a marine fungus, Penicillium fellutanum isolated from coastal mangrove sediment. *Colloids Surf. B: Biointerfaces*, 71, 1, 133–137, 2009.
124. Sathishkumar, M., Sneha, K., Kwak, I.S. *et al.*, Phyto-crystallization of palladium through reduction process using Cinnamom zeylanicum bark extract. *J. Hazard. Mater.*, 171, 1-3, 400–404, 2009.
125. Dubey, S.P., Lahtinen, M., Sillanpää, M., Tansy fruit mediated greener synthesis of silver and gold nanoparticles. *Process Biochem.*, 45, 7, 1065–1071, 2010.
126. Andreescu, D., Eastman, C., Balantrapu, K. *et al.*, A simple route for manufacturing highly dispersed silver nanoparticles. *J. Mater. Res.*, 22, 9, 2488–2496, 2007.

127. Dwivedi, A.D. and Gopal, K., Biosynthesis of silver and gold nanoparticles using Chenopodium album leaf extract. *Colloids Surf. A: Physicochem. Eng. Asp.*, 369, 1-3, 27–33, 2010.
128. Ghodake, G.S., Deshpande, N.G., Lee, Y.P. et al., Pear fruit extract-assisted room-temperature biosynthesis of gold nanoplates. *Colloids Surf. B: Biointerfaces*, 75, 2, 584–589, 2010.
129. Rai, A., Singh, A., Ahmad, A., Sastry, M., Role of halide ions and temperature on the morphology of biologically synthesized gold nanotriangles. *Langmuir*, 22, 2, 736–741, 2006.
130. Bilal, M., Rasheed, T., Iqbal, H.M.N. et al., Silver nanoparticles: Biosynthesis and antimicrobial potentialities. *Int. J. Pharmacol.*, 13, 7, 832–845, 2017.
131. Cruz, D., Falé, P.L., Mourato, A. et al., Preparation and physicochemical characterization of Ag nanoparticles biosynthesized by Lippia citriodora (Lemon Verbena). *Colloids Surf. B: Biointerfaces*, 81, 1, 67–73, 2010.
132. Dwivedi, A.D. and Gopal, K., Biosynthesis of silver and gold nanoparticles using Chenopodium album leaf extract. *Colloids Surf. A: Physicochem. Eng. Asp.*, 369, 1-3, 27–33, 2010.
133. Lin, L., Wang, W., Huang, J. et al., Nature factory of silver nanowires: Plant-mediated synthesis using broth of Cassia fistula leaf. *Chem. Eng. J.*, 162, 2, 852–858, 2010.
134. Gericke, M. and Pinches, A., Biological synthesis of metal nanoparticles. *Hydrometallurgy*, 83, 1-4, 132–140, 2006.
135. Makarov, V.V., Love, A.J., Sinitsyna, O.V. et al., "Green" nanotechnologies: Synthesis of metal nanoparticles using plants. *Acta Naturae (англоязычная версия)*, 6, 1 (20), 35–44, 2014.
136. Aziz, Z.A.A., Mohd-Nasir, H., Ahmad, A. et al., Role of nanotechnology for design and development of cosmeceutical: Application in makeup and skin care. *Front. Chem.*, 7, 739, 2019.
137. Ankamwar, B., Kirtiwar, S., Shukla, A.C., Plant-mediated green synthesis of nanoparticles, in: *Advances in Pharmaceutical Biotechnology*, J. Patra, A. Shukla, G. Das (Eds.), pp. 221–234, Springer, Singapore, 2020.
138. Chung, I.M., Park, I., Seung-Hyun, K. et al., Plant-mediated synthesis of silver nanoparticles: Their characteristic properties and therapeutic applications. *Nanoscale Res. Lett.*, 11, 1, 1–14, 2016.
139. Liu, H., Zhang, X., Xu, Z. et al., Role of polyphenols in plant-mediated synthesis of gold nanoparticles: Identification of active components and their functional mechanism. *Nanotechnology*, 31, 41, 415601, 2020.
140. Park, Y., Hong, Y.N., Weyers, A. et al., Polysaccharides and phytochemicals: A natural reservoir for the green synthesis of gold and silver nanoparticles. *IET nanobiotechnol.*, 5, 3, 69–78, 2011.
141. Elavazhagan, T. and Arunachalam, K.D., Memecylon edule leaf extract mediated green synthesis of silver and gold nanoparticles. *Int. J. Nanomed.*, 6, 1265–1278, 2011.

142. Aromal, S.A. and Philip, D., Green synthesis of gold nanoparticles using Trigonella foenum-graecum and its size-dependent catalytic activity. *Spectrochim. Acta Part A: Mol. Biomol. Spectrosc.*, 97, 1–5, 2012.
143. Song, J.Y., Jang, H.K., Kim, B.S., Biological synthesis of gold nanoparticles using Magnolia kobus and Diopyros kaki leaf extracts. *Process Biochem.*, 44, 10, 1133–1138, 2009.
144. Li, X., Xu, H., Chen, Z.S., et al., Biosynthesis of nanoparticles by microorganisms and their applications. *J. Nanomater.*, 2011, 270974, 2011.
145. Simkiss, K. and Wilbur, K.M., *Biomineralization*, Elsevier, Netherlands, 2012.
146. Hasan, M., Hossain, E., Balasubramaniam, S. et al., Social behavior in bacterial nanonetworks: Challenges and opportunities. *IEEE Network*, 29, 1, 26–34, 2015.
147. Singh, P., Kim, Y.J., Zhang, D. et al., Biological synthesis of nanoparticles from plants and microorganisms. *Trends Biotechnol.*, 34, 7, 588–599, 2016.
148. Poinern, G.E.J., *A laboratory course in nanoscience and nanotechnology*, CRC Press, Boca Raton, Florida, 2014.
149. Zhang, X., Yan, S., Tyagi, R.D. et al., Synthesis of nanoparticles by microorganisms and their application in enhancing microbiological reaction rates. *Chemosphere*, 82, 4, 489–494, 2011.
150. Diallo, M. and Brinker, C.J., Nanotechnology for sustainability: Environment, water, food, minerals, and climate, in: *Nanotechnology Research Directions for Societal Needs in 2020*, pp. 221–259, Springer, Dordrecht.Diallo, 2011.
151. Wright, G.D., Bacterial resistance to antibiotics: Enzymatic degradation and modification. *Adv. Drug Delivery Rev.*, 57, 10, 1451–1470, 2005.
152. Fadeel, B. and Garcia-Bennett, A.E., Better safe than sorry: Understanding the toxicological properties of inorganic nanoparticles manufactured for biomedical applications. *Adv. Drug Delivery Rev.*, 62, 3, 362–374, 2010.
153. Chaloupka, K., Malam, Y., Seifalian, A.M., Nanosilver as a new generation of nanoproduct in biomedical applications. *Trends Biotechnol.*, 28, 11, 580–588, 2010.
154. Emerich, D.F. and Thanos, C.G., The pinpoint promise of nanoparticle-based drug delivery and molecular diagnosis. *Biomol. Eng.*, 23, 4, 171–184, 2006.
155. Mandhata, C.P., Sahoo, C.R. and Padhy, R.N., Biomedical applications of biosynthesized gold nanoparticles from cyanobacteria: An overview. *Biol. Trace Elem. Res.*, 200, 12, 5307–5327, 2022.
156. Singh, R., Tiwari, P., Kumari, N. et al., Biomedical applications of green synthesized nanoparticles, in: *Advances in Pharmaceutical Biotechnology*, pp. 235–245, Springer, Singapore, 2020.
157. Jones, M.E., Karlowsky, J.A., Draghi, D.C. et al., Rates of antimicrobial resistance among common bacterial pathogens causing respiratory, blood, urine, and skin and soft tissue infections in pediatric patients. *Eur. J. Clin. Microbiol. Infect. Dis.*, 23, 6, 445–455, 2004.
158. Rai, M., Yadav, A., Gade, A., Silver nanoparticles as a new generation of antimicrobials. *Biotechnol. Adv.*, 27, 1, 76–83, 2009.

159. Prabhu, S. and Poulose, E.K., Silver nanoparticles: Mechanism of antimicrobial action, synthesis, medical applications, and toxicity effects. *Int. Nano Lett.*, 2, 1, 1–10, 2012.
160. Dawadi, S., Katuwal, S., Gupta, A., *et al.*, Current research on silver nanoparticles: Synthesis, characterization, and applications. *J. Nanomater.*, 2021, 6687290, 2021.
161. Yan, X., He, B., Liu, L. *et al.*, Antibacterial mechanism of silver nanoparticles in Pseudomonas aeruginosa: Proteomics approach. *Metallomics*, 10, 4, 557–564, 2018.
162. Burduşel, A.C., Gherasim, O., Grumezescu, A.M. *et al.*, Biomedical applications of silver nanoparticles: An up-to-date overview. *Nanomaterials*, 8, 9, 681, 2018.
163. Sondi, I. and Salopek-Sondi, B., Silver nanoparticles as antimicrobial agent: A case study on E. coli as a model for Gram-negative bacteria. *J. Colloid Interface Sci.*, 275, 1, 177–182, 2004.
164. Ravichandran, V., Vasanthi, S., Shalini, S. *et al.*, Results in physics photocatalytic activity of Parkia speciosa leaves extract mediated silver nanoparticles. *Results Phys.*, 15, 102565–102573, 2019.
165. Chandrasekaran, R., Seetharaman, P., Krishnan, M. *et al.*, Carica papaya (Papaya) latex: A new paradigm to combat against dengue and filariasis vectors Aedes aegypti and Culex quinquefasciatus (Diptera: Culicidae). *3 Biotech.*, 8, 2, 1–10, 2018.
166. Fahimmunisha, B.A., Ishwarya, R., AlSalhi, M.S. *et al.*, Green fabrication, characterization and antibacterial potential of zinc oxide nanoparticles using Aloe socotrina leaf extract: A novel drug delivery approach. *J. Drug Delivery Sci. Technol.*, 55, 101465, 2020.
167. Azam, A., Ahmed, A.S., Oves, M. *et al.*, Antimicrobial activity of metal oxide nanoparticles against Gram-positive and Gram-negative bacteria: A comparative study. *Int. J. Nanomed.*, 7, 6003, 2012.
168. Khezerlou, A., Alizadeh-Sani, M., Azizi-Lalabadi, M. *et al.*, Nanoparticles and their antimicrobial properties against pathogens including bacteria, fungi, parasites and viruses. *Microb. Pathogen.*, 123, 505–526, 2018.
169. Li, Y., Zhang, W., Niu, J. *et al.*, Mechanism of photogenerated reactive oxygen species and correlation with the antibacterial properties of engineered metal-oxide nanoparticles. *ACS Nano*, 6, 6, 5164–5173, 2012.
170. Nikolova, M.P. and Chavali, M.S., Metal oxide nanoparticles as biomedical materials. *Biomimetics*, 5, 2, 27, 2020.
171. Dobrucka, R. and Długaszewska, J., Biosynthesis and antibacterial activity of ZnO nanoparticles using Trifolium pratense flower extract. *Saudi J. Biol. Sci.*, 23, 517–523, 2016.
172. Lok, C.N., Ho, C.M., Chen, R. *et al.*, Proteomic analysis of the mode of antibacterial action of silver nanoparticles. *J. Proteome Res.*, 5, 4, 916–924, 2006.
173. Dobrucka, R. and Długaszewska, J., Biosynthesis and antibacterial activity of ZnO nanoparticles using Trifolium pratense flower extract. *Saudi J. Biol. Sci.*, 23, 517–523, 2015.

174. Beyth, N., Houri-Haddad, Y., Domb, A., et al., Alternative antimicrobial approach: Nano-antimicrobial materials. *Evid. based Complement. Altern. Med.*, 2015, 246012, 2015.
175. Gomathi Devi, L. and Nagaraj, B., Disinfection of E scherichia coli gram negative bacteria using surface modified TiO2: Optimization of Ag metallization and depiction of charge transfer mechanism. *Photochem. Photobiol.*, 90, 5, 1089–1098, 2014.
176. Pelgrift, R.Y. and Friedman, A.J., Nanotechnology as a therapeutic tool to combat microbial resistance. *Adv. Drug Delivery Rev.*, 65, 13-14, 1803–1815, 2013.
177. Pasha, A., Kumbhakar, D.V., Sana, S.S., et al., Role of biosynthesized Ag-NPs using *Aspergillus niger* (MK503444. 1) in antimicrobial, anti-cancer and anti-angiogenic activities. *Front. Pharmacol.*, 12, 812474, 2022.
178. Nazeruddin, G.M., Prasad, N.R., Waghmare, S.R. et al., Extracellular biosynthesis of silver nanoparticle using Azadirachta indica leaf extract and its anti-microbial activity. *J. Alloys Compd.*, 583, 272–277, 2014.
179. Sarkar, R., Kumbhakar, P., Mitra, A.K., Green synthesis of silver nanoparticles and its optical properties. *Dig. J. Nanomater. Biostruct.*, 5, 2, 491–496, 2010.
180. Vlachou, E., Chipp, E., Shale, E. et al., The safety of nanocrystalline silver dressings on burns: A study of systemic silver absorption. *Burns*, 33, 8, 979–985, 2007.
181. Al-Halifa, S., Gauthier, L., Arpin, D. et al., Nanoparticle-based vaccines against respiratory viruses. *Front. Immunol.*, 10, 22, 2019.
182. Balan, K., Qing, W., Wang, Y. et al., Antidiabetic activity of silver nanoparticles from green synthesis using Lonicera japonica leaf extract. *RSC Adv.*, 6, 46, 40162–40168, 2016.
183. Prabhu, S., Vinodhini, S., Elanchezhiyan, C. et al., Evaluation of antidiabetic activity of biologically synthesized silver nanoparticles using Pouteria sapota in streptozotocin-induced diabetic rats. *World J. Diabetes*, 10, 1, 28–42, 2018.
184. Saratale, G.D., Saratale, R.G., Benelli, G. et al., Anti-diabetic potential of silver nanoparticles synthesized with Argyreia nervosa leaf extract high synergistic antibacterial activity with standard antibiotics against foodborne bacteria. *J. Cluster Sci.*, 28, 3, 1709–1727, 2017.
185. Mukherjee, S., Chowdhury, D., Kotcherlakota, R. et al., Potential theranostics application of bio-synthesized silver nanoparticles (4-in-1 system). *Theranostics*, 4, 3, 316, 2014.
186. Patra, C.R., Mukherjee, S., Kotcherlakota, R., Biosynthesized silver nanoparticles: A step forward for cancer theranostics? *Nanomedicine*, 9, 10, 1445–1448, 2014.
187. Jadhav, D.A., Chendake, A.D., Ghosal, D. et al., Advanced microbial fuel cell for biosensor applications to detect quality parameters of pollutants, in: *Bioremediation, Nutrients, and Other Valuable Product Recovery*, pp. 125–139, Elsevier, Netherlands, 2021.

188. Qiao, J. and Qi, L., Recent progress in plant-gold nanoparticles fabrication methods and bio-applications. *Talanta*, 223, 121396, 2021.
189. Sawant, S.N., Development of biosensors from biopolymer composites, in: *Biopolymer Composites in Electronics*, pp. 353–383, Elsevier, Netherlands, 2017.
190. Kawamura, A. and Miyata, T., Biosensors, in: *Biomaterials Nanoarchitectonics*, pp. 157–176, William Andrew Publishing, Norwich, 2016.
191. Sawant, S.N., Development of biosensors from biopolymer composites, in: *Biopolymer Composites in Electronics*, Elsevier, Netherlands, 2017.
192. Bollella, P., Schulz, C., Favero, G. *et al.*, Green synthesis and characterization of gold and silver nanoparticles and their application for development of a third generation lactose biosensor. *Electroanalysis*, 29, 1, 77–86, 2017.
193. Santhosh, A., Theertha, V., Prakash, P. *et al.*, From waste to a value added product: Green synthesis of silver nanoparticles from onion peels together with its diverse applications. *Mater. Today: Proc.*, 46, 4460–4463, 2021.
194. Zamarchi, F. and Vieira, I.C., Determination of paracetamol using a sensor based on green synthesis of silver nanoparticles in plant extract. *J. Pharm. Biomed. Anal.*, 196, 113912, 2021.
195. Caliman, F.A., Robu, B.M., Smaranda, C. *et al.*, Soil and groundwater cleanup: Benefits and limits of emerging technologie. *Clean Technol. Environ. Policy*, 13, 2, 241–268, 2011.
196. Imran, K., Mohd, F., Pratichi, S. *et al.*, Nanotechnology for environmental remediation. *Res. J. Pharm. Biol. Chem. Sci.*, 5, 3, 1916–1927, 2014.
197. Sharma, K., Singh, G., Kumar, M. *et al.*, Silver nanoparticles: Facile synthesis and their catalytic application for the degradation of dyes. *RSC Adv.*, 5, 33, 25781–25788, 2015.
198. Zhao, J., Wu, T., Wu, K. *et al.*, Photoassisted degradation of dye pollutants. 3. Degradation of the cationic dye rhodamine B in aqueous anionic surfactant/ TiO2 dispersions under visible light irradiation: Evidence for the need of substrate adsorption on TiO2 particles. *Environ. Sci. Technol.*, 32, 16, 2394–2400, 1998.
199. Kharissova, O.V., Dias, H.R., Kharisov, B.I. *et al.*, The greener synthesis of nanoparticles. *Trends Biotechnol.*, 31, 4, 240–248, 2013.
200. Xiao, C., Li, H., Zhao, Y. *et al.*, Green synthesis of iron nanoparticle by tea extract (polyphenols) and its selective removal of cationic dyes. *J. Environ. Manage.*, 275, 111262, 2020.
200. Das, S.K., Das, A.R., Guha, A.K., Gold nanoparticles: Microbial synthesis and application in water hygiene management. *Langmuir*, 25, 14, 8192–8199, 2009.
201. Yu, G., Jiang, P., Fu, X. *et al.*, Phytoextraction of cadmium-contaminated soil by Celosia argentea Linn.: A long-term field study. *Environ. Pollut.*, 266, 115408, 2020.
202. Caliman, F.A., Robu, B.M., Smaranda, C. *et al.*, Soil and groundwater cleanup: Benefits and limits of emerging technologies. *Clean Technol. Environ. Policy*, 13, 2, 241–268, 2011.

203. Imran, K., Mohd, F., Pratichi, S. et al., Nanotechnology for environmental remediation. *Res. J. Pharm. Biol. Chem. Sci.*, 5, 3, 1916–1927, 2014.
204. Salvadori, M.R., Ando, R.A., Nascimento, C.A.O. et al., Dead biomass of Amazon yeast: A new insight into bioremediation and recovery of silver by intracellular synthesis of nanoparticles. *J. Environ. Sci. Health, Part A*, 52, 11, 1112–1120, 2017.
205. Saravanan, A., Kumar, P.S., Karishma, S. et al., A review on biosynthesis of metal nanoparticles and its environmental applications. *Chemosphere*, 264, 128580, 2021.
206. Calaf, G.M. and Roy, D., Cancer genes induced by malathion and parathion in the presence of estrogen in breast cells. *Int. J. Mol. Med.*, 21, 2, 261–268, 2008.
207. Das, S.K., Das, A.R., Guha, A.K., Gold nanoparticles: Microbial synthesis and application in water hygiene management. *Langmuir*, 25, 14, 8192–8199, 2009.
208. de Jesus, R.A., de Assis, G.C., de Oliveira, R.J. et al., Environmental remediation potentialities of metal and metal oxide nanoparticles: Mechanistic biosynthesis, influencing factors, and application standpoint. *Environ. Technol. Innovation*, 24, 101851, 2021.
209. Pandian, C.J., Palanivel, R., Dhananasekaran, S., Green synthesis of nickel nanoparticles using Ocimum sanctum and their application in dye and pollutant adsorption. *Chin. J. Chem. Eng.*, 2, 8, 1307–1315, 2015.
210. Vidya, C., Prabha, M.C., Raj, M.A., Green mediated synthesis of zinc oxide nanoparticles for the photocatalytic degradation of Rose Bengal dye. *Environ. Nanotechnol. Monit. Manage.*, 6, 134–138, 2016.
211. Khodadadi, B., Bordbar, M., Nasrollahzadeh, M., Green synthesis of Pd nanoparticles at Apricot kernel shell substrate using Salvia hydrangea extract: Catalytic activity for reduction of organic dyes. *J. Colloid Interface Sci.*, 490, 1–10, 2017.
212. Ahmed, S.F., Mofijur, M., Parisa, T.A. et al., Progress and challenges of contaminate removal from wastewater using microalgae biomass. *Chemosphere*, 286, 131656, 2022.
213. Ahmed, S.F., Mofijur, M., Nuzhat, S. et al., Recent developments in physical, biological, chemical, and hybrid treatment techniques for removing emerging contaminants from wastewater. *J. Hazard. Mater.*, 416, 125912, 2021.
214. Das, P., Ghosh, S., Ghosh, R. et al., Madhuca longifolia plant mediated green synthesis of cupric oxide nanoparticles: A promising environmentally sustainable material for wastewater treatment and efficient antibacterial agent. *J. Photochem. Photobiol. B: Biol.*, 189, 66–73, 2018.
215. Luo, F., Yang, D., Chen, Z. et al., The mechanism for degrading Orange II based on adsorption and reduction by ion-based nanoparticles synthesized by grape leaf extract. *J. Hazardous Mater.*, 296, 37–45, 2015.
216. Batool, M., Qureshi, Z., and Basir, A., Removal of melachite green dye by using zinc oxide nanoparticles prepared by the green synthesis by using

Camellia Sinensis (Green Tea) Leafs Extract. *Arch. Nanomed: Open Access J.*, 1, 4, 2018.
217. Devatha, C.P., Thalla, A.K., Katte, S.Y., Green synthesis of iron nanoparticles using different leaf extracts for treatment of domestic wastewater. *J. Cleaner Prod.*, 139, 1425–1435, 2016.
218. Ahmed, S.F., Mofijur, M., Rafa, N. *et al.*, Green approaches in synthesising nanomaterials for environmental nanobioremediation: Technological advancements, applications, benefits and challenges. *Environ. Res.*, 204, 111967, 2022.
219. Mukhopadhyay, S.S., Nanotechnology in agriculture: Prospects and constraints. *Nanotechnol. Sci. Appl.*, 7, 63–7, 2014.
220. Sharon, M., Choudhary, A.K. and Kumar, R., Nanotechnology in agricultural diseases and food safety. *J. Phytol.*, 2, 4, 83–92, 2010.
221. Prasad, R., Kumar, V., Prasad, K.S., Nanotechnology in sustainable agriculture: Present concerns and future aspects. *Afr. J. Biotechnol.*, 13, 6, 705–713, 2014.
222. Rostamizadeh, E., Iranbakhsh, A., Majd, A. *et al.*, Green synthesis of Fe2O3 nanoparticles using fruit extract of Cornus mas L. and its growth-promoting roles in Barley. *J. Nanostruct. Chem.*, 10, 125–130, 2020.
223. Shaik, A.M., David Raju, M., Rama Sekhara Reddy, D., Green synthesis of zinc oxide nanoparticles using aqueous root extract of Sphagneticola trilobata Lin and investigate its role in toxic metal removal, sowing germination and fostering of plant growth. *Inorg. Nano-Metal Chem.*, 50, 7, 569–579, 2020.
224. Ndaba, B., Roopnarain, A., Rama, H. *et al.*, Biosynthesized metallic nanoparticles as fertilizers: An emerging precision agriculture strategy. *J. Integr. Agric.*, 21, 5, 1225–1242, 2022.

5

Green Conversion Methods to Prepare Nanoparticle

Pradip Kumar Sukul and Chirantan Kar*

Department of Chemistry, Amity University Kolkata, West Bengal, India

Abstract

The synthesis of nanoparticles has become the matter of great interest in recent times due to its various advantageous properties and applications in various fields. Nanomaterials are synthesized following two general approaches: the top-down, in which process, a bigger particle is broken down into tiny particles using mechanical and chemical forces. On the other approach, the bottom-up approach, a process where various interactions, such as physical, chemical and biological forces, are utilized to create self-assembled larger architecture, nanomaterials. The physical and chemical methods are very common to prepare nanostructured materials. The toxic chemicals utilized in the bottom-up approach are responsible for the induction of cancer, cytotoxicity, and environmental contamination. The toxicity problems become more serious when organic solvents, reducing agents, and capping agents are the major components of the nanomaterials. The capping agents are used to prevent the agglomeration of colloid particles. In common solution methods, different toxic chemicals, such as sodium borohydride, tetrakis(hydroxymethyl)phosphonium chloride and poly-N-vinyl pyrrolidone, are generally used. The application of nanomaterials in the biomedical and medicinal fields is limited due to the presence of toxic chemicals and organic solvents utilized during the preparation of nanoparticles, whereas in dry methods, lasers, UV light, lithography technique, and aerosol are not considered as eco-friendly techniques. The biological and environmental safety is now a major concern for the nanoparticle synthesis although physical and chemical synthesis methods for nanoparticle are common. These two popular methods are costly as well as not safe for environment. In nanomaterials, "green" synthesis concept is introduced to overcome the shortcoming of the physical and chemical methods. "Green synthesis" is a substitute, biocompatible, and environmentally safe protocol. The green

---

Corresponding author: chirantankar2000@gmail.com

Mousumi Sen and Monalisa Mukherjee (eds.) *Bioinspired and Green Synthesis of Nanostructures: A Sustainable Approach,* (115–140) © 2023 Scrivener Publishing LLC

synthesis method has created the platform to grow simple, cost-effective, and safe procedures. Several limitations also exist in green method although it is considered as safe, cost-effective, sustainable, and biocompatible. These limitations are mainly the handling of microbes and the control of dimension of the nanoparticles, crystallinity, and morphology. In addition to these limitations, the nanoparticles synthesized by green synthesis are mostly polydisperse and takes a long time to complete the reaction. However, the shortcoming could be addressed by tuning the reaction parameters, such as time, temperature, pH, and the amount of biological components taken to reduce to metal ions. This chapter is designed to understand the method of green synthesis, the effects of various parameters on the size, morphology, and amount of metal nanoparticles produced.

Keywords: Green synthesis, microorganism, biosorption, bioreduction, metal nanoparticles

5.0 Introduction

Nanotechnology deals with the strategy for the development of very tiny particles having sizes from 1 to 100 nm. The formulations of the nanostructures could be attained with the knowledge of combinations of all the natural sciences fields like chemistry, biology, physics, engineering, computer sciences and others [1].

There are two general strategies for the synthesis of nanomaterials: the top-down approach and bottom-up approach. In top-down approach, a big size particle is cut down into very small size objects using physical, chemical and biological forces (Figure 5.1). The bottom-up approach follows the synthesis of nanoparticle from the atomic level. The driving forces for the assembly of atoms into nanomaterials are chemical, physical and biological interactions [2].

Varieties of nanoparticles are developed via modifications in their size, shape and surface properties for the applications in biomedical and healthcare field. Targeted delivery of drugs or therapeutic agents can be achieved by means of different functionalized nanoparticles. Several nanoparticles are now utilizing in imaging, disease detection, sensor for biological molecules and treatment of advance diseases such as cancer, cardiovascular disease and central nervous system diseases [3, 4].

The major biomedical and healthcare fields where nanoparticles are playing pivotal role are as below in Figure 5.2.

In the medicine and diagnostic area, nanoparticles can be used for diagnosis and imaging purpose. In diagnosis applications, they can be used as luminescent probes for the detection of biomolecules and pathogens. They

Green Conversion Methods to Prepare Nanoparticle 117

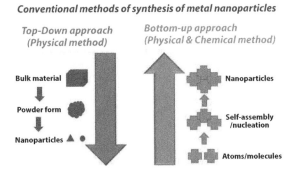

Figure 5.1 Schematic representation of the conventional synthesis technique of nanoparticles. Top-down approach which also known as physical method. Bottom-up approach which also known as physical & chemical method.

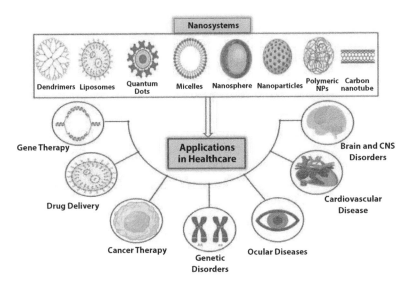

Figure 5.2 Schematic representation of the applications of various nanomaterials in biomedical sciences [with permission from Anjum *et al.* Pharmaceuticals 2021, 14, 707].

are also used as contrast agents in Magnetic Resonance Imaging due to their excellent magnetic properties. Very recently nanoparticles are using in photodynamic therapy.

Nanoparticles have also been used in designing biosensors, including those based on carbon nanotubes for identifying bacterial cells [5], measuring the glucose level [6] and detecting specific DNA fragments and regions [7].

However, the most common methods used in the synthesis of nanoparticles are physical and chemical methods. But the use of harmful chemicals could initiate serious toxicity at cellular level and environmental level [8]. The hazardous problems are now points of concern to the human civilizations. The chemicals used for the synthesis of nanoparticles such as organic solvents, reducing agents, stabilizers are toxic. These harmful effects of the utilized chemicals limit the applications of the nanoparticles in health and biomedical field.

Therefore, it is essential to develop a green, biologically safe and environmental-friendly method for the synthesis of nanomaterials [9, 10].

The attractive and alternative route towards the green synthesis of nanoparticles would be the biological synthesis of nanoparticles. Different biological entities are responsible for the synthesis of the nanoparticles which includes bacteria [11, 12], fungi [13, 14], and yeasts [15, 16]. The biological entities may create the pattern of the assembly, which will help in organization of the nanometer scale particles.

The green synthesis of can be classified according to their path of formation as: (a) Bioreduction: The chemical reduction of the metal ion by means of biological entities, such as enzyme, to form more stable zero valent metal nanoparticles. The enzyme is oxidized, and metal ion is reduced [17]; (b) Biosorption: This is common in case soil samples or water-soluble metal ions. Bacteria, fungi, or some plants synthesized peptides on their cell wall, which attracts the metal ions. These peptides helped in forming assembly of metal nanoparticles [18].

In this chapter, we cover the use of biological routes for the synthesis of metal oxide and metal nanoparticles, and various factors affecting their synthesis, and possible mechanisms employed along with likely applications of nanoparticles formed using biological entities.

The biological synthesis of the nanoparticles is described according to the biological entities responsible for the production of the metal nanoparticles as below:

5.1 Bacteria

Beveridge *et al.* [19] reported the *Pseudomonas aeruginosa* bacteria mediated synthesis of gold nanoparticles. The quasi-steady state biofilms were allowed to react with high concentration of $AuCl_3$ (i.e., 0.5–5 mM) for 30 min at 20°C. The intracellular and extracellular gold nanoparticles were formed due to the reduction of the Au^{3+} ion. The size of the nanoparticle was confirmed by transmission electron microscopy (TEM).

The concentration dependency of the nanoparticle formation inside cells and on the surfaces was also tested where they found that the treatment of 0.1 mM $AuCl_3$, was not enough to produce nanoparticles. Whereas 1 mM $AuCl_3$ produced nanoparticles predominantly at the outer surface of the cells (Figure 5.3). But when the concentration was high (e.g., 5 mM), equal amounts of gold nanoparticle was deposited on the surfaces and inside the cells (Figure 5.3c).

Magnetic nanoparticles show various kind of diagnostic and drug delivery applications. It is essential to synthesize magnetic nanoparticles of higher saturation magnetization. The commercially available magnetic nanoparticles show low saturation magnetic saturation. It was reported that bacteria can synthesize magnetic nanoparticles of smaller size and higher magnetic saturation. The cost of the synthesis via bacteria mediated was also low. But the only concern was the slower rate of conversion to nanoparticles. So many approaches are developed to make the rate faster

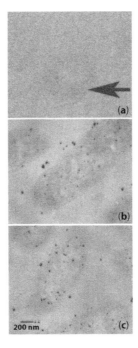

Figure 5.3 Transmission electron micrographs showing the binding of gold to the biofilm at different concentrations. (a) At 0.1 mM concentration, little binding occurred conferring only a slight contrast to cells. (b and c) At concentrations of 1 mM and 5 mM, both show contrast image of particle to the cells [with permission from Beveridge *et al.* Environmental Microbiology (2002) 4(11), 667–675].

with different strains of bacteria. The metal ions are reduced by mainly the extracellular enzymes as those enzymes show excellent redox properties. In some cases, the hydroquinones released from the bacterial cell wall can act as reducing agent [20]. It was reported that bacteria, like *Geothrix fermentans, Mycobacterium paratuberculosis, Shewanella oneidensi*, produced small and diffusible redox compounds, which reduce Fe^{3+} ions to iron oxide nanoparticles.

The extracellular synthesis of iron oxide nanoparticles was successfully demonstrated by *Bacillus subtilis* strains isolated from rhizosphere soil [21]. The supernatant of *Bacillus subtilis* was used for the nanoparticle synthesis. In a typical method, 50 mL of 2 mM Fe^{3+} solution was treated with *Bacillus subtilis* supernatant solution, and the pH was adjusted to 8.5. The mixture was incubated at 35°C for 5 days under a dark condition. The formation of the nanoparticles was confirmed by the presence of plasmon band at 250 to 350 nm in the UV-Vis spectrum of the aqueous solution (Figure 5.4).

In another approach, silver nanoparticles were synthesized using different types of bacteria strains, such as *Escherichia coli* (PTCC 1399),

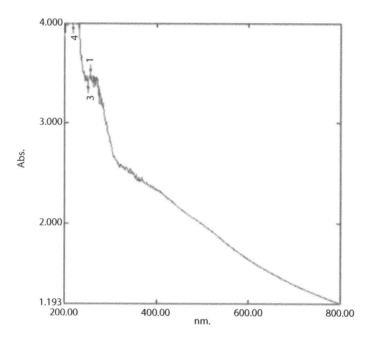

Figure 5.4 UV-Vis spectrum of Fe_3O_4 nanoparticles [with permission from Kannan *et al.*, Biotechnol. and Bioproc. E. 2012. 17: 835-840].

Lactobacillus acidophilus (PTCC 1608), *Bacillus subtilis* (PTCC 1023), *Klebsiella pneumoniae* (PTCC 1053), *Candida albicans* (PTCC 5011), *Enterobacter cloacae* (PTCC 1238), and *Staphylococcus aureus* (PTCC 1112). They have first prepared different cultured test strains by incubating at 37°C for 24 hours, and only albicans were incubated at 30°C. The supernatants were collected after centrifugation at 12,000 rpm after incubation. For the synthesis of silver nanoparticles, 10^{-3} M silver nitrate solutions were separately added to the reaction vessels containing different supernatants (1% v/v). The reaction mixtures were allowed to react under dark, as well as bright condition. The formation of the silver nanoparticles was monitored by taking aliquot of the solution frequently and measure the UV-Vis spectra. Typical surface plasmon resonance (SPR) absorbance peak at 420 to 430 nm confirmed the presence of silver nanoparticles (Figure 5.5a). The nanoparticles were further confirmed from the SEM micrograph where 50 to 100 nm particles were observed [22] (Figure 5.5b).

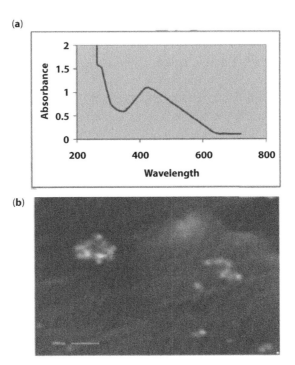

Figure 5.5 (a) The absorbance spectrum of silver nanoparticles synthesized by Klebsiella pneumoniae and (b) the SEM micrograph of silver nanoparticles [with permission from Nohi *et al.*, Winter 2008. J. Sci. I. A. U. (JSIAU), Vol 17, No. 66].

5.2 Fungi

Fungus is another important bio-species, which are used as green method for the synthesis of nanoparticles. One such interesting fungal strain, fusarium oxysporum was separated and successfully applied to produce platinum nanoparticles the intercellularly and extracellularly [23]. The growth of the nanoparticle was optically monitored by the change of color of the solution from yellow to dark brown after the reaction with fungal strain. The amount of residual hexachloroplatinic acid was measured to trace the change in concentration from a standard curve at 456 nm. It was found that the size of the nanoparticles formed was 10 to 100 nm when prepared extracellularly (Figure 5.6). The shape of the nanoparticles was different starting from circles, squares, rectangles, pentagons, to hexagons at both extracellular and intercellular compartments. In a typical procedure, first 1 g of the biomass was dissolved in 10 ml water and covered with aluminium foil to incubate at 28°C for 72 hours. Then, the fungal mycelia were separated by centrifugation from the extracellular solution. Finally, hexachloroplatinate (IV) solution was mixed with the cell-free solution. Then, the mixture was equilibrated at fixed temperature, the pH values checked at regular interval and allowed for 72 hours. The absorption band of pure hexachloroplatinic acid at 456 nm and the platinum nanoparticles at 261 nm were monitored to track the growth of the nanoparticles.

Eukaryotic organism was used to synthesize magnetic nanoparticles. Sastry *et al.* extracted two fungi, *F. oxysporum* and *Verticillium* sp. for the green synthesis of nanoparticles [24]. They have mixed the fungi with the aqueous solution of $K_3[Fe(CN)_6]$ and $K_4[Fe(CN)_6]$ in a 2:1 molar ratio and incubated for 24 hours to results in the formation of magnetic nanoparticles extracellularly. TEM micrographs confirm the presence of 20 to 50 nm particles of irregular shapes (Figure 5.7). A cationic protein with a molecular weight of 55 kDa induces the hydrolysis of the anionic complexes, which was further transformed into magnetite particles.

In a detail synthetic method, first *Fusarium oxysporum* and *Verticillium* sp. were cultured. Then, the fungal mycelia were harvested and washed under sterile conditions. In a 500-mL Erlenmeyer flask, 20 g of each of the fungal biomasses were added to 100 ml aqueous solutions of $K_3[Fe(CN)_6]$ and $K_4[Fe(CN)_6]$ in a 2:1 molar ratio. The pH of the suspension was maintained at 3.1. Then, the Erlenmeyer flask was shaken at 200 rpm at 27°C for 24 hours.

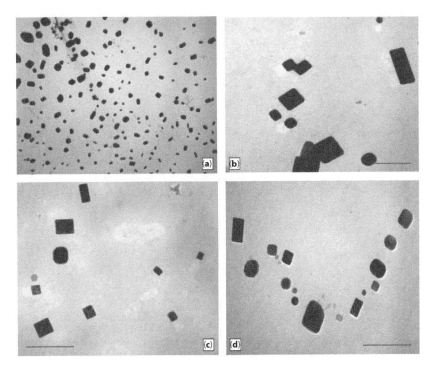

Figure 5.6 TEM images of (a) samples of nanoparticles after 72 hours incubation with fungal cells. (b) The sample after 2 hours incubation with extracellular fungal solution. (c) Sample showing square and rectangular nanoparticles after 2 hours of incubation with the extracellular fungal solution. (d) Sample showing irregular nanoparticles of widely varying shapes and sizes in the 0 hour incubation [with permission from Whiteley et al., Nanotechnology. 2006. 17, 3482–3489].

Silver nanoparticles were synthesized by Sanghi et al. using a white rot fungus C. versicolor [25]. They have succeeded to synthesize silver nanoparticles intracellularly as well as extracellularly.

The fungal mycelium was incubated and grown at 37°C for 7 days at pH 5.5 to 6 with continuous agitation at 200 rpm in an incubator agitator. Typically, in a 150 ml Erlenmeyer flask, 1 mM $AgNO_3$ solution was mixed with 7 g of fungal biomass. The flask was kept at 37°C in the dark. The color of the mixture was turned to light brown after 48 hours and attained dark brown after 72 hours which indicated the formation of the silver nanoparticles. The formation of the silver nanoparticles was further confirmed by UV-Vis study. UV-Vis spectra show the onset of surface plasmon band for silver nanoparticles at around 430 to 440 nm (Figure 5.8).

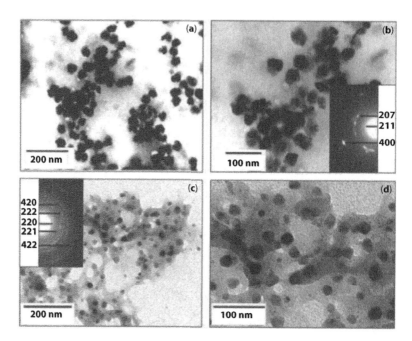

Figure 5.7 TEM images of iron oxide nanoparticles synthesized using *Fusarium oxysporum* before (a, b) and after (c, d) calcination at 400°C for 3 hours after exposure on iron cyanide precursors [with permission from Sastry *et al.*, Small. 2006. 2, 1, 135–141].

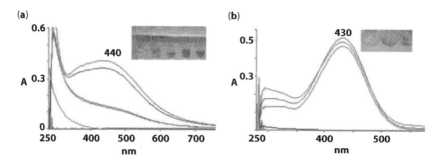

Figure 5.8 Absorbance spectra of (a) Ag/media and (b) Ag/mycelium solutions at normal pH depicting the increase in color intensity with increased absorbance with time [with permission from Verma *et al.*, Bioresource Technology. 2009. 100, 501–504].

Extracellular green synthesis of silver nanoparticles was developed by D'Souza *et al.* using fungal strain, *A. fumigatus* [26]. The biomass required for the synthesis of the nanoparticle was harvested in a liquid media containing 7 g/l KH_2PO_4, 2 g/l K_2HPO_4, 0.1 g/l $MgSO_4 \cdot 7H_2O$, 1 g/l $(NH_4)_2SO_4$, 0.6 g/l yeast extract, and 10 g/l glucose. The container was incubated at 25°C on an orbital shaker at 150 rpm. The biomass was extracted after 72 hours of growth and washed before using further. 20 g of the biomass was taken in an Erlenmeyer flask. 200 ml of Milli Q-deionized water was added to the flask and stirred for 72 hours at 25°C. After the incubation, the cell extract was filtered through Whatman filter no. 1 paper. 50 ml of the cell filtrate was taken in a 250-ml Erlenmeyer flask with 1 mM $AgNO_3$ solution and agitated at 25°C. The growth of the nanoparticles was monitored by UV-Vis study where the typical band of silver nanoparticles appeared at 420 nm (Figure 5.9a, b). Further, the size of the particles was measured by TEM micrograph (Figure 5.9c). The size of the particles was typically 5 to 25 nm. The X-ray diffraction of the film prepared from the nanoparticles showed the presence of crystalline peak for silver nanoparticles (Figure 5.9d).

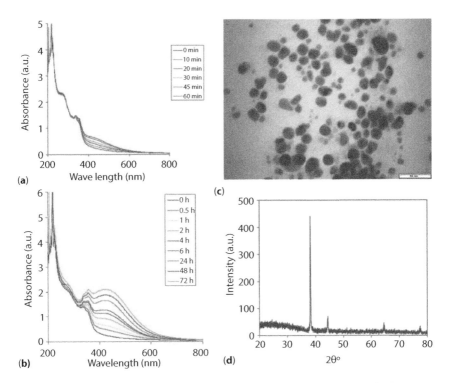

Figure 5.9 Absorbance spectra of aqueous solution containing cell filtrate and silver ion (1 mM) (a) short incubation and (b) long incubation time. (c) TEM micrograph of the silver nanoparticles synthesized by A. fumigatus. (d) XRD pattern of nanoparticle film on Si surface [with permission from D'Souza et al., Colloids and Surfaces B: Biointerfaces. 2006. 47, 160–164.].

Extracellular synthesis of CdS nanoparticles was developed where the process controlled purely secreted enzyme. The fungus, *Fusarium oxysporum* was used for the synthesis of CdS nanoparticles [27]. First, *Fusarium oxysporum* was grown on potato-dextrose agar slants at 25°C in a 500-ml Erlenmeyer flask. 100 ml MGYP media (a mixture of yeast extract [0.3%], malt extract [0.3%], glucose [1.0%], and peptone [0.5%]) was taken for the incubation and agitated at 200 rpm at 25°C to 28°C for 96 hours. Then centrifugation at 5000 rpm at 10°C was done to collect the mycelia from the culture. The mycelia were then washed with sterile water under aseptic condition. 10 g of the biomass was then resuspended in a 500-ml conical flask with 10^{-3} M aqueous $CdSO_4$ solution at pH 5.5 to 6. The whole mixture was finally agitated at 28°C and the aliquot was taken frequently to monitor the growth of the nanoparticles (Figure 5.10).

The role of the reductase enzymes releases by the fungus was established by performing control experiment. The control experiments were conducted by taking *Fusarium oxysporum* biomass with (a) a 10^{-3} M aqueous solution of $CdNO_3$ and (b) an aqueous solution containing a mixture of the salts $CdCl_2$ and Na_2SO_4 (concentration ratio of Cd^{2+}/SO_4^{2-} = 1:1 in solution) of 10^{-3} M.

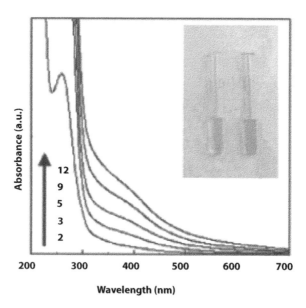

Figure 5.10 Absorbance spectra of aqueous solution of 10^{-3} M $CdSO_4$ solution as a function of time (in days) after addition of fungal biomass. The inset shows test tubes containing $CdSO_4$ solution before (no color) and after the reaction with fungal biomass for 12 days (yellow color) [with permission from Sastry *et al.*, J. Am. Chem. Soc. 2002, 124, 12108-12109].

It was interesting observation that CdS nanoparticles were formed in case of the later mixed salt solution just after 4 days. Whereas in a former case, even after 21 days, there was no evidence of the formation of nanoparticles.

The entrapment of the Cd^{2+} ion first, then the reaction with sulfide ions was ruled out by doing the control experiment. In the control experiment, the fungal biomass was suspended in sterile water and allowed to react for 12 days. Then, the formed precipitate was filtered and 10^{-3} M $CdSO_4$ solution was allowed to react with filtrate. The appearance of the bright yellow color of the reaction mixture after 6 days indicated the formation of the cadmium sulfide nanoparticles. This control experiment confirmed the release of the reductase enzyme from *Fusarium oxysporum*, which was responsible for the production of CdS nanoparticles.

5.3 Yeast

Non-conventional yeast, such as *Yarrowia lipolytica*, is an important biological species shows application in biotechnological area [28]. *Yarrowia*

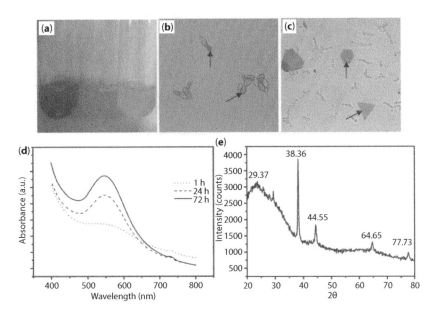

Figure 5.11 Photos of Y. lipolytica cells incubated with 1 mM HAuCl4, (a) image of tubes at pH 9.0, 7.0, and 2.0 (left to right). Optical images of cells at (b) pH 9.0 and (c) pH 2.0, (d) UV-Visible spectra of *Y. lipolytica* cells at pH 9.0 as function of time, (e) XRD pattern of *Y. lipolytica* biofilm [with permission from Kulkarni *et al.*, Materials Letters. 2009. 63, 1231–1234].

lipolytica is also used for waste treatment. Gold nanoparticle can be synthesised using *Y. lipolytica* as reported by Kulkarni et al. [29].

Typically, *Y. lipolytica* was harvested in YNB glucose (7 g yeast nitrogen base; 10 g l^{-1} glucose in distilled water) for 24 hours at 30°C, agitated at 130 rpm and centrifuged at 6000 g for 10 min at 4°C. Cells were made free from residual glucose by washing with distilled water.

The biomass (1010 ml^{-1}) were re-suspended in aqueous solution containing $HAuCl_4$ solutions. The mixture was incubated at 30°C for 72 hours and agitated at 130 rpm. The pH was maintained at 7.0 or 9.0. The UV-Vis spectra were recorded frequently to trace the growth of the nanoparticles. The characteristic surface plasmon resonance band at 545 nm confirms the presence of gold nanoparticles (Figure 5.11 a-d). The X-ray diffraction pattern of the films prepared from the nanoparticles exhibits peak due to (111), (200), (220), and (311) Bragg reflection at 2θ= 38.36°, 45.55°, 64.65°, and 77.73°, respectively, confirmed the presence of gold nanoparticles (Figure 5.11 e).

Amorphous iron phosphate nanoparticles were synthesized by Yan et al. using yeast cells [30]. For the synthesis of the nanoparticles, 1 g yeast cells were grown in 80 ml glucose solution in distilled water for 30 min at 36°C. Separately, two metal salt solutions were prepared.

First one, 540-mg $FeCl_3.6H_2O$ was dissolved in 20-ml distilled water and second, aqueous solution of Na_3PO_4 (0.12 mol/l). The $FeCl_3$ solution

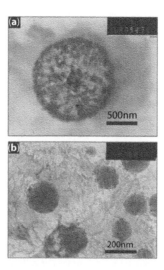

Figure 5.12 TEM images of iron phosphate nanoparticles biomineralized in a yeast cell (a) a complete yeast cell and (b) the magnification part of yeast cells [with permission from Yan et al., Materials Science and Engineering C. 2009. 29, 1348–1350].

was added dropwise into the solution of the yeast cells and stirred vigorously. The resulting mixture was continuously stirred at 25°C for 3 hours. Then, the previously prepared Na_3PO_4 solution was added to the mixture of $FeCl_3$-yeast cells suspension under constant stirring. The pH of the final solution was adjusted to 5.5 and incubated for 4 days. The nanoparticles were recovered from the final solution by centrifugation and washing with distilled water and ethanol. The size of the nanoparticles was observed by TEM micrograph. TEM micrograph shows the 50- to 200-nm particle size distribution of the iron phosphate aggregations (Figure 5.12).

5.4 Viruses

A virus is an "infectious microbe consisting of a segment of nucleic acid (either DNA or RNA) surrounded by a protein coat." The coating surrounding the nucleic acid can work as a platform to synthesize nanoparticles. Recently, Górzny et al. [31] has used a template of a tobacco mosaic virus to synthesize platinum nanotubes with high surface area and high aspect ratio. The virus provides a tubular frame with aspect ratio (16:9) and the inner and outer surface charge is dependent on the pH of the environment and hence can be controlled easily. These interesting properties make the tobacco mosaic virus a suitable scaffold for inorganic assembly. Additionally, virus fragments also tend to self-assemble into long filament like structure which leads to the development of nanoelectronic component. Usually, wet-lab synthetic chemistry is applied for the metallization of the virus. In this technique, a metal salt is added to the virus, and the pH of the medium is controlled to get the desirable charge on the virus surface, this will allow the metal ions to stick to the surface of the virus particle and then the reduction of the metal ions is carried out using reducing agents, such as hydrazine, dimethylamine borane (DMAB), or sodium hypophosphite.

The nanotubes reported by Górzny et al. are synthesized by alcohol-mediated reduction of the platinum salt where the virus particles worked as a template. The Platinum (II) metal ions stick to the virus surface due to electrostatic interaction, and the reduction from platinum (II) to platinum (0) happens on the surface of the virus particles (Figure 5.13). The procedure of transition metal nanoparticle synthesis using aqueous alcohol was first reported by Hirai et al. [32]. Generally, alcohols like ethanol, methanol, or glycol work simultaneously as the solvent and reductant. The alcohol converts to aldehyde or ketone in the process. The process is very simple and can be used for preparation of nanotubes with many useful functions due to their high porous nature, good uniformity, and excellent aspect ratio.

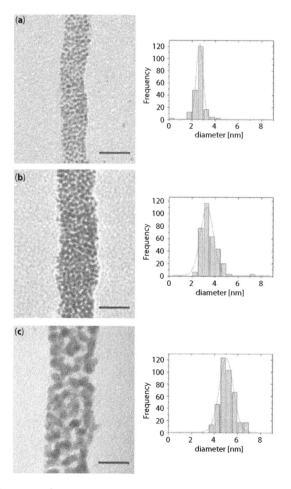

Figure 5.13 Transition electron microscope images of Tobacco mosaic virus with the platinum nanoparticles embedded on its surface. Inset images show the relationship between the size of platinum nanoparticle with varying concentration of platinum salt. (a–c) 0.05 mM, 0.125 mM, and 0.25 mM, Pt salt concentrations, respectively. The average diameters of platinum nanoparticles are 2.6 nm (a), 3.5 nm (b), and 5 nm (c). Scale bar in all images is 20 nm. Adapted and reproduced in part from Górzny et al. [31]; with permission from Wiley.

Like tobacco mosaic virus, the surface of cowpea mosaic virus can also be used as a template for synthesizing nanoparticles [33]. Cowpea mosaic virus is a plant RNA virus containing a single RNA strand. The three-dimensional shape of the virus is icosahedral with a diameter of 28 nm is a bipartite, single-stranded RNA plant virus with icosahedral symmetry and a diameter of 28 nm. The capsid of the virus is made up of two different proteins, the large one is around 41 kDa and the small one is around 24 kDa. The virus particles are found to be quite stable within a wide pH range (3.5–10). They are very stable at room temperature and can withstand 60°C temperature for an hour. The robust properties make the virus an efficient template for chemical, as well as genetic modification. In recent literature, many scientists have reported new peptide molecules that promote specific mineralization on the surface loops of cowpea mosaic virus. They have also described various processes for chemically coupling appropriate peptide to the virus surface without genetically modifying the capsid. The peptide coupled virus surface can work as an efficient surface for monodispersing silica and iron-platinum nanoparticles.

Aljabali et al. [33] have used the virus capsid for the synthesis of metallic nanoparticles without being genetically modifying the virus or performing any chemical change of the surface. They have used electroless deposition method, which is an autocatalytic redox process for chemically reducing metal ions without the use of external current. This process is commonly used for the formation of thin metallic layers with uniform composition and thickness. The metallization of the virus particles is done by preactivating the surface of the capsid with palladium (II) followed by electroless deposition of platinum, iron, nickel, cobalt, iron–nickel, and platinum–cobalt at room temperature, which leads to the formation of monodisperse metallic nanoparticles. The method is reported to be completely environmentally benign as well as green.

Along with significant development in the use of groups present on cowpea mosaic virus surface for metal deposition, genetic moderation has also been considered to improve the properties of the surface and enhance fabrication with metals and metalloids. For instance, Steinmetz et al. [34] have developed a process of genetically modifying the virus to display a dodecapeptide on the surface, which helps it in silica binding. They have incubated the virus particles with tetraethylorthosilicate and aminopropyltriethoxysilane precursors to generate a 2-nm layer of silica on the virus surface. In another study by Shah et al. [35], the virus was genetically modified to express specific peptides on their surface, which binds metal ions of iron and platinum and can be converted to metal nanosheets with a thickness of 1 nm by exposing it to typical reducing agents. The virus can be genetically modified to produce cysteine residues at specific sites on their surface to anchor metal ions. In this regard, Blum et al. [36] have reported a technique to synthesize gold nanoparticle by anchoring gold ion on the surface using the thiol-metal interaction of the cysteine residue (Figure 5.13). These bound gold nanoparticles now form a network structure by interlinking using di-thiolated linkers (Figure 5.14). Due to this network

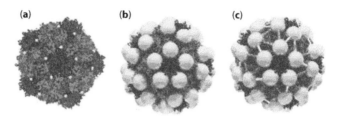

Figure 5.14 (a) Cowpea mosaic virus surface genetically modified with cysteine residues (white spots). (b) Formation of gold nanoparticles over the cysteine residue. (c) Network formation between the gold nanoparticles with di-thiolated Pt linkers. Adapted and reproduced in part from Blum *et al.* [36]; with permission from Elsevier.

structure, they are quite conductive in nature, and the conductivity can be modified by chemical techniques; thus, an effective substance for designing biosensors.

5.5 Algae

In case of viruses, we have seen that the surface of the virus capsid can work as the template for nanoparticle synthesis, but algae have the ability to absorb heavy metal ions in its cytoplasm. As they accumulate metal ions, so they can be utilized for the biosynthesis of metallic nanoparticles.

In cellular systems, metal ions are allowed to enter the cellular environment as soluble metal salt and then they are reduced to for the metal particles. Hence, the effectiveness of reduction depends on various factors of the system like the reduction potential and capacity. Compared to standard hydrogen electrode, reduction potential in higher plant cells is 0 V, which indicates that they can be utilized as a nanoparticle production vehicle for only precious and semiprecious metals, like platinum, palladium, rhodium, iridium, and ruthenium [37]. Like higher plants microalgae could also be a favorable platform for producing nanoparticle, in many ways, using algae could be even advantageous. Compared to higher plants, microalgae grow extremely rapidly and can grow up to ten times faster than normal higher plant [38]. New methods are developed to produce a large scale of microalgae in a controlled manner to provide a regular supply of the biomass. Moreover, algal cells are easy to handle, a large portion of aqueous slurries containing algae can be prepared for infusion of metal ions, and after the synthesis of the nanoparticles, the algal biomass can be transferred in large volume [39] for disruption of the cells and commercial extraction [40] of the intercellular nanoparticle.

Luangpipat *et al.* [41] has come up with a new technique using the live cells of the green microalga *chlorella vulgaris*, which allows easy recovery of the nanoparticles. The gold salt was infused in the cell by incubation of the microalga in a solution of gold chloride and the gold nanoparticles are finally recovered by centrifugation of the cells. As shown in Figure 5.15, the intercellular nanoparticles are easily detected using transmission electron microscopy and confirmed by electron microscopy. The metallic character of the particles is verified by synchrotron-based X-ray powder diffraction and X-ray absorption spectroscopy. The average diameter of the intercellular gold nanoparticles is around 50 nm with a recovery rate of 97% at a concentration of 1.4% Au in the algae. The same method was also applied with other metal ions, e.g., salts of palladium, ruthenium, and rhodium, which results in the formation of similar types of nanoparticle for the corresponding metal ions. Although efficiency of *C. vulgaris* on reducing different metal ions is not the same, it has been found that *C. vulgaris* can

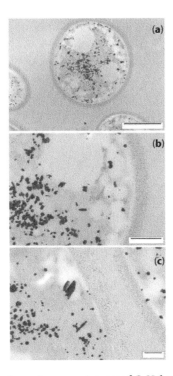

Figure 5.15 Transition electron microscope images of C. Vulgaris cells, where the black dots are the gold nanoparticles. The scale bars denoted in the images are 2000 nm (a), 500 nm (b), and 200 nm (c), respectively. Adapted and reproduced in part from Luangpipat *et al.* [41]; with permission from Springer.

readily reduce Gold (III) to metallic gold nanoparticles but less effective for other noble metal ions.

5.6 Plants

One of the major advantages for using plant as a medium for green and environmentally friendly production of nanoparticles is their non-pathogenic nature, hence different routes can be studied without any potential risk. A large variety of nanoparticles using different kind of metals is already produced using different plants [42–45]. These nanoparticles have a wide range of physical, thermal, optical, and electrical properties. Shankar *et al.* [46] have reported the use of Geranium plant extract (leaf broth) for biological reduction of gold ions to nanosized metallic gold particles. In this case, the alkaloids present in the leaf extract behave as the reducing agent, during the reduction of the gold ions the alcohol groups of these alkaloids is oxidized to carbonyl groups. The proteins present in the plant extract also behave as the stabilizing agent as the amide groups of the protein bind with the metal surface. Transmission electron microscopic studies confirmed the formation of spherical shaped nanoparticles with an average diameter of 14 nm [47]. Broad X-ray diffraction pattern also supports the crystalline nature and nanodimension of the particles. Like the leaf extract, Geranium root extract were also used for synthesizing round-shaped gold nanoparticles with an average diameter of 20 nm.

Indian gooseberry or amla is a native of tropical and southern Asia, the herbal benefit of this plant is well known. Recently [48], the extract of these berries is used for synthesizing highly stable gold nanoparticles with a diameter range of 12 to 25 nm. Like Indian gooseberry, another important tree with herbal benefit is tamarind, the extract of the tamarind leaf is also used for synthesizing gold nanoparticles [49]. TEM analysis confirmed that the shape of these nanoparticles (where tamarind extract is used as the reducing agent) is triangular with an edge length of 100 to 500 nm and thickness of 20 to 40 nm. The high stability of the synthesized nanoparticles is probably due to the presence of tartaric acid in these plant extracts. The presence of carbonyl groups is confirmed by FTIR analysis of the nanoparticles, indicating the presence of tartaric acid as the stabilizing molecule.

In another article, Shankar *et al.* [50] have reported the formation silver nanoparticles by using an extract obtained from the leaf of neem tree. The rate of reduction is also quite efficient as 90% of the silver nitrate is converted to metal nanoparticles within just 4 hours of time, which is confirmed by

SPR analysis showing a peak at 450 nm. Although the reduction process for silver is slower compared to gold because of the lower redox potential of metallic silver. Similar to Geranium, the alkaloids present in the extract of neem leaves help in the reduction process. TEM analysis shows that the shape of the silver nanoparticles is round with a diameter range of 5 to 35 nm. Aloe vera extract is also used for the synthesis of silver nanoparticles, but in this case, the reduction of silver ions is facilitated by the addition of ammonia, as it helps in solubilizing the Ag(I) by complex formation. The shape of these nanoparticles is again spherical, having an average diameter of 17 nm [51].

Rubber a natural polymer is used in many healthcare products due to its nontoxic and biocompatible nature. The presence of silver in rubber could be beneficial due to its antimicrobial and electrical property. It has been found that the proteins present in natural rubber can absorb silver ions and can act as a template for reducing them to form metallic silver. A group of scientists has used this technique for absorbing silver ions and use UV-light for the reduction of the silver salts [52]. They have used different natural rubber-based template for example de-proteinized natural rubber latex, natural rubber latex containing sodium dodecyl sulfate and aqueous solutions of bovine serum albumin. TEM images show the presence of globular natural rubber-silver nanocomposite with an average size of 5.7 nm.

Apart from these, there are many other examples of silver nanoparticle produced from *Jatropha curcas* extract, [53] the leaf extracts of *Acalypha indica* [54] and from the seed of *Medicago sativa*.

These naturally sources of reducing agent and stabilizers are chemically less toxic and mostly biodegradable. Hence, the abovementioned techniques are an effective way for green synthesis of nanocomposites with different metals.

5.7 Conclusion and Perspectives

The application of the nanomaterials in the field of biomedical and healthcare system has been grown exponentially within the last few decades. However, their applications are still associated with several limitations, which are mostly toxicity issues. There is a need of identifying potential hazards of these nanomaterials to avoid any unforeseen circumstances. Their unpredictable toxicity on the human body is challenging to overcome. The chemical composition of the metal nanoparticles and the shell part of the particles are the controlling factor for the toxicity.

Green synthesis strategy furnishes safe, clean, and environment-friendly methodology to produce metallic nanoparticles. There is a large demand of developing new protocols for making cost-effective and high yield production of nanoparticle in comparable to conventional methods. The significant step will be the improvement of the eco-friendly processes for the creation of metallic nanoparticles. There are tremendous efforts manifested toward the synthesis of metallic nanoparticles by microorganism. But the limitation is there for the efficiency of the methods and the control of the morphology. The most important challenge is the reaction time in microorganism-based production of the nanoparticles. These methods are very slow which takes in general several days to be completed compared to the conventional method of synthesis. The decomposition of the nanoparticles with time formed by microorganism is another issue needs to be addressed.

The challenges may overcome by the variation of the physicochemical parameters of the substrate and the type of microorganism to create nanoparticles of controlled shape, size, and long stability in the biological medium. Owing to the rich biodiversity of microbes, their potential as biological materials for nanoparticle synthesis is still to be explored.

References

1. Medvedeva, N.V., Ipatova, O.M., Ivanov, Y.D., Drozhzhin, A.I., Archakov, A.I., Nanobiotechnology and nanomedicine. *Biochem. (Moscow) Suppl. B Biomed. Chem.*, 1, 114–124, 2007.
2. Anastas, P.T. and Horvath, I.T., *Green chemistry for a sustainable future*, Wiley-Blackwell, Hoboken, 2011.
3. Shiku, H., Wang, L., Ikuta, Y., Okugawa, T., Schmitt, M., Gu, X., Akiyoshi, K., Sunamoto, J., Nakamura, H., Development of a cancer vaccine: Peptides, proteins, and DNA. *Cancer Chemother. Pharmacol.*, 46, S77–S82, 2000.
4. Saul, J.M., Annapragada, A.V., Bellamkonda, R.V., A dual-ligand approach for enhancing targeting selectivity of therapeutic nanocarriers. *J. Controlled Release*, 114, 277–287, 2006.
5. Timur, S., Anik, U., Odaci, D., Gorton, L., Development of a microbial biosensor based on carbon nanotube (CNT) modified electrodes. *Electrochem. Commun.*, 9, 1810–1815, 2007.
6. Muguruma, H., Matsui, Y., Shibayama, Y., Carbon nanotube–plasma polymer-based amperometric biosensors: Enzyme-friendly platform for ultrasensitive glucose detection. *Jpn. J. Appl. Phys.*, 46, 6078–6082, 2007.
7. Clendenin, J., Kim, J.-W., Tung, S., An aligned carbon nanotube biosensor for DNA detection. *2nd IEEE Int. Conf. Nano/Micro Eng. Mol. Syst.*, IEEE, pp. 1028–1033, 2007.

8. Gupta, R. and Xie, H.J., Nanoparticles in daily life: Applications, toxicity and regulations. *Environ. Pathol. Toxicol. Oncol.*, 37, 209–230, 2018.
9. Kulkarni, N. and Muddapur, U., Biosynthesis of metal nanoparticles: A review. *J. Nanotechnol.*, 2014, 1–8, 2014.
10. Jain, N., Bhargava, A., Majumdar, S., Tarafdar, J., Panwar, J., Extracellular biosynthesis and characterization of silver nanoparticles using aspergillus flavus NJP08: A mechanism perspective. *Nanoscale*, 3, 635–641, 2010.
11. Roh, Y., Lauf, R.J., McMillan, A.D., Zhang, C., Rawn, C., Bai, J. *et al.*, Microbial synthesis and the characterization of metal-substituted magnetites. *Solid State Commun.*, 118, 529–534, 2001.
12. Husseiny, M., El-Aziz, M., Badr, Y., Mahmoud, M., Biosynthesis of gold nanoparticles using Pseudomonas aeruginosa. *Spectrochim. Acta A*, 67, 1003–1006, 2007.
13. Mukherjee, P., Ahmad, A., Mandal, D., Senapati, S., Sainkar, S., Khan, M. *et al.*, Fungus-mediated synthesis of silver nanoparticles and their immobilization in the mycelial matrix: A novel biological approach to nanoparticle synthesis. *Nano Lett.*, 1, 515–519, 2001.
14. Bhainsa, K. and D'Souza, S., Extracellular biosynthesis of silver nanoparticles using the fungus Aspergillus fumigatus. *Colloids Surf. B Biointerfaces*, 47, 160–164, 2006.
15. Dameron, C.T., Reese, R.N., Mehra, R.K., Kortan, A.R., Carroll, P.J., Steigerwald, M.L. *et al.*, Biosynthesis of cadmium sulphide quantum semiconductor crystallites. *Nature*, 338, 596–597, 1989.
16. Gericke, M. and Pinches, A., Microbial production of gold nanoparticles. *Gold Bull.*, 39, 22–28, 2006.
17. Deplanche, K., Caldelari, I., Mikheenko, I., Sargent, F., Macaskie, L., Involvement of hydrogenases in the formation of highly catalytic Pd(0) nanoparticles by bioreduction of Pd(II) using Escherichia coli strains. *Microbiology*, 156, 2630–2640, 2010.
18. Yong, P., Rowson, N.A., Farr, J.P.G., Harris, I.R., Macaskie, L.E., Bioaccumulation of palladium by Desulfovibrio desulfuricans. *J. Chem. Technol. Biotechnol.*, 77, 593–601, 2002.
19. Karthikeyan, S. and Beveridge, T.J., Pseudomonas aeruginosa biofilms react with and precipitate toxic soluble gold. *Environ. Microbiol.*, 4, 11, 667–675, 2002.
20. Baker, R.A. and Tatum, J.H., Novel anthraquinones from stationary cultures of Fusarium oxysporum. *J. Ferment. Bioeng.*, 85, 359–361, 1998.
21. Sundaram, P.A., Augustine, R., Kannan, M., Extracellular biosynthesis of iron oxide nanoparticles by Bacillus subtilis strains isolated from rhizosphere soil. *Biotechnol. Bioprocess. Eng.*, 17, 835–840, 2012.
22. Minaeian, S., Shahverdi, A.R., Nohi, A.S., Extracellular biosynthesis of silver nanoparticles by some bacteria, Winter 2008. *J. Sci. I. A. U. (JSIAU)*, 17, 1–4, 66.
23. Riddin, T.L., Gericke, M., Whiteley, C.G., Analysis of the inter- and extracellular formation of platinum nanoparticles by Fusarium oxysporum f. sp.

Lycopersici using response surface methodology. *Nanotechnology*, 17, 3482–3489, 2006.
24. Bharde, A., Rautaray, D., Bansal, V., Ahmad, A., Sarkar, I., Yusuf, S.M., Sanyal, M., Sastry, M., Extracellular biosynthesis of magnetite using fungi. *Small*, 2, 1, 135–141, 2006.
25. Sanghi, R. and Verma, P., Biomimetic synthesis and characterisation of protein capped silver nanoparticles. *Bioresour. Technol.*, 100, 501–504, 2009.
26. Bhainsa, K.C. and D'Souza, S.F., Extracellular biosynthesis of silver nanoparticles using the fungus Aspergillus fumigatus. *Colloids Surf. B: Biointerfaces.*, 47, 160–164, 2006.
27. Ahmad, A., Mukherjee, P., Mandal, D., Senapati, S., Khan, M.I., Kumar, R., Sastry, M., Enzyme mediated extracellular synthesis of CdS nanoparticles by the fungus, Fusarium oxysporum. *J. Am. Chem. Soc.*, 124, 12108–12109, 2002.
28. Barth, G. and Gaillardin, C., Physiology and genetics of the dimorphic fungus Yarrowia lipolytica. *FEMS Microbiol. Rev.*, 19, 219–237, 1997.
29. Agnihotri, M., Joshi, S., Ameeta, R.K., Zinjarde, S., Kulkarni, S., Biosynthesis of gold nanoparticles by the tropical marine yeast Yarrowia lipolytica NCIM 3589. *Mater. Lett.*, 63, 1231–1234, 2009.
30. He, W., Zhou, W., Wang, Y., Zhang, X., Zhao, H., Li, Z., Yan, S., Biomineralization of iron phosphate nanoparticles in yeast cells. *Mater. Sci. Eng. C*, 29, 1348–1350, 2009.
31. Górzny, M.Ł., Walton, A.S., Evans, S.D., Synthesis of high-surface-area platinum nanotubes using a viral template. *Adv. Funct. Mater.*, 20, 1295–1300, 2010.
32. Hirai, H., Nakao, Y., Toshima, N., Adachi, K., Colloidal rhodium in polyvinyl alcohol as hydrogenation catalyst of olefins. *Chem. Lett.*, 5, 905–910, 1976.
33. Aljabali, A.A., Barclay, J.E., Lomonossoff, G.P., Evans, D.J., Virus templated metallic nanoparticles. *Nanoscale*, 2, 2596–2600, 2010.
34. Steinmetz, N.F., Lin, T., Lomonossoff, G.P., Johnson, J.E., Structure-based engineering of an icosahedral virus for nanomedicine and nanotechnology, in: *Viruses and Nanotechnology*, pp. 23–58, 2009.
35. Shah, S.N., Steinmetz, N.F., Aljabali, A.A., Lomonossoff, G.P., Evans, D.J., Environmentally benign synthesis of virus-templated, monodisperse, iron-platinum nanoparticles. *Dalton Trans.*, 40, 8479–8480, 2009.
36. Blum, A.S., Soto, C.M., Sapsford, K.E., Wilson, C.D., Moore, M.H., Ratna, B.R., Molecular electronics based nanosensors on a viral scaffold. *Biosens. Bioelectron.*, 26, 2852–2857, 2011.
37. Haverkamp, R.G. and Marshall, A.T., The mechanism of metal nanoparticle formation in plants: Limits on accumulation. *J. Nanopart. Res.*, 11, 1453–1463, 2009.
38. Chisti, Y., Fuels from microalgae. *Biofuels*, 1, 233–235, 2010.
39. Grima, E.M., Belarbi, E.H., Fernández, F.A., Medina, A.R., Chisti, Y., Recovery of microalgal biomass and metabolites: Process options and economics. *Biotechnol. Adv.*, 20, 491–515, 2003.

40. Chisti, Y. and M.-Young, M., Disruption of microbial cells for intracellular products. *Enzyme Microb. Technol.*, 8, 194–204, 1986.
41. Luangpipat, T., Beattie, I.R., Chisti, Y., Haverkamp, R.G., Gold nanoparticles produced in a microalga. *J. Nanopart. Res.*, 13, 6439–6445, 2011.
42. Narayanan, K.B. and Sakthivel, N., Green synthesis of biogenic metal nanoparticles by terrestrial and aquatic phototrophic and heterotrophic eukaryotes and biocompatible agents. *Adv. Colloid Interface Sci.*, 169, 59–79, 2011.
43. Iravani, S. and Zolfaghari, B., Green synthesis of silver nanoparticles using Pinus eldarica bark extract. *BioMed. Res. Int.*, 2013, 1–5, 2013.
44. Mittal, A.K., Chisti, Y., Banerjee, U.C., Synthesis of metallic nanoparticles using plant extracts. *Biotechnol. Adv.*, 31, 346–356, 2013.
45. Das, R.K., Pachapur, V.L., Lonappan, L., Naghdi, M., Pulicharla, R., Maiti, S., Cledon, M., Dalila, L.M., Sarma, S.J., Brar, S.K., Biological synthesis of metallic nanoparticles: Plants, animals and microbial aspects. *Nanotechnol. Environ. Eng.*, 2, 1–21, 2017.
46. Shankar, S.S., Ahmad, A., Pasricha, R., Sastry, M., Bioreduction of chloroaurate ions by geranium leaves and its endophytic fungus yields gold nanoparticles of different shapes. *J. Mater. Chem.*, 13, 1822–1826, 2003.
47. Shankar, S.S., Rai, A., Ahmad, A., Sastry, M., Biosynthesis of silver and gold nanoparticles from extracts of different parts of the geranium plant. *Appl. Nanosci.*, 1, 69–77, 2004.
48. Ankamwar, B., Damle, C., Ahmad, A., Sastry, M., Biosynthesis of gold and silver nanoparticles using Emblica officinalis fruit extract, their phase transfer and transmetallation in an organic solution. *J. Nanosci. Nanotechnol.*, 5, 1665–1671, 2005.
49. Ankamwar, B., Chaudhary, M., Sastry, M., Gold nanotriangles biologically synthesized using tamarind leaf extract and potential application in vapor sensing. *Synth. React. Inorg. Met.-Org. Nano-Metal Chem.*, 35, 19–26, 2005.
50. Shankar, S.S., Rai, A., Ahmad, A., Sastry, M., Rapid synthesis of Au, Ag, and bimetallic Au core-Ag shell nanoparticles using Neem (Azadirachta indica) leaf broth. *J. Colloid Interface Sci.*, 275, 496–502, 2004.
51. Chandran, S.P., Chaudhary, M., Pasricha, R., Ahmad, A., Sastry, M., Synthesis of gold nanotriangles and silver nanoparticles using Aloevera plant extract. *Biotechnol. Progr.*, 22, 2, 577–83, 2006.
52. Bakar, N.A., Ismail, J., Bakar, M.A., Synthesis and characterization of silver nanoparticles in natural rubber. *Mater. Chem. Phys.*, 104, 276–83, 2007.
53. Bar, H., Bhui, D.K., Sahoo, G.P., Sarkar, P., De, S.P., Misra, A., Green synthesis of silver nanoparticles using latex of Jatropha curcas. *Colloids Surf. A: Physicochem. Eng. Asp.*, 339, 134–139, 2009.
54. Krishnaraj, C., Jagan, E.G., Rajasekar, S., Selvakumar, P., Kalaichelvan, P.T., Mohan, N.J., Synthesis of silver nanoparticles using Acalypha indica leaf extracts and its antibacterial activity against water borne pathogens. *Colloids Surf. B: Biointerfaces*, 76, 50–56, 2010.

6

Bioinspired Green Synthesis of Nanomaterials From Algae

Reetu[1], Monalisa Mukherjee[1,2] and Monika Prakash Rai[1]*

[1]*Amity Institute of Biotechnology, Amity University, Uttar Pradesh, Noida, India*
[2]*Amity Institute of Click Chemistry and Research, Amity University, Uttar Pradesh, Noida, India*

Abstract

In this period of nanotechnology, various engineered nanomaterials have been manufactured by a greener technique. Green nanomaterials synthesis has evolved as an eco-friendly method of producing nanomaterials with a wide range of physical, chemical, and biological properties. It has emerged as a viable alternative, offering an eco-friendly, economical, and energy-efficient method for producing a wide spectrum of nanomaterials. Algae are one of the most prevalent autotrophic biological units, accounting for almost half of all photosynthesis worldwide. Algae acts as a living cell factory for the efficient synthesis of nanomaterials because of the presence of numerous reducing, capping, and stabilizing substances in the algal extract that could be employed to transform metal ions to nanoforms such as carbohydrates, phenolics, polypeptides, alkaloids, and terpenoids. Additionally, these nanomaterials offer a wide range of applications in the realms of medicine, reaction catalysis, and various other fields. The goal of this chapter is to provide a brief overview of the variety of algal strains used in this booming field and factors affecting along with the disparate nanocomposites synthesized.

Keywords: Algae, biomass, nanomaterial, green synthesis, cell factory, eco-friendly

6.1 Introduction

Globally scientific community has recognized the environmental catastrophe, and providing clean air, food, water, and sustainable source

---

Corresponding author: mprai@amity.edu

Mousumi Sen and Monalisa Mukherjee (eds.) Bioinspired and Green Synthesis of Nanostructures: A Sustainable Approach, (141–156) © 2023 Scrivener Publishing LLC

of energy in an equal way as a matter of major concern. The worldwide increasing population is rapidly transitioning to a resource exhausting lifestyle; hence, the dynamics of resource usage are alarming, leading to climate change and differences in land use pattern. In next 30 years, the desire for sustainable green alternatives is anticipated to double; therefore, the interdisciplinary holistic approaches pushing the idea of turning waste into profit require special emphasis [1]. Significant research on using microbes for monitoring the pollution and obtaining the clean energy, food, and water is of utmost interest [2]. Numerous studies have been conducted on the efficiency of microbes and produced metabolites for the photocatalytic decomposition of various complex organic molecules, antibiotics, dyes, and also administering them as biosensors for aquatic contaminants [3, 4].

Algae is a notable biological organism that has demonstrated an exceptional performance as a biofactory in the present circumstances for the treatment of wastewater, biofuel synthesis, and the mass production of value added products of economic importance [5]. One such significant product is algae-derived nanomaterials (NMs) with potential uses in numerous fields. Nowadays, the field of nanotechnology has received a remarkable prominence and the materials with a diameter between 1 and 100 nm fall within the category of NMs [6]. In the past decades, the synthesis of NMs has steadily increased due to their distinct features and wide variety of applications. Various types of pollutants entering the environment are more likely now because of the rise in NM production and uses. Recently, scientists have adopted the trend toward greener, more environmentally friendly procedures within the traditional physicochemical pathways for synthesizing NMs [7]. Many studies have been done to explore their ecotoxicological consequences due to the dearth of knowledge regarding their long-term effects on environmental health. So, by developing more sustainable and eco-friendly NM production techniques is one option to deal with the associated consequences [8]. Physical and chemical procedures, which were previously employed, produce harmful by products, require expensive instrumentation, and need energy-intensive experimental processes, high temperatures, poisonous solvents, or unwanted chemical as by products, hence, restricting their utility. Whereas, the biogenic synthesis fulfil the regulations of "green chemistry," which is described as the process and synthesis of chemicals that minimize the usage and production of harmful matter, preventing the generation of waste and by-products; avoiding the use of potentially toxic solvents, reagents and energy efficient experimentation typically conducted at ambient temperature using natural resources [9]. The hydrophilic surface groups like sulphate, carboxyl, and hydroxyl that are present on algae-mediated NMs give them a special range of applications [8]. This chapter examines the variations, benefits, and drawbacks of algal-mediated biosynthetic pathways. Additionally, a number of important process variables that affect the biosynthesis process are identified and covered in detail, including the pH, temperature, incubation duration, precursor concentration, and light intensity. The broad range of uses for NMs as they are now created are also highlighted. The chapter's overall goal is to develop this highly promising biosynthetic technique into a scalable platform.

6.2 Algal System-Mediated Nanomaterial Synthesis

Algae constitutes a group of single or multicellular organisms that exists in marine or freshwater and damp surfaces. Due to the greater growth rate, significantly higher rate of CO_2 sequestration, hyperaccumulation of heavy metals, absence of harmful by-products, low energy input, and employment, algae are potential candidates for NM synthesis [10]. Algae have already been used in commercial and industrial applications as food, cosmetics, nutraceuticals, and fertiliser, but the paradigm is now changing to algae based production of NM. Algal cells may ingest heavy metals and use them as a nutrition source to build biomass via physiological and biological ways that regulate their metabolic activities. Algae-mediated NMs have a special applicability due to the presence of hydrophilic surface groups such as sulphate, carboxyl, and hydroxyl [11]. On the basis of the size, algae can be classified as microalgae (observed under the microscope) or macroalgae (visible to naked eyes). Various micro- and macroalgae are abundant in active chemicals, making them an interesting platform for biorefineries to produce a wide range of high-value products other than fuels, in addition to their antioxidant, anticancer, antibacterial, and heavy metal bioaccumulator roles.

The three main divisions of algae are categorized as chlorophyta (green algae), phaeophyta (brown algae), and rhodophyta (red algae), all of which have the property of possessing chlorophyll in abundance. In addition to xylans, cellulose, alginic acid, and galactans, respectively, are found in the cell walls of all the algae. Green algae, which are representatives of the Cladophorales group, are widely used in a diverse range of industrial, medical, and biotechnological applications. Various secondary metabolites, carbohydrates, and respective functional groups that could act as reducing and stabilizing agents in algae based biosynthesis of nanoparticles (NPs) are just a few of the many critical components it possesses [12].

Table 6.1 Green algae-mediated biosynthesis of various nanoparticles.

Algae	Nanoparticles	Size	References
Ulva intestinalis (Macroalga)	Ag, Au	Ag-17.8 nm Au-14.2 nm	[13]
Chaetomorpha linum (Macroalgae)	Ag	3-44 nm	[14]
Gracilaria edulis (Macroalgae)	Ag	36 nm	[15]
Lessonia flavicans (Macroalgae)	ZnO	50-200 nm	[16]
Sargassum Horneri (Macroalgae)	Graphene	2.2-3.2 nm	[17]
Oscilatoria sancta (Microalgae)	Ag	24-34 nm	[18]
Chlorella vulgaris (Microalgae)	SnO_2	32 nm	[19]
Chlorella K01 (Microalgae)	Fe_3O_4	20-200 nm	[20]
Chlorococcum humicola (Microalgae)	Ag	16 nm	[21]
Scencedesmus sp. (Microalgae)	Ag	15-20 nm	[22]

As indicated in Table 6.1, Ag is the most often produced NMs from diverse species of the green algae. Brown algae belongs to the order Fucales and family Sargassaceae. Cholesterols, fucosterols, and associated functional moieties, like vinyl, muramic, and alginic acid, that serve as capping and reducing agent for the formation of NMs make up the majority of the components of fucales [23]. Currently, multiple species of brown algae have been used to synthesize metal oxide (zinc oxide and titanium oxide) and metallic (silver and gold) NPs, as stated in Table 6.2.

Table 6.2 Brown algae mediated biosynthesis of various nanoparticles.

Algae	Nanoparticles	Size	References
Sargassum bovinum	Pd	5–10 nm	[24]
Padina sp.	Ag	25–60 nm	[25]
Sargassum myriocystum	Ag	20 nm	[26]
Sargassum cymosum	Au	7–10 nm	[27]
Sargassum muticum	ZnO	30–57 nm	[28]
Sargassum longifolium	CuO	40–60 nm	[29]

Table 6.3 Red algae mediated biosynthesis of various nanoparticles.

Algae	Nanoparticles	Size	References
Halymenia dilatata	Pt	15 nm	[30]
Kappaphycus alvarezii	Ag	80 nm	[31]
Gelidium corneum	Ag	20–50 nm	[32]
Gracilaria verrucosa	Au	73 nm	[33]
Galaxaura elongate	Au	3–77 nm	[34]

Due to the distinct flavor, high protein and vitamin content, red algae, which are members of the Rhodophyta family, are largely consumed as food in many countries. The metabolites are effective for reduction and stabilization in the algae-mediated production of NMs. However, because to self-aggregation, sluggish crystallization growth, and stability difficulties, the synthesis of NMs from seaweed red algae is still in the nascent phase [12]. Many red algae strains have been studied for the biosynthesis of NMs, like *Palmaria decipiens*, *Kappaphycus* sp., *Gelidiella acerosa*, *Gracilaria dura*, and many more, as summarized in Table 6.3.

6.3 Factors Affecting the Green Synthesis of Nanomaterials

The method for synthesizing NM using algae is influenced by a variety of parameters. Temperature, medium pH, incubation period, precursor ion concentration, and light are a few of the important variables. These variables have a significant impact on the stability and architecture of NMs [23]. Further notably, scaling of as-produced NMs is largely dependent on the optimization of these parameters. As a result, a comprehensive discussion of the most significant elements influencing algae-mediated NMs production is as follows.

6.3.1 Light

Light acts as a catalyst for biosynthetic reactions that are carried out by algae for NM synthesis. Despite the few instances where algae enhance NM synthesis in the dark, the process of biosynthesis is primarily driven by light. It is essential to mention that NM production is frequently controlled by photosynthetic pigments found in the algal cells. Like, the conversion of Ag into AgNPs is attributed to fucoxanthin, a photosynthetic pigment of the diatom Amphora-46 [35]. Similar phenomenon was observed, when the extracellular polysaccharides of the *Scenedesmus* sp. failed to produce AgNPs in the absence of light [36].

6.3.2 Temperature

The temperature during the process is considered as the major factor affecting the rate of NM production, and determining the stability and size of NM. The thermodynamics hypothesis states that raising the reaction temperature speeds up the conversion of metal ions into NPs. The fact that the synthetic reaction can typically proceed below 100 °C or even at ambient temperature is one of the key benefits of green synthesis. Thus, according to a report on CuO green synthesis, temperature between 25°C and 100°C were typically used for the synthesis of the NM [37]. Similarly, the, hematite NPs synthesis can be carried out between 30°C and 40°C, but the reaction stops above 40°C [38]. As a result, these findings suggest that temperature plays a key role in the bioactivity of active enzymes or biomolecules. It also has an impact on the rate of NM synthesis, in addition to their size, shape, and environmental stability [23].

6.3.3 Incubation Period

Algae are generally known to synthesize NPs faster than the other microorganisms. *Laminaria japonica* within 10 to 20 minutes can convert 90% to 95% of the $HAuCl_4$ precursor (at 2 mM) to AuNPs intracellularly. In about 12 hours, brown algae *Sargassum wightii*, and green freshwater algae

Prasiola crispa reach their maximum levels of AuNP production, whereas *Sargassum longifolium*-based optimal synthesis of AuNPs requires 64 h [39]. Therefore, effective time control is essential to enhance the NP synthesis pathways mediated by algae.

6.3.4 pH

The physiochemical and optoelectronic properties of the synthesized NMs are directly correlated with the morphology of the NMs. pH is recognized as the key parameter, pH of the medium can modify the nucleation centres so it greatly influences the size, shape, and rate of synthesis of NMs. Typically, the process of NP production by algae occurs at a pH of 7, for instance, *Spirulina platensis* produced AgNPs at pH 7 and *Sargassum myriocystum* produced ZnO NPs at pH 8, respectively. While majority of the algae produces NPs at pH values that are neutral, there are those that carries the synthesis at basic as well as acidic pH. Like, AuNPs *Fucus vesiculosus* produces when the pH is in the range of 4 and 9 (the ideal range) [8]. Hence, pH should be monitored carefully in the synthesis process involving algae.

6.3.5 Precursor Concentration and Bioactive Catalyst

The yield of NMs is significantly influenced by the bioactive catalyst and the precursor concentration. According to a recent study by Costa *et al.*, the quantity of *Sargassum cymosum* extract used in conjunction with the right biocatalyst-to-metallic precursor ratio had a substantial impact on both the reaction rate and the quality of the NPs [27]. Another report found that addition of 1mM $AgNO_3$, the biocatalyst (algal extract)-to-metallic precursor (Ag ion) ratio of 1:30 was found most beneficial stable Ag-NP synthesis because particle agglomeration was supported by higher precursor concentration, slowing the process [40]. Collectively, the choice of algae strain, catalyst, and their ratio are the important elements impacting the yield and properties of the obtained NMs, even reaction rates and processes may vary with the biomolecules and precursors, as a result, these aspects need to be carefully evaluated [23].

6.4 Applications of the Green Synthesized Nanomaterials

Due to their distinct physiological, chemical, and electrical characteristics, NPs have a wide range of applications several disciplines. Their surface

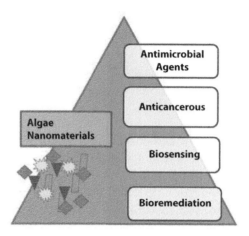

Figure 6.1 Applications of algae based nanomaterials in various fields.

chemistry and reactivity and the general make them suitable for numerous applications. Nevertheless, hydroxyl, sulphate, carboxyl, and other hydrophilic surface groups are readily available over the algal-mediated NPs attributing the unique applications [41]. Since these do not require any external capping or reducing agents during the synthesis, therefore NMs made producing various green technologies are typically biocompatible and free of harmful compounds on their surfaces and exhibit less toxicity than NMs synthesized using chemical processes [42]. In the following sections, different applications of algae-mediated NMs are discussed in detail (Figure 6.1).

6.4.1 Antimicrobial Agents

Antibiotics are being commonly used to treat bacterial ailments, which has caused the rise of bacterial strains that are multi-drug resistant. A significant global health concern is developing a safe and effective way to treat drug-resistant bacterial pathogens. Therefore, the use of NMs as an alternative antibacterial agent has changed because they have proven to have effective and better bactericidal activity. NMs exhibit broad-spectrum antibacterial efficacy against both gram-positive and gram-negative bacteria because they kill the bacteria cell by rupturing the cell membrane and generating reactive oxygen species (ROS) [43]. In the recent times, to inhibit the bacterial colonization, Ag-NPs are applied to surgical instruments including catheters and bandages [44]. The heating, ventilation, and air-conditioning system in a humid environment now uses a silver-activated carbon filter to remove bio-aerosols, which can lead to chronic diseases [45]. AgCl-NPs produced using *Scenedesmus* sp., exhibited antibacterial activity against *Escherichia coli* and *Escherichia coli* (DH5-α), *Bacillus subtilis*, *Bacillus sphaericus*, and *Bacillus pasteurii* [46]. The aqueous extracts of *Sargassum vulgare* and *Ulva fasciata*, were used to biosynthesize iron oxide NPs that had an antibacterial effect on marine biofilm bacteria [47].

6.4.2 Anticancerous

Aberrant cell growth manifesting in any organ of the body is considered as cancerous. The present day treatments are expensive along with having negative health effects on humans, hence finding appropriate medications to treat different types of cancer is necessary. Algae based NPs can be employed for the cancer treatments as these are economical with less biohazard effects [48]. The extracts of *Ulva rigida* were used for synthesizing the Ag-NPs and these NPs inhibited breast cancer cell lines (IC_{50} = 13 µg/ml) [49]. Another study investigated the anticancer property of *Ulva lactuca*–derived Fe NPs (30-40 nm) against HeLa cell lines [50].

6.4.3 Biosensing

The NMs possess electronic and optical properties, therefore biosensors applications are influenced by the combination between NPs and biomolecules. AgNPs synthesized from *Noctiluca scintillans* biomass were recently tested for colorimetric sensing of hydrogen peroxide, an antibacterial that is recommended for minor skin scrapes; in the mouth, gums, and toothache. It was demonstrated that the hydrogen peroxide breakdown on the catalytic AgNPs was pH, temperature, and time-dependent. The experiment exhibited a brown color to colorless change as well, with hydrogen peroxide showing the greatest observable color shift [51]. Platinum NPs (PtNPs) produced from *Sargassum myriocystum* also function as biosensors for measuring the amount of adrenaline in the human body [12].

6.4.4 Bioremediation

The basis of life is water, but water pollution is jeopardizing aquatic ecosystems, human lives, and the economy. Algae-based NMs have the capacity to serve as disinfectants due to the antibacterial capabilities, heavy metal mitigation, nitrogen and phosphorus removal, antifouling, and biosensing of contaminants, therefore can be used for the purification purposes [52]. *Chlorella* sp. and *Gracilaria vermicTulophylla* were used to synthesize

ZnO-NPs and based on the photocatalytic activities they could degrade dibenzothiophene and aqueous phenol, respectively [53]. In a report, lipid-cadmium sulphide NPs were produced by the green alga *Scenedesmus obliquus*, it was discovered that the adsorption kinetics of Cd^{2+} ions changed dramatically during synthesis. The Cd^{2+} monolayer (which was enlarged and chemisorbed) became permanently attached to the algal biomass. Hence, due to high retention capability *Scenedesmus* proved to be an efficient model alga for the bioremediation of Cd^{2+} ions [12]. This approach of producing different algae-mediated NPs can be considered beneficial contributing to the remediation of different kinds of heavy metals, organic, aromatic chemicals, and dyes.

6.5 Future Perspectives

Algal-mediated biosynthesis of NMs is an eco-friendly, readily available, non-toxic, economical, and energy-efficient process. However, research is needed to examine the mechanisms and processes by which algae produce NMs. Identification of bio-active molecules that operate as reducing and capping agents shall be carried out, artificial intelligence and genetics may play a crucial role in the modulation and optimization of the dimensions of these algae-NPs. For a successful commercial production scale, monitoring and control of factors affecting phyco-NPs synthesis must be advanced on a wide scale. It is necessary to study new algae strains and methodologies for the synthesis of organic and inorganic NPs. In addition, the effects of phyco-NPs need to be studied in various fields.

6.6 Conclusion

The utilization of energy-efficient, safe technologies with minimal ecological and biological risks has become the norm in modern science. As a result, algae are now useful raw materials that have a variety of uses, particularly in nanotechnology. This chapter focuses on the function of algae as phyco-nano factories and provides illustrations of several synthesis processes using cyanobacteria, microalgae, and macroalgae. Polysaccharides, phenols, proteins, and enzymes are essential chemicals found both in microalgae and macroalgae that can be employed in the production of NPs. The synthesis of several NPs, including cellulose, and chitosan, which have significance in pharma, agricultural and food industries, can also be carried out using some macroalgae. The use of green NMs is done for numerous applications, such as drug insemination, tissue engineering, tumor dissolution, biosensors, as antimicrobial agents, and water treatments to enhance human life and make it more secure.

References

1. Jacob, J.M., Ravindran, R., Narayanan, M., Samuel, S.M., Pugazhendhi, A., Kumar, G., Microalgae: A prospective low cost green alternative for nanoparticle synthesis. *Curr. Opin. Environ. Sci. Health*, 20, 100163, 2021, https://doi.org/10.1016/j.coesh.2019.12.005.
2. Pathak, J., Rajneesh, H.A., Singh, D.K., Pandey, A., Singh, S.P., Sinha, R.P., Recent developments in green synthesis of metal nanoparticles utilizing cyanobacterial cell factories, in: *Nanomaterials in Plants, Algae and Microorganisms: Concepts and Controversies*, vol. 2, 2018, https://doi.org/10.1016/B978-0-12-811488-9.00012-3.
3. Javed, F., Aslam, M., Rashid, N., Shamair, Z., Khan, A.L., Yasin, M., Fazal, T., Hafeez, A., Rehman, F., Rehman, M.S.U., Khan, Z., Iqbal, J., Bazmi, A.A., Microalgae-based biofuels, resource recovery and wastewater treatment: A pathway towards sustainable biorefinery. *Fuel*, 255, 115826, 2019, https://doi.org/10.1016/j.fuel.2019.115826.
4. Jeffryes, C., Agathos, S.N., Rorrer, G., Biogenic nanomaterials from photosynthetic microorganisms. *Curr. Opin. Biotechnol.*, 33, 23–31, 2015, https://doi.org/10.1016/j.copbio.2014.10.005.
5. Kumar, B.R., Mathimani, T., Sudhakar, M.P., Rajendran, K., Nizami, A.S., Brindhadevi, K., Pugazhendhi, A., A state of the art review on the cultivation of algae for energy and other valuable products: Application, challenges, and opportunities. *Renewable Sustainable Energy Rev.*, 138, 110649, 2021, https://doi.org/10.1016/j.rser.2020.110649.
6. Yap, Y.H., Azmi, A.A., Mohd, N.K., Yong, F.S.J., Kan, S.Y., Thirmizir, M.Z.A., Chia, P.W., Green synthesis of silver nanoparticle using water extract of onion peel and application in the acetylation reaction. *Arab. J. Sci. Eng.*, 45, 4797–4807, 2020, https://doi.org/10.1007/s13369-020-04595-3.
7. Palem, R.R., Ganesh, S.D., Kronekova, Z., Sláviková, M., Saha, N., Saha, P., Green synthesis of silver nanoparticles and biopolymer nanocomposites: A comparative study on physico-chemical, antimicrobial and anticancer activity. *Bull. Mater. Sci.*, 41, 55, 2018, https://doi.org/10.1007/s12034-018-1567-5.
8. Rahman, A., Kumar, S., Nawaz, T., Biosynthesis of nanomaterials using algae, in: *Microalgae Cultivation for Biofuels Production*, 2019, https://doi.org/10.1016/B978-0-12-817536-1.00017-5.
9. Dahoumane, S.A., Mechouet, M., Wijesekera, K., Filipe, C.D.M., Sicard, C., Bazylinski, D.A., Jeffryes, C., Algae-mediated biosynthesis of inorganic nanomaterials as a promising route in nanobiotechnology-a review. *Green Chem.*, 19, 552–587, 2017, https://doi.org/10.1039/c6gc02346k.

10. Ankit, N., Bordoloi, Tiwari, J., Kumar, S., Korstad, J., Bauddh, K., Efficiency of algae for heavy metal removal, bioenergy production, and carbon sequestration, In: Bharagava, R. (eds) *Emerging Eco-friendly Green Technologies for Wastewater Treatment. Microorganisms for Sustainability*, vol 18. Springer, Singapore, 2020, https://doi.org/10.1007/978-981-15-1390-9_4.
11. Kumar, N., Balamurugan, A., Balakrishnan, P., Vishwakarma, K., Shanmugam, K., Biogenic nanomaterials: Synthesis and its applications for sustainable development, in: *Biogenic Nano-Particles and Their Use in Agro-Ecosystems*, 2020, https://doi.org/10.1007/978-981-15-2985-6_7.
12. Chaudhary, R., Nawaz, K., Khan, A.K., Hano, C., Abbasi, B.H., Anjum, S., An overview of the algae-mediated biosynthesis of nanoparticles and their biomedical applications. *Biomolecules*, 10, 1498, 2020, https://doi.org/10.3390/biom10111498.
13. González-Ballesteros, N., Diego-González, L., Lastra-Valdor, M., Rodríguez-Argüelles, M.C., Grimaldi, M., Cavazza, A., Bigi, F., Simón-Vázquez, R., Immunostimulant and biocompatible gold and silver nanoparticles synthesized using the: Ulva intestinalis L. aqueous extract. *J. Mater. Chem. B*, 7, 4677–4691, 2019, https://doi.org/10.1039/c9tb00215d.
14. Kannan, R.R.R., Arumugam, R., Ramya, D., Manivannan, K., Anantharaman, P., Green synthesis of silver nanoparticles using marine macroalga Chaetomorpha linum. *Appl. Nanosci. (Switzerland)*, 3, 229–233, 2013, https://doi.org/10.1007/s13204-012-0125-5.
15. Madhiyazhagan, P., Murugan, K., Kumar, A.N., Nataraj, T., Subramaniam, J., Chandramohan, B., Panneerselvam, C., Dinesh, D., Suresh, U., Nicoletti, M., Alsalhi, M.S., Devanesan, S., Benelli, G., One pot synthesis of silver nanocrystals using the seaweed Gracilaria edulis: Biophysical characterization and potential against the filariasis vector Culex quinquefasciatus and the midge Chironomus circumdatus. *J. Appl. Phycol.*, 29, 649–659, 2017, https://doi.org/10.1007/s10811-016-0953-x.
16. Berneira, L.M., Poletti, T., de Freitas, S.C., Maron, G.K., Carreno, N.L.V., de Pereira, C.M.P., Novel application of sub-Antarctic macroalgae as zinc oxide nanoparticles biosynthesizers. *Mater. Lett.*, 320, 132341, 2022, https://doi.org/10.1016/J.MATLET.2022.132341.
17. Ai, N., Lou, S., Lou, F., Xu, C., Wang, Q., Zeng, G., Facile synthesis of macroalgae-derived graphene adsorbents for efficient CO_2 capture. *Process Saf. Environ. Prot.*, 148, 1048–1059, 2021, https://doi.org/10.1016/J.PSEP.2021.02.014.
18. Elumalai, D., Hemavathi, M., Rekha, G.S., Pushpalatha, M., Leelavathy, R., Vignesh, A., Ashok, K., Babu, M., Photochemical synthesizes of silver nanoparticles using Oscillatoria sancta micro algae against mosquito vectors Aedes aegypti and Anopheles stephensi. *Sens. Biosens. Res.*, 34, 100457, 2021, https://doi.org/10.1016/J.SBSR.2021.100457.
19. Nagajyothi, P.C., v. Prabhakar Vattikuti, S., Devarayapalli, K.C., Yoo, K., Shim, J., Sreekanth, T.V.M., Green synthesis: Photocatalytic degradation of textile dyes using metal and metal oxide nanoparticles-latest trends and

advancements. *Crit. Rev. Environ. Sci. Technol.*, 50, 2617–2723, 2020, https://doi.org/10.1080/10643389.2019.1705103.
20. Win, T.T., Khan, S., Bo, B., Zada, S., Fu, P.C., Green synthesis and characterization of Fe3O4 nanoparticles using Chlorella-K01 extract for potential enhancement of plant growth stimulating and antifungal activity. *Sci. Rep.*, 11, 1–11, 2021, https://doi.org/10.1038/s41598-021-01538-2.
21. Jena, J., Pradhan, N., Nayak, R.R., Dash, B.P., Sukla, L.B., Panda, P.K., Mishra, B.K., Microalga Scenedesmus sp.: A potential low-cost green machine for silver nanoparticle synthesis. *J. Microbiol. Biotechnol.*, 24, 522–533, 2014, https://doi.org/10.4014/jmb.1306.06014.
22. Vigneshwaran, N., Ashtaputre, N.M., v. Varadarajan, P., Nachane, R.P., Paralikar, K.M., Balasubramanya, R.H., Biological synthesis of silver nanoparticles using the fungus Aspergillus flavus. *Mater. Lett.*, 61, 1413–1418, 2007, https://doi.org/10.1016/j.matlet.2006.07.042.
23. Li, S.N., Wang, R., Ho, S.H., Algae-mediated biosystems for metallic nanoparticle production: From synthetic mechanisms to aquatic environmental applications. *J. Hazard. Mater.*, 420, 126625, 2021, https://doi.org/10.1016/J.JHAZMAT.2021.126625.
24. Momeni, S. and Nabipour, I., A Simple Green Synthesis of Palladium Nanoparticles with Sargassum Alga and Their Electrocatalytic Activities Towards Hydrogen Peroxide. *Appl. Biochem. Biotechnol.*, 176, 1937–1949, 2015, https://doi.org/10.1007/s12010-015-1690-3.
25. Bhuyar, P., Rahim, M.H.A., Sundararaju, S., Ramaraj, R., Maniam, G.P., Govindan, N., Synthesis of silver nanoparticles using marine macroalgae Padina sp. and its antibacterial activity towards pathogenic bacteria. *Beni Suef Univ. J. Basic Appl. Sci.*, 9, 1–15, 2020, https://doi.org/10.1186/s43088-019-0031-y.
26. Balaraman, P., Balasubramanian, B., Kaliannan, D., Durai, M., Kamyab, H., Park, S., Chelliapan, S., Lee, C.T., Maluventhen, V., Maruthupandian, A., Phyco-synthesis of silver nanoparticles mediated from marine algae Sargassum myriocystum and its potential biological and environmental applications. *Waste Biomass Valorization*, 11, 5255–5271, 2020, https://doi.org/10.1007/s12649-020-01083-5.
27. Costa, L.H., v. Hemmer, J., Wanderlind, E.H., Gerlach, O.M.S., Santos, A.L.H., Tamanaha, M.S., Bella-Cruz, A., Corrêa, R., Bazani, H.A.G., Radetski, C.M., Almerindo, G.I., Green synthesis of gold nanoparticles obtained from Algae Sargassum cymosum: Optimization, characterization and stability. *Bionanoscience*, 10, 1049–1062, 2020, https://doi.org/10.1007/s12668-020-00776-4.
28. Azizi, S., Ahmad, M.B., Namvar, F., Mohamad, R., Green biosynthesis and characterization of zinc oxide nanoparticles using brown marine macroalga Sargassum muticum aqueous extract. *Mater. Lett.*, 116, 275–277, 2014, https://doi.org/10.1016/j.matlet.2013.11.038.
29. Rajeshkumar, S., Nandhini, N.T., Manjunath, K., Sivaperumal, P., Krishna Prasad, G., Alotaibi, S.S., Roopan, S.M., Environment friendly synthesis

copper oxide nanoparticles and its antioxidant, antibacterial activities using Seaweed (Sargassum longifolium) extract. *J. Mol. Struct.*, 1242, 130724, 2021, https://doi.org/10.1016/J.MOLSTRUC.2021.130724.
30. Sathiyaraj, G., Vinosha, M., Sangeetha, D., Manikandakrishnan, M., Palanisamy, S., Sonaimuthu, M., Manikandan, R., You, S.G., Prabhu, N.M., Bio-directed synthesis of Pt-nanoparticles from aqueous extract of red algae Halymenia dilatata and their biomedical applications. *Colloids Surf. A Physicochem. Eng. Asp.*, 618, 126434, 2021, https://doi.org/10.1016/J.COLSURFA.2021.126434.
31. Khan, M.S., S, R., S, H., Synthesis and characterization of Kappaphycus alvarezii derived silver nanoparticles and determination of antibacterial activity. *Mater. Chem. Phys.*, 282, 125985, 2022, https://doi.org/10.1016/J.MATCHEMPHYS.2022.125985.
32. Yılmaz Öztürk, B., Yenice Gürsu, B., Dağ, İ., Antibiofilm and antimicrobial activities of green synthesized silver nanoparticles using marine red algae Gelidium corneum. *Process Biochem.*, 89, 208–219, 2020, https://doi.org/10.1016/J.PROCBIO.2019.10.027.
33. Chellapandian, C., Ramkumar, B., Puja, P., Shanmuganathan, R., Pugazhendhi, A., Kumar, P., Gold nanoparticles using red seaweed Gracilaria verrucosa: Green synthesis, characterization and biocompatibility studies. *Process Biochem.*, 80, 58–63, 2019, https://doi.org/10.1016/J.PROCBIO.2019.02.009.
34. Abdel-Raouf, N., Al-Enazi, N.M., Ibraheem, I.B.M., Green biosynthesis of gold nanoparticles using Galaxaura elongata and characterization of their antibacterial activity. *Arabian J. Chem.*, 10, S3029–S3039, 2017, https://doi.org/10.1016/J.ARABJC.2013.11.044.
35. Khanna, P., Kaur, A., Goyal, D., Algae-based metallic nanoparticles: Synthesis, characterization and applications. *J. Microbiol. Methods*, 163, 105656, 2019, https://doi.org/10.1016/j.mimet.2019.105656.
36. Patel, V., Berthold, D., Puranik, P., Gantar, M., Screening of cyanobacteria and microalgae for their ability to synthesize silver nanoparticles with antibacterial activity. *Biotechnol. Rep.*, 5, 112–119, 2015, https://doi.org/10.1016/j.btre.2014.12.001.
37. Akintelu, S.A., Folorunso, A.S., Folorunso, F.A., Oyebamiji, A.K., Green synthesis of copper oxide nanoparticles for biomedical application and environmental remediation. *Heliyon*, 6, e04508, 2020, https://doi.org/10.1016/j.heliyon.2020.e04508.
38. Rajendran, K. and Sen, S., Optimization of process parameters for the rapid biosynthesis of hematite nanoparticles. *J. Photochem. Photobiol. B*, 159, 82–87, 2016, https://doi.org/10.1016/j.jphotobiol.2016.03.023.
39. Parial, D., Patra, H.K., Dasgupta, A.K.R., Pal, R., Screening of different algae for green synthesis of gold nanoparticles. *Eur. J. Phycol.*, 47, 22–29, 2012, https://doi.org/10.1080/09670262.2011.653406.
40. Khoshnamvand, M., Ashtiani, S., Chen, Y., Liu, J., Impacts of organic matter on the toxicity of biosynthesized silver nanoparticles to green microalgae

Chlorella vulgaris. *Environ. Res.*, 185, 109433, 2020, https://doi.org/10.1016/J.ENVRES.2020.109433.
41. Rahman, A., Kumar, S., Bafana, A., Lin, J., Dahoumane, S.A., Jeffryes, C., A mechanistic view of the light-induced synthesis of silver nanoparticles using extracellular polymeric substances of Chlamydomonas reinhardtii. *Molecules*, 24, 3506, 2019, https://doi.org/10.3390/molecules24193506.
42. Bhattacharya, P., Swarnakar, S., Ghosh, S., Majumdar, S., Banerjee, S., Disinfection of drinking water via algae mediated green synthesized copper oxide nanoparticles and its toxicity evaluation. *J. Environ. Chem. Eng.*, 7, 102867, 2019, https://doi.org/10.1016/j.jece.2018.102867.
43. Anwar, A., Masri, A., Rao, K., Rajendran, K., Khan, N.A., Shah, M.R., Siddiqui, R., Antimicrobial activities of green synthesized gums-stabilized nanoparticles loaded with flavonoids. *Sci. Rep.*, 9, 1–12, 2019, https://doi.org/10.1038/s41598-019-39528-0.
44. Nakamura, S., Sato, M., Sato, Y., Ando, N., Takayama, T., Fujita, M., Ishihara, M., Synthesis and application of silver nanoparticles (Ag nps) for the prevention of infection in healthcare workers. *Int. J. Mol. Sci.*, 20, 3620, 2019, https://doi.org/10.3390/ijms20153620.
45. Pascariu, P., Airinei, A., Iacomi, F., Bucur, S., Suchea, M.P., Electrospun TiO2-based nanofiber composites and their bio-related and environmental applications, in: *Functional Nanostructured Interfaces for Environmental and Biomedical Applications*, 2019, https://doi.org/10.1016/B978-0-12-814401-5.00012-8.
46. Kashyap, M., Samadhiya, K., Ghosh, A., Anand, V., Lee, H., Sawamoto, N., Ogura, A., Ohshita, Y., Shirage, P.M., Bala, K., Synthesis, characterization and application of intracellular Ag/AgCl nanohybrids biosynthesized in Scenedesmus sp. as neutral lipid inducer and antibacterial agent. *Environ. Res.*, 201, 111499, 2021, https://doi.org/10.1016/j.envres.2021.111499.
47. Salem, D.M.S.A., Ismail, M.M., Tadros, H.R.Z., Evaluation of the antibiofilm activity of three seaweed species and their biosynthesized iron oxide nanoparticles (Fe3O4-NPs). *Egypt. J. Aquat. Res.*, 46, 333–339, 2020, https://doi.org/10.1016/j.ejar.2020.09.001.
48. El-Sheekh, M.M., Morsi, H.H., Hassan, L.H.S., Ali, S.S., The efficient role of algae as green factories for nanotechnology and their vital applications. *Microbiol. Res.*, 263, 127111, 2022, https://doi.org/10.1016/J.MICRES.2022.127111.
49. Algotiml, R., Gab-Alla, A., Seoudi, R., Abulreesh, H.H., El-Readi, M.Z., Elbanna, K., Anticancer and antimicrobial activity of biosynthesized red sea marine algal silver nanoparticles. *Sci. Rep.*, 12, 333–339, 2022, https://doi.org/10.1038/s41598-022-06412-3.
50. Bensy, A.D.V., Christobel, G.J., Muthusamy, K., Alfarhan, A., Anantharaman, P., Green synthesis of iron nanoparticles from Ulva lactuca and bactericidal activity against enteropathogens. *J. King Saud Univ. Sci.*, 34, 101888, 2022, https://doi.org/10.1016/j.jksus.2022.101888.

51. Elgamouz, A., Idriss, H., Nassab, C., Bihi, A., Bajou, K., Hasan, K., Haija, M.A., Patole, S.P., Green synthesis, characterization, antimicrobial, anti-cancer, and optimization of colorimetric sensing of hydrogen peroxide of algae extract capped silver nanoparticles. *Nanomaterials*, 10, 1861, 2020, https://doi.org/10.3390/nano10091861.
52. Khan, F., Shahid, A., Zhu, H., Wang, N., Javed, M.R., Ahmad, N., Xu, J., Alam, M.A., Mehmood, M.A., Prospects of algae-based green synthesis of nanoparticles for environmental applications. *Chemosphere*, 293, 133571, 2022, https://doi.org/10.1016/j.chemosphere.2022.133571.
53. Khalafi, T., Buazar, F., Ghanemi, K., Phycosynthesis and enhanced photocatalytic activity of zinc oxide nanoparticles toward organosulfur pollutants. *Sci. Rep.*, 9, 1–10, 2019, https://doi.org/10.1038/s41598-019-43368-3.

7
Interactions of Nanoparticles with Plants: Accumulation and Effects

Indrajit Roy

Department of Chemistry, University of Delhi, Delhi, India

Abstract

Several factors, such as global climate change, rapid urbanization, growing food demands, uncontrolled disposal of non-biodegradable waste, overuse of harmful agrochemicals such as fertilizers and pesticides, etc., have contributed to a steady decline in crops, plants and trees worldwide. These factors have led to disturbance in the oxygen-carbon dioxide balance, crop loss, soil erosion, enhanced pollution, etc., that has harmed both the ecology and the economy. Therefore, we must seek novel scientific and technological interventions for the restoration of the well-being and productivity of the flora.

Advances in nanotechnology promise to usher unprecedented developments in several critical areas such as energy storage/conversion, healthcare, agriculture, environmental remediation, etc. It has been observed that several nanoparticles are efficiently uptaken by plants through their roots, leaves and seeds, followed by their rapid translocation to other plant parts and cellular accumulation, with negligible harmful effects at moderate dosages. This provides ample opportunities where the composition, tailored dimensions, high surface area, unique physical properties and stimuli-responsiveness of nanoparticles and nanoparticle-based devices can be exploited for the benefit of plants.

Recently, we have witnessed a surge in the application of several nanoparticles, without or with loaded fertilizers, pesticides, or genetic drugs, in plants and trees for a specific purpose. Nanoparticles of selenium, silver, copper, ceria, etc., having antioxidant and antimicrobial activities, can rescue plants from various abiotic and biotic stress. Nanoparticles of zinc oxide and iron oxide have been shown to enhance the nutrient content in plants without the necessity of traditional fertilizers. NPK-based fertilizers and other nutrients can be delivered in plants with nanoparticles of silica, lipids, polymers, etc., with an efficiency higher than

---

Email: indrajitroy11@gmail.com

Mousumi Sen and Monalisa Mukherjee (eds.) *Bioinspired and Green Synthesis of Nanostructures: A Sustainable Approach*, (157–188) © 2023 Scrivener Publishing LLC

that observed using free fertilizers. These nanoparticles can also deliver harmful agrochemicals such as pesticides and herbicides to plants without environmental leakage and ecological poisoning. In addition, nanoparticles can introduce functional genes, such as DNA, RNA, CRISPR/Cas gene editing agents, etc., for providing genetic stimulus for enhanced growth and function, and producing resistant crops. Concurrently, the unique physical properties of several nanoparticles have been explored to fabricate sensing platforms (nanosensors) for the detection of physical, chemical and biochemical markers that indicate the state of growth, development and productivity of plants. Moreover, fluorescent or Raman-active nanoprobes have served as internal sensors for monitoring physiological events and molecular changes occurring in plants. This chapter shall critically discuss such developments, and underline the future prospects of this very exciting research area.

Keywords: Nanotechnology, nanosensors, agrochemicals, translocation, foliar spray, CRISPR/Cas gene editing

7.1 Introduction

Over the last few decades, the understanding and applications of nanotechnology in several areas have witnessed unprecedented advancements. On one hand, the discovery of unique optical, electronic and magnetic aspects of nanomaterials have facilitated the fabrication of several nanoscale systems and devices with unprecedented advantages in technological areas, such as energy harvesting and conversion, data storage, and communications [1, 2]. On the other hand, biocompatible nanoparticles have been interfaced with active molecules, such as image-contrast agents, therapeutic drugs, functional genes, growth factors, etc., for advancing the diagnostic, prophylactic, therapeutic, and regenerative capabilities in biomedical science [3–5]. In parallel, the possible deleterious effects of engineered nanoparticles in the ecological and environmental milieu have also been actively explored as a research area [6–8]. It is quite natural, therefore, to investigate the effect of nanomaterials and nanoscale devices in the flora, not only from the point of view of reducing the collateral ecological exposure of such nanomaterials but also to exploit the unique features of such materials for enhancing the growth and development in plants.

Our planet was created with such conditions that established a delicate balance and interdependence between the flora and fauna. However, a combination of factors, such as climate change, rapid urbanization and industrialization, rampant deforestation, overcultivation, excess use of agrochemicals and pesticides, etc., have perturbed this balance and thus created significant stress for the flora. Therefore, mankind must seek immediate solutions to restore this balance; nanotechnology is expected to play a major role in this pursuit. The botanical world has inspired several nanotechnology-based functions/phenomena, such as artificial photosynthesis/light harvesting, superhydrophobic surfaces, stimuli-responsive drug delivery, etc. [9, 10]. Conversely, various nanotechnology-based applications are expected to interface with and bring about unique benefits to the crops, plants and trees. At the same time, it is critical to investigate the potential ill effects of nanoparticles on plants following their accidental or intentional treatment.

Although nanotechnology applications in plants is a relatively recent area of research, already several publications have appeared related to the uptake pathways and translocation mechanism of various nanomaterials in plants [11]. Concurrently, it has been also reported that several nanoparticles, such as those of metals, metal oxides, hybrid metal-organic frameworks, lipids, polymers, etc., are not only well-tolerated by plants at moderate dosages, but also bring about beneficial functions such as nutrient enrichment, reactive oxygen species (ROS) scavenging, stress and biofouling resistance, growth stimulation, enhanced productivity, etc. [12, 13]. In addition, various active components, such as fertilizers, nutrients, pesticides, growth factors, etc., can be delivered to specific parts, cells and subcellular organelles within plants for improving their structure and function, even in stress conditions, for producing resilient plants and crops [14, 15]. Genetic engineering and gene editing in plants using nanoparticles incorporating functional genes for producing resistant crops is another active area of research for smart agricultural applications [16, 17]. Moreover, the unique physical properties shown by several nanoparticles, such as localized surface plasmon resonance (LSPR), narrow band-edge photoluminescence, upconversion luminescence, etc., are crucial towards the development of nanoparticle-based sensing platforms that relay various chemical, electrical and molecular plant-based signals in real-time, thus allowing the continuous monitoring of plant growth and development [18, 19]. This chapter attempts to summarize these aspects that make nanoparticles as promising probes for improving the flora in the world. The various applications of nanoparticles in plants covered in this chapter are summarized in Figure 7.1. Owing to the vast literature already available in a very short time in this rapidly expanding research area, we are able to provide only selected examples from the various applications.

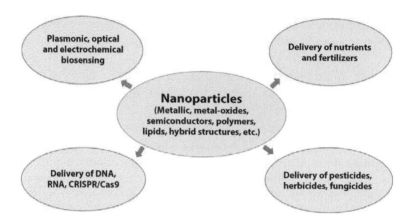

Figure 7.1 Summary of the various applications of nanoparticles covered in this chapter.

7.2 Uptake and Translocation of Nanoparticles and Nanocarriers in Plants

Nanoparticles, without or with incorporated active molecules (e.g., fertilizers, pesticides), first needs to be efficiently uptaken by the plants, followed by their distribution (translocation) to other parts within the plant body, often by traversing across various biological barriers. For efficient delivery and distribution, it is essential that the nanoparticles remain both structurally and colloidally stable, resist premature leakage of loaded active molecules, and protect the loaded molecules from enzymatic degradation. Nanoparticles are delivered in the plants via three major routes, namely absorption through roots (using the conventionally used soil absorption or the recent technique of hydroponics where plants are grown without soil or mineral-rich water), infiltration through the leaves (foliar delivery), and seed coating (also known as nanopriming of seeds) [20]. While soil absorption is the preferred and easiest route of administration, it suffers from the risk of ecological leakage and diffusion to non-target areas. In contrast, foliar infiltration is a more direct way of delivery; however, this necessitates the aerosolization of the nanoparticles for uniform uptake, as well as proper visco-rheological characteristics for prolonged adhesion and retention. Nanopriming of seeds is also a direct method of nanoparticle application, although it is limited to young plants/crops only. Once successfully administered, the nanoparticles must efficiently traverse across biological barriers (e.g. plant cell wall) and reach target cells and critical subcellular organelles (e.g., chloroplasts). Along with the route of administration, the physical parameters of nanoparticles, such as their size, shape, charge, porosity, surface chemistry, etc., play critical roles in directing their uptake, translocation and targeting potential. In the subsequent section, we shall underline some examples of nanoparticle delivery in plants and their translocation to desired sites.

In an early example of root-mediated delivery, Le *et al.* have investigated the uptake, transport, and functional effects of silica (SiO_2) nanoparticles on Bt-transgenic cotton. Analysis of biomass, nutrients, hormone levels, xylem sap and enzymes was carried out over a 3-week period following delivery of various concentrations (0, 10, 100, 500, and 2000 mg·L^{-1}) of the nanoparticles. Transmission electron microscopy (TEM) experiment revealed that the nanoparticles could efficiently translocate from the roots to the shoots via the xylem sap, as shown in Figure 7.2. Significant functional effects, such as reduction in plant height, biomass and nutrient content, and alteration in levels of superoxide dismutase (SOD) enzyme were observed [21]. Root-mediated delivery can also be carried out hydroponically. Prasad *et al.* probed the uptake and translocation of fluorescently tagged cationic zein nanoparticles (ZNPs) in sugar cane plants grown in hydroponic culture. Fluorescence analysis of digested plant tissues revealed that most of the ZNPs remained adhered to the roots, with little diffusion in the leaves. Although poor translocation of the nanoparticles was observed, it was suggested that prolonged adhesion with the roots can potentially allow the sustained release of active molecules encapsulated within these nanoparticles. The cationic charge of these nanoparticles can be a contributing factor towards their poor translocation ability [22]. In a recent example, Tombuloglu *et al.* have sonochemically synthesized

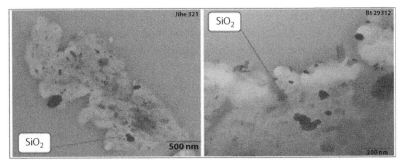

Figure 7.2 TEM images showing nanoparticles present in the xylem sap of (left panel) non-transgenic cotton, and (right panel) Bt-transgenic cotton. Reprinted with permission from Reference [20].

terbium-substituted cobalt ferrite nanoparticles and investigated their effect on barley (*Hordeum vulgare* L.) plants following hydroponic delivery. Although a significant reduction in germination rate was observed, plant growth and biomass enhancement were also seen. Moreover, efficient root-to-leaf translocation was recorded, as evidenced by observation of significant magnetic signals in both the root and leaf samples using vibrating-sample magnetometry (VSM) analysis. Nanoparticle translocation also altered the photoluminescence signal and thus possible photosynthesis characteristics of the leaves [23].

Several groups have studied the translocation of nanoparticles following foliar delivery. In an early demonstration, Wang *et al.* have developed an aerosol process for foliar delivery of nanoparticles in watermelon plant. The aerosolized nanoparticles were found to efficiently enter the leaf stomata via gas uptake, without straying into the soil. Using techniques, such as TEM and inductively coupled plasma mass spectrometry (ICPMS), it was observed that following entry through the stomatal pathway, the nanoparticles could translocate to the roots via passing through the stem. In a similar demonstration, the same group has shown that aerosolized gold nanostructures (30–80 nm) could be efficiently taken up by watermelon plants through direct penetration and translocate to the roots via phloem transport mechanism following foliar delivery [24]. In a recent example, Hu *et al.* systematically analyzed the role of hydrodynamic diameter and surface charge (zeta-potential) of hydrophilic nanoparticles for dictating their interaction with specific leaf cells and subcellular organelles [25]. They probed the delivery of several nanomaterials, such as carbon dots (CDs), cerium oxide (CeO_2), and SiO_2 nanoparticles, on both maize (monocotyledon) and cotton (dicotyledon) leaves. Confocal fluorescence microscopy was used for real-time *in planta* tracking of the translocation and distribution of the CDs. Low size, positive surface charge, and low-surface tension following surfactant coating were found to be key factors associated with rapid foliar uptake via stomata and cuticle pathways, as well as efficient distribution in the guard cells, extracellular space, and chloroplasts.

Nanoparticle delivery can yet be improved using more direct delivery methods, such as petiole feeding and trunk injection, particularly in trees. In a recent report, Su *et al.* have probed the effect of surface coating with citrate, polyvinylpyrrolidone, and gum Arabic, on the mobility and distribution of silver nanoparticles following petiole feeding and trunk injection on Mexican lime citrus trees. Steric repulsion between nanoparticles and conducting tube surfaces can facilitate the efficient translocation of nanoparticles. It was found that coating with gum Arabic resulted in reduced aggregation and enhanced mobility of the nanoparticles in trees. Over a 7-day observation period, about 80% of the silver was found throughout the trunk, followed by their presence in the roots, branches, and leaves [26]. This result shows that appropriate surface coating can play a crucial role in the uptake and translocation of nanoparticles in plants.

Coating of nanoparticles on seeds (nanopriming) is another promising method for the delivery in plants. Using x-ray fluorescence spectroscopic analysis of seedlings, Lau *et al.* have concluded that surface coating of iron oxide nanoparticles has a profound effect on their translocation in tomato (*Solanum lycopersicum*) plants following seed coating [27]. In another work, Afzal *et al.* have synthesized biocompatible and "green" iron oxide nanoparticles using extracts of *Cassia occidentalis* L. flower, and investigated their effect on germination of *Pusa basmati* rice seeds following nanopriming. Much enhanced germination and seedling vigor were observed in case of the nanoprimed seeds, as compared to that observed in seeds primed with ferrous sulphate ($FeSO_4$) or hydro-primed control. Much higher stimulation was observed in biophysical parameters (root length and dry weight), as well as sugar and amylase content, along with enhancement in antioxidant enzyme activity in case of nanoprimed seeds, when compared to the other controls [28]. In a similar demonstration, Mahakham *et al.* have shown that upon nanopriming of rice seeds with green-synthesized silver nanoparticles (AgNPs), significant improvement in germination performance (Figure 7.3), seedling vigor, α-amylase activity, aquaporin gene expression, ROS production, etc., were observed in comparison to those observed using controls (no priming, $AgNO_3$ priming, and conventional hydropriming) [29].

Overall, it can be concluded that several nanoparticles can be efficiently uptaken by plants using various routes, and translocate to other parts. Physicochemical characteristics as well as surface coating of the nanoparticles are essential factors that control the extent of uptake, retention and translocation. In case of foliar uptake, the viscosity, elasticity, surface tension and other rheological parameters are important determinants of their surface coverage, adhesion, and retention. Although this facile uptake and translocation indicates a promising landscape for the various applications of nanoparticles intended for botanical benefits, the collateral diffusion of nanoparticulate wastes, especially those containing toxic heavy metals, may lead to deleterious effects in plants.

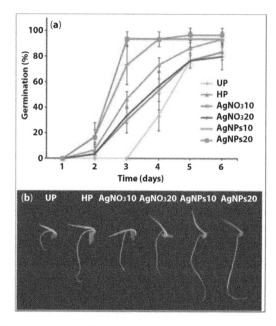

Figure 7.3 (a) Germination rate of rice seeds after priming with different priming agents, and (b) phenotype of rice seedlings at 6-day post-germination. AgNPs, silver nanoparticles; $AgNO_3$, silver nitrate control; UP, non-primed control; HP, hydroprimed control. Reprinted with permission from Reference [29].

7.3 Nanoparticle-Mediated Sensing and Biosensing in Plants

Sensing is essential for continuous monitoring of the health and well-being of the flora. A number of analytes can be sensed, ranging from the pH, humidity, salinity, content of agrochemicals, such as fertilizers and pesticides, etc., in the soil where the plants grow, to various harmful pathogens that impede plant growth [30]. In addition, global climate change, pollution, and other stress factors can lead to the generation of various chemical and electrical signals by plants and trees, that can be measured either externally (using wearable sensors), or internally by delivering sensing probes within the plants [31]. Internal sensing can also lead to the monitoring of functional parameters within the plants, such as ionic and molecular dynamics, which may lead to the fundamental understanding of physiological processes within normal as well as diseased (or stressed) plants. Expression of various functional genes essential for plants are also studied using sensing methods [32]. Therefore, it is necessary to interface various sensing platforms with the flora that can reliably transduce the analyte expressions within plants into quantitative and digitized output in real time.

Nanoparticles, owing to their unique physical properties, are associated with several functional effects that make them ideal candidates in sensing and biosensing applications [33]. For example, the collective oscillation of free surface electrons (plasmons) in nanoparticles of noble metals such as gold and silver, which is termed as localized surface plasmon resonance (LSPR), can be exploited in plasmonic imaging and sensing applications. LSPR phenomena also facilitates surface enhanced Raman spectroscopy (SERS)-based sensing in conjunction with Raman-active molecules. On the other hand, the tunable band-gap in nanoparticles of semiconductors (quantum dots), such as silicon, cadmium sulphide, and titanium dioxide, results in highly photostable and narrow-band photoluminescence, that are very useful in multimodal optical sensing applications. Upconversion nanophosphors (UCNPs) are another unique class of rare-earth doped nanoparticles, which show the phenomenon of photon upconversion, i.e., the emission of a higher energy photon following the sequential absorption of two lower energy photons. Owing to multiple emission peaks in the ultraviolet-visible region of the electromagnetic spectrum following absorption of photons in the near infra-red (NIR) region, UCNPs can serve as luminescent probes for ratiometric sensing. Porous nanostructures, such as nanosilica and nanoscale metal organic frameworks (NMOFs), can also serve as optical sensing nanoprobes following incorporation of fluorescent molecules within their well-defined pore network. Several carbon-based nanoparticles, such as carbon dots, nanographene oxide, etc., also show the phenomenon of photoluminescence and therefore serves as non-toxic probes in optical sensing. Carbon nanotubes and nanographene are also useful in electrochemical sensing owing to their unique conductive and optoelectronic behaviour. Iron oxide and ferrite nanoparticles, owing to their unique magnetic behavior, are also promising sensing nanoprobes. In addition to their functional features, the tunable physical parameters of nanoparticles, such as size, shape, porosity, etc., makes them amenable to be integrated within a variety of sensing devices that can be used in both internal and external (wearable) sensing applications. Furthermore, the facile digitization of plasmonic, optical, electrochemical, and magnetic signals allows their remote monitoring in real-time, thus facilitating a variety of sensing platforms. Furthermore, incorporation of more than one type of nanoprobe within a single platform can facilitate multimodal sensing applications.

Glucose is produced via photosynthesis in plants, and serves as a key energy source. Therefore, glucose sensing is an important indicator of the well-being of plants. Li *et al.* have fabricated an optical sensing platform based on fluorescent quantum dots (QDs) and have quantified glucose levels in wild-type plants using a ratiometric approach. The ratiometric probe is based on a QD pair: where the fluorescence of one probe (boronic acid-conjugated QDs) is quenched in presence of glucose in a dose-dependent manner while the other probe (thioglycolic acid-capped QDs) acted as an internal standard by remaining unresponsive to changing glucose levels. The probe was also found to be highly selective for glucose among all common sugars in plants. The detection limit of this sensing probe was 500 µM in planta, which falls within the physiological range. Confocal microscopy revealed that glucose could be detected in single chloroplast within algal cells (*Chara zeylanica*), as well as in leaf tissues (*Arabidopsis thaliana*). Thus, this method provided a sensitive, selective and reliable avenue for detection of glucose levels in photosynthetic plants [34]. Recently, Mou *et al.* have reported a glucose sensor based on DNA-aptamer nanostructures which were internalized into Arabidopsis and tobacco leaf cells using a thiol-mediated approach [35]. Continuous monitoring of the pH levels within plants can also be carried out using nanoparticle-based sensors. Ruiz-Gonzalez *et al.* have fabricated an implantable electrochemical sensor based on ruthenium oxide nanofilms for direct pH monitoring within the xylem sap of tomato plants (Figure 7.4). The thin-films were fabricated with a thickness of 80 nm, along with roughness

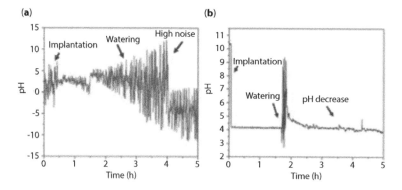

Figure 7.4 (a) Measurement of pH inside the tomato stems using a pristine electrode without the incorporation of a cellulose-based coating. An increasing noise was recorded, with a change upon watering the plants. (b) Plot obtained after the recording of pH in a tomato stem using cellulose-based device. The changes in pH upon watering the tomato plant could be determined. In particular, a stable decrease in pH after the watering was obtained. Reprinted with permission from Reference [36].

below 3 nm. The output signal can be relayed online wirelessly through an inexpensive Wio Terminal device. A lignified layer was found to form between the sensing film and xylem [36].

Nanosensors can be employed for *in situ* hormone detection in plants. Wang *et al.* have developed a microneedle array sensor using microelectrodes of Gold@Tin oxide-vertical graphene (VG)/Tantalum for detection and quantification of the plant hormone abscisic acid (ABA) in plants. Direct electrocatalytic oxidation method was employed on the microarray sensor electrodes for quantification of ABA, which was facilitated by the presence of several catalytic sites on the gold/Tin oxide nanoparticle surface, assisted by the excellent conductivity of graphene nanosheets. The detection limit of the analyte varied between 0.002 and 0.005 µM, within the pH range of 4 to 7, with a wide linear concentration range [37].

Deficiency of critical nutrients in growing plants is a fundamental reason for poor crop development and low yield. Nutritional deficiency leads to the production of several associated biomarkers, which can be quantitatively monitored in order to prevent crop loss. Giust *et al.* developed a sensor that can quantitatively detect biomarkers associated with zinc-deficiency, which is an essential nutrient in plants, using graphene oxide and UCNPs. The biomarker chosen was an mRNA that encodes zinc-regulated, iron-regulated transporter-like protein (ZIP), which is a family of membrane transport proteins that are up-regulated during zinc deficiency. These mRNA biomarkers were detected based on changes on optical output as a result of their specific interaction with complementary oligonucleotide-coated UCNP and graphene oxide, which can be relayed using a smartphone camera [38].

Another key reason for poor plant growth and plant loss is the infiltration of contaminants in the plants. Therefore, monitoring of harmful chemical and biochemical contaminants in plants is of paramount importance in smart agricultural practices. Hao *et al.* have developed an assay based on fluorescence resonance energy transfer (FRET) for the sensitive and fast detection of the contaminant Ochratoxin A (OTA) in corn plants, using gold nanorod-based aptasensors. Herein, gold nanorods attached with DNA tetrahedron containing the OTA aptamer served as the energy acceptor, with a complementary sequence of the OTA aptamer fluorescently labelled with Cy5 serving as the energy donor. Therefore, dose dependent hybridization of the OTA aptamers with Cy5-modified complementary sequences lead to FRET, the intensity of which could be linearly correlated with the logarithmic concentration of OTA, in the concentration range of 0.01 to 10 ng/mL. Ultralow detection limit of 0.005 ng/mL of the contaminant was detected using this sensor, thus highlighting its potential in sensitive, specific, reliable and fast detection of contaminant-induced stress biomarkers [39].

Gene expression analysis within plants provides several critical information, such as their state, levels of nutrients, possible stress condition, etc. However, conventional analytical techniques of gene expression in plants, such as polymerase chain reaction (PCR), RNase protection assay, *in situ* hybridization, etc., are time-consuming and inefficient [40]. Nanoparticle-mediated *in situ* biosensing and continuous monitoring of gene expression profiling in plants can lead to reliable and efficient data. Crawford *et al.* devised a multimodal method for biosensing of microRNA targets within leaves of live plants using plasmonic nanostars. Specifically, by integrating the techniques of SERS and X-ray fluorescence (XRF) with plasmonic-enhanced luminescence, the authors have demonstrated multimodal, quantitative biosensing of microRNA, with high spatial (200 μm) and temporal resolution (30 mins) [41]. In another report, Ghazi *et al.* have studied the transformation and expression of the reporter GUS transgene in Zabol melon plants using gold nanoparticles modified with complementary sequences. The GUS gene was first introduced in melon seeds using Agrobacterium tumefaciens. Using gold nanoparticles attached to DNA, RNA and cDNA probes complementary to the target gene, the expression of the transgene was correlated with color changes due to plasmonic interactions. A linear relationship between signal change and analyte concentration, along with a low detection limit of 0.25 ng/μL indicated the suitability of that this method for in planta gene expression analysis [42].

Overall, it can be concluded that nanoparticles based on plasmonic, optical, and electrochemical signals can provide inexpensive, reliable, reusable, transmissible and highly sensitive sensing platforms for the continuous monitoring of various essential chemical and biochemical analytes in plants. Nanoparticles can also facilitate simultaneous multimodal sensing of several analytes using a same platform.

7.4 Tolerance Versus Toxicity of Nanoparticles in Plants

The widespread use of nanoparticles worldwide in a variety of medical, industrial, and technological applications has led to their copious release in the environment. The facile uptake and translocation of a variety of nanoparticles in plants following intentional or accidental exposure provokes the question about their potential toxicity in the hosts. A number of studies have already addressed this concern, and the results yet obtained are rather surprising. In most of the cases, it has ben observed that the nanoparticles have a beneficial effect on the treated plants; this, however relies on the composition of the nanoparticle. Some nanoparticles, such as those of zinc, has behaved as a plant nutrient itself without the incorporation of any additional agrochemical [43]. Several other nanoparticles have shown a protective as well as growth stimulating effect on treated plants. A detailed understanding of the effects of nanoparticles on treated plants shall pave the way for the selection of appropriate nanoparticles for advanced applications. Below, we shall highlight some select examples of the effects of placebo nanoparticles on treated plants.

Selenium nanoparticles are known for their safety and antioxidant nature. Quiterio-Gutiérrez et al. have investigated the effects of selenium and copper nanoparticles on tomato plants infected with the fungal pathogen *Alternaria solani*, which is known to cause Blight disease and subsequent plant loss. Foliar application of the nanoparticles led to attenuation of the severity of the infection in the plants. Moreover, several key enzymes, such as superoxide dismutase (SOD), glutathione peroxidase (GPX) and phenylalanine ammonia lyase in the leaves, along with chlorophyll a and b in the leaves, as well as other vital biochemicals such as vitamin C, glutathione, phenols, and flavonoids in the fruits, were found to be increased upon nanoparticle treatment. As a result of their effect on the upregulation of critical enzymatic and non-enzymatic compounds, as well as protection from pathogen-induced biotic stress, it can be concluded that these nanoparticles have profound beneficial effect on the tomato plants [44]. In a report by Ma *et al.*, copper-based nanostructures were found to rescue soybean plants from biotic stress mediated by pathogenic infection and subsequent sudden death syndrome. These nanoparticles were delivered to the infected plants via the foliar route, which resulted in significant reversal of attenuated biomass and photosynthesis. Concurrently, other disease biomarkers, such as antioxidant enzymes and fatty acid, were alleviated upon nanoparticle treatment. Gene expression profiling revealed that several genes associated with the defence and health of the plants were upregulated upon nanotreatment [45]. Recently, Dev Sarkar *et al.* showed that green-synthesized selenium nanoparticles can rescue Brassica campestris (mustard) plants from salt-induced abiotic stress. These nanoparticles, with an average size of about 60 nm, were formed via the reduction of selenium dioxide using *Allamanda cathartica* L. flower extract in aqueous system. The seed germination percentage, shoot and root lengths, and total chlorophyll content of the nanoparticle-treated plants, grown under salt (200 mM NaCl) stress, were found to be significantly improved as compared to the untreated controls [46].

Abiotic stress, caused by the exposure to extreme heat, light, soil salinity, etc., leads to the elevation in the levels of reactive oxygen species (ROS) in plants, which in turn are responsible for extensive oxidative damage to plants metabolites, enzymes, lipids, and DNA, along with reduction in photosynthesis [47]. Therefore, the introduction of ROS scavengers is critical to rescue the afflicted plants. Cerium dioxide nanoparticles (nanoceria) are well known for their antioxidant and ROS scavenging activity. Wu *et al.* prepared anionic (zeta potential of -16.9 mV), ultra-low sized (below 11 nm) spherical, polyacrylic acid-coated nanoceria and delivered them via the foliar route to Arabidopsis thaliana plants exposed to excess light and extreme temperature conditions. The nanoparticles were found to infiltrate the chloroplasts through nonendocytic pathways, and reduce the levels of several ROS, such as hydrogen peroxide and hydroxyl radicals. The photosynthesis levels in the nanoparticle-treated plants were also upregulated, as evidenced by enhancement in the quantum yield of photosystem II and carbon-assimilation [48]. Recently, Liu *et al.* have shown that upon application of poly acrylic acid-coated nanoceria (PNC) on cotton plants exposed to high salinity, significantly enhanced phenotypic performance and chlorophyll contents (Figure 7.5), as well as biomass and carbon assimilation were observed, when compared to that in non-nanoparticle treated control. Conversely, the levels of several other chemical markers, such as malondialdehyde, as well as ROS such as hydrogen peroxide, were found to be significantly reduced upon nanoceria treatment. Fluorescence microscopic studies revealed that treatment with the nanoparticles led to significant enhancement (about 84%) in the levels of cytosolic K^+ and concurrent reduction (about 77 %) in cytosolic Na^+, in the plants under salt stress. These results implicate that nanoceria can significantly improve cotton salt tolerance by ROS scavenging and maintaining cytosolic K^+/Na^+ homeostasis [49].

Nanoparticles with antimicrobial properties can also rescue plants from biotic stress and related diseases. Using a microwave-assisted 'green' formulation of silver nanoparticles from the leaf extract of Melia azedarach, Ashraf *et al.* have shown that tomato plants can be prevented from infection by Fusarium oxysporum, which is responsible for wilt disease in plants. The spherical nanoparticles with particle size in the range of 12 to 46 nm were shown to inhibit fungal mycelial growth by about 98% *in vitro*, when compared to that without nanoparticle-treated control. Growth parameters of F. oxysporum-infected tomato seedlings were also significantly enhanced upon nanoparticle treatment. Electron microscopic analysis showed extensive damage to the fungal hyphae and spores after nanoparticle treatment. Additionally, intracellular ROS

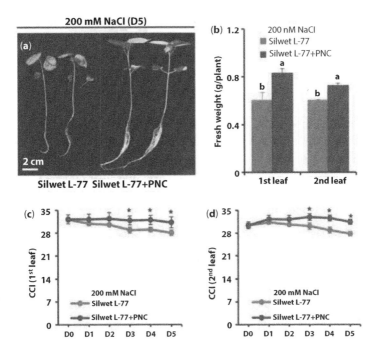

Figure 7.5 The phenotypic performance and chlorophyll content of salt stressed (200 mM NaCl, 5 days) cotton plants with and without PNC treatment. (a) Phenotypic performance of salt stressed cotton plants with and without PNC treatment. (b) The fresh weight of salt stressed cotton plants with and without PNC treatment. (c, d) CCI (chlorophyll content index) readout in the first (c) and second (d) true leaves of salt stressed cotton plants with and without PNC treatment (mean±SE (n=6)) Different lower-case letters indicate the significance level at 0.05. (*Means P<0.05). The comparison was made between PNC treated cotton plants and the NNP control at each day. Silwet L-77: surfactant used for enhancing foliar delivery. Reprinted with permission from Reference [49].

enhancement and altered membrane permeability of the fungal cells were also observed [50].

Owing to the rampant assimilation of industrial wastes in groundwater and agricultural lands, several plants and crops are exposed to high levels of toxic heavy metals such as mercury, lead and cadmium, leading to severe abiotic stress related loss of their quality and growth. Zou *et al.* have probed the application of iron oxide and zinc oxide nanoparticles on tobacco plants under cadmium stress. Upon foliar spraying of the nanoparticles, the repressed growth of the afflicted plants were found to be reversed, as evidenced from the enhancement in physical parameters such as plant height, root length, shoot weight, etc. Furthermore, dissolution of the nanoparticles within the plant cells led to higher content of essential nutrients such as Zn, K and Mn, which promoted seedling growth. Cadmium stress severely affected the metabolic profiling of the plants, which was substantially rescued following nanoparticle treatment. Since these metabolites are associated with several essential molecules, such as amino acids, nicotinate, nicotinamide, arginine, proline, flavone, flavonol, 6-hydroxynicotinic acid, farrerol and quercetin-3-O-sophoroside, their restoration upon nanoparticle treatment are essential for the plant growth under abiotic stress [51].

As carbon-based nanomaterials are widely used in several applications, Velikova *et al.* have probed the effects of single-walled carbon nanotubes (SWCNTs) on structural and functional parameters of pea plants following foliar delivery. They observed that lower dosages (10 mg L^{-1}) of the nanotubes did not have any deleterious effects on the plants. However, upon increasing the dosage beyond 10 mg L^{-1}, several harmful effects, such as swelling in granal and stromal regions, impaired photosynthesis, slower non-photochemical quenching, etc., could be observed. This suggests that SWCNTs do show dose-dependent toxicity on treated plants. However, at lower doses (10 mg L^{-1}), they can be used as safe nanocarriers of active agents, such as fertilizers, for delivery in plants [52]. In a recent work involving targeted carbon-based nanomaterials, Santana *et al.* have reported the use of carbon dots and SWCNTs for delivery of active components within the chloroplasts in plants. Since the chloroplasts are major intracellular sites for photosynthesis, nutrient assimilation and agrochemical delivery, they were targeted using a chloroplast targeting peptide. No significant damage to the leaves and chloroplasts were observed at the dosages tested, though moderate enhancement in leaf H_2O_2 levels were observed. This work shows that biotargeted carbon-based nanomaterials can be used as a safe delivery vehicle in plant chloroplasts [53].

Most of the reports are related to studying the effects of inorganic-based nanoparticles in plants, while those of organic nanoparticles based on polymers or lipids are scarce. Salinas *et al.* have addressed this issue by investigating the effects of hydroponically delivered biodegradable polymeric nanoparticles based on the proteins lignin (LNP) and zein (ZNP) on soybean plants. The treatment dosages of both the anionic LNPs (114 ± 3.4 nm) and cationic ZNPs (142 ± 3.9 nm) were varied (0.02, 0.2, and 2 mg/ml), and plants were harvested at 1, 3, 7 and 14 days post-treatment. Analysis of several physical and biochemical parameters post-harvest, such as length and dry biomass of both root and stem, chlorophyll content, nutrient uptake, ROS levels, etc., revealed that although no discernible effect was observed upon LNP (at all dosages) and ZNP (at low dosage) treatment, some effect, such as enhanced ROS content, was observed for high dose (2 mg/ml) ZNP treatment groups. Overall, the authors concluded that these polymeric nanoparticles are largely non-toxic to the plants at low and medium treatment dosages, and can be used as a delivery vehicle in plants [54].

Overall, it can be concluded that nanoparticles composed of metals, metal oxides, cerium dioxide, polymers, etc., do not exert deleterious effects on plants at low and moderate dosages [55]. Surprisingly, several of these nanoparticles served as nutrients, antioxidants, antimicrobials, etc., and have exerted a beneficial effect on the plants. These observations, however, do not automatically lead to the conclusion that engineered nanoparticles are safe for plants. Rather, extensive research efforts have to be devoted to probe in detail the effects of size, shape, crystallinity, composition, stability, etc., on plants. For example, the interaction of heavy-metal containing nanoparticles, such as that of cadmium sulfide, with plants are required. The mode of synthesis can also play a critical role in this regard. For example, iron oxide nanoparticles synthesized using green, bioinspired, aqueous-based approaches can have an altogether different compatibility/toxicity profile in plants from the same nanoparticles synthesized using hot colloidal method in organic solvents. Nevertheless, the general perspective of nanoparticle-plant interactions is quite positive; however, it requires vigilance from regulatory agencies before widespread applications, similar to regulations that exist with regards to nanoparticle applications in biomedicine.

7.5 Nanoparticle-Mediated Delivery of Fertilizers, Pesticides, Other Agrochemicals in Plants

Nanoparticle mediated delivery of drugs and other active components have been extensively investigated in animal models. Nanoparticles can stably incorporate drug molecules, protect them from premature release and degradation, guide them to target tissues, cells and subcellular organelles by transport across complex biological barriers, and release them in a controlled manner. These benefits make them ideal candidates for delivery of essential agrochemicals and biochemicals in plants as well. In particular, delivery of toxic agrochemicals such as pesticides and herbicides are desirable in a nanoencapsulated form rather than in their free form as in the former the major risks of ecological leakage and subsequent deleterious effects on non-target flora and fauna are reduced, while their functional effects on the target plants are preserved. Additionally, nanoencapsulation protects these agrochemicals from degradation induced by sunlight and stray chemicals. As discussed earlier, nanoparticles can be introduced in plants via several routes, from where they can translocate to other parts in the plants using phloem transport and other mechanisms. Delivery of active reagents in plants can accelerate plant growth, enhance their productivity, rescue them from nutrient deficiency, protect them from various biotic and abiotic stresses, present crop loss, and prevent ecological leakage and off-target toxicity, thereby paving the way for smart farming practices [56].

Around the world, millions of metric tonnes of fertilizers are applied daily for agricultural purposes, out of which a small fraction is absorbed by the plants/crops owing to inefficient absorption and retention pathways. This not only leads to massive economic loss, but also significant environmental fouling and off-target side effects. This necessitates the development of technologically driven platforms for optimized delivery of fertilizers in plants. Nanoparticles are capable of incorporating and efficiently delivering fertilizers to plants with optimal uptake and retention, thus preventing economic and environmental losses. In one of the early examples, Abdel-Aziz *et al.* used nanoparticles of the natural, biocompatible polymer chitosan for the delivery of nitrogen, phosphorus and potassium (NPK) in wheat plants. Following foliar delivery, the nanoparticle-loaded fertilizers entered the stomata through gas uptake, while avoiding soil contact. Electron microscopy data showed that the nanoparticles could translocate through the phloem tissues to other parts in the plant. When grown in nutrient-depleted soil, much higher harvest index and crop production, along with accelerated growth, were observed in the plants treated with these fertilizer-loaded nanoparticles, in comparison to the plants treated with free NPK-fertilizer [57]. Liposomes are one of the most common nanostructured drug delivery systems owing to their unique bilayered geometry and structural flexibility. Karny *et al.* composed 'green' liposomal nanoparticles from plant-derived lipids, that can translocate to other leaves and roots following foliar delivery in tomato plants. Fe and Mg-loaded liposomes could rescue the treated tomato plants from acute nutrient deficiency. These liposomes were composed in such a manner that they remained stable only over short spraying distances (about 2 meters), thus minimizing their potential for environmental leakage and contamination [58]. Similarly, Ramírez-Rodríguez *et al.* prepared multinutrient nanofertilizers by employing biomimetic calcium phosphate (CaP) nanoparticles doped with calcium (Ca), potassium (K), phosphorous (P) and nitrogen (N). These nanoparticles showed slow release of the nutrients (NPK) in durum wheat, resulting in reduction in the necessity of nitrogen supply in the plants (40% reduction in comparison with conventional treatment), while retaining the final kernel weight per plant [59].

The overuse of harmful agrochemicals such as pesticides, herbicides, antibiotics, etc. in agriculture and horticulture are not only associated with acute ecological poisoning, but also leads to the evolution of resistant pests, weeds, microbes, etc. that are refractive to conventional agents. Therefore, optimization of the dosage of these toxic chemicals through nanoparticle-mediated targeting and controlled delivery in plants is urgently needed. Oliveira et al. have investigated the use of poly(epsilon-caprolactone) (PCL) nanocapsules encapsulating the herbicide atrazine on mustard (*Brassica juncea*) plants. They reported that upon treatment with 1 mg mL^{-1} of the herbicide encapsulated within the PCL nanoparticles, several harmful effects could be observed, such as impaired photosynthesis, low photosystem II quantum yield (Figure 7.6), enhanced lipid peroxidation, inhibition of shoot growth, etc. However, the herbicidal activity was enhanced in the nanoencapsulated atrazine formulation, when compared to that of free atrazine-treated control. This result shows that atrazine dosage in plants can be reduced using nanoparticles without

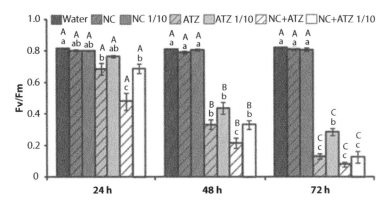

Figure 7.6 Maximum photosystem II quantum yields of B. juncea plants treated with the formulations. Chlorophyll fluorescence parameters were evaluated 24, 48, and 72 h after the plants were sprayed with 3.1 mL of water, empty PCL nanocapsules (NC), commercial atrazine (ATZ), or PCL nanocapsules containing atrazine (NC+ATZ). The formulations containing atrazine at 1 mg mL^{-1} were used undiluted or after 10-fold dilution in water (1/10), resulting in atrazine dosages equivalent to 2000 or 200 g ha^{-1}, respectively. Different letters indicate significantly different values according to two-way ANOVA followed by the Tukey test ($p < 0.05$). Lowercase letters indicate comparison among all treatments at each time point, while capital letters indicate comparison of the same treatment along its individual time course analysis. Data are shown as means ± SE (n = 5). Reprinted with permission from Reference [60].

compromising with the post-emergence herbicidal effect [60]. Recently, Rashidipour *et al.* have demonstrated that upon encapsulation of the herbicide paraquat within nanoparticles of pectin, chitosan, and sodium tripolyphosphate (PEC/CS/TPP), the herbicide was found to have a delayed release effect, was did not cause discernible toxicity in normal mammalian cell lines, remained minimally mutagenic, and had reduced deep soil penetration to curtail systemic leakage, with respect to its free form. At the same time, the herbicidal activity on the target crops (maize and mustard) was found to be enhanced, possibly due to higher bioavailability owing to the controlled release of the herbicide following nanoencapsulation [61].

In one of the early examples of nanoparticle-mediated pesticide delivery, Torre-Roche *et al.*, investigated the uptake of the insecticide and pesticide dichlorodiphenyldichloroethylene (p,p'-DDE) in several plant species, such as zucchini, soybean, and tomato, which were exposed to different dosages of fullerene (C60). It was found that the p,p'-DDE uptake in all these plants were enhanced with varying degrees, as evidenced from comparing the root and shoot p,p'-DDE levels in fullerene treated plants [62]. This report paved the way for several works in the area of nanoparticle-mediated enhancement of pesticide uptake in plants. Tong *et al.* have prepared polymeric nanoparticles of mPEG-PLGA for the stable aqueous dispersion of the hydrophobic pesticide metolachlor. The nanoencapsulated metolachlor was found to be efficiently uptaken by *Oryza sativa* and *Digitaria sanguinalis* plants [63]. In a recent report, Wang *et al.* have prepared porous silica nanoparticles incorporating the pesticide acetamiprid. Upon treatment with tea seedlings (*Camellia sinensis* L.), the nanopesticide formulation was found to be well uptaken via the roots and transported to the leaves through the stem, where it was retained for over a month. Upon metabolomic analysis, it was found that several metabolites, such as salicylic acid, ribonic acid, glutamine, naringenin diglucoside, and epiafzelechin, were differently expressed in the plants treated with both the nanopesticide formulation and a commercially available sample of the pesticide [64]. In yet another recent report, Luo *et al.* have prepared nanogels from block-copolymers of Polyethylene glycol (PEG) and 4,4-methylenediphenyl diisocyanate, encapsulating hydrophobic pesticides. Upon foliar spraying, these flexible nanogels can cover high leaf surface area and enhance protection against pests. Moreover, these viscous, gel-like polymeric nanopesticides could be retained in the leaves for much longer time than free pesticides, even after repeated washing. Furthermore, it could be seen that these polymeric nanogels could protect the leaves from degradation mediated by aquatic pests and UV light [65].

Nanoparticles can also deliver protease inhibitors as nanopesticides in plants. Bapat *et al.* prepared surface tunable silica nanoparticles incorporating soybean trypsin inhibitor (STI), and have investigated their effects on tomato plants and the pest *Helicoverpa armigera*. The nanopesticides could be uptaken by the plants via both roots and leaves, and infiltrate in the vasculature. This formulation could inhibit the gut proteinase (HGP) activity in *Helicoverpa armigera* by 50%, and retard growth of the insect larvae. Choice assays revealed that the nanopesticide-applied leaf discs were refractory to the insect larvae [66].

Nanoparticles also serve as efficient carriers of fungicides for protection of plants against fungal infections. Liang *et al.* have incorporated the fungicide pyraclostrobin within pH-responsive mesoporous silica nanoparticles coated with a copper-tannic acid coordination complex. The coating prevented the encapsulated fungicide from light-induced degradation, and also facilitated fungicide release at low pH. Tannic acid on the nanoparticle surface was responsible for enhancing their adhesion and retention in crop foliage. This fungicide nanoformulation was shown to effectively protect rice plants from infection by the fungus *Rhizoctonia solani* [67]. In another recent report, polysuccinimide nanoparticles conjugated with glycine methyl ester incorporating the fungicide fludioxonil was prepared by Wu *et al.* When applied to banana plants via foliar delivery, it was observed that these nanoparticles could travel to the rhizomes and roots via phloem transport, and preferentially release the fungicide at alkaline pH in the plant phloem. Significant reduction in the severity of fungus-mediated Fusarium Wilt disease was observed in banana plants treated with this nanofungicide formulation. An active amino-acid transporter-mediated cellular diffusion of the nanoparticles was observed [68].

Overall, it is evident that nanoparticles can be used for the delivery of several active agents in plants for a protective/beneficial intent. Traits such as slow release, pH-dependent release, etc., shown by these nanoparticles can help in optimizing the dosage and targeting potential of these nanoparticles to specific regions of the plants. In combination with nanoparticle-mediated sensing platforms, which were discussed previously in this chapter, co-incorporation of active molecules can also facilitate on-demand delivery guided by sensing signals.

7.6 Nanoparticle-Mediated Non-Viral Gene Delivery in Plants

The term gene therapy represents any therapeutic or prophylactic effect brought about in a living species via the introduction of a functional gene or genetic tool. Following introduction, these genes most travel to target cells and even specific intracellular organelles (e.g. nucleus) where they exert their functional effect. However, since genes such as DNA and RNA are inherently negatively charged, and are extremely vulnerable to enzymatic digestion within a living being, they require a carrier for their protection, targeting, and controlled release. As viruses are natural gene carriers, engineered viruses were developed as promising vectors for gene delivery. However, their toxicity, immunogenicity, mutagenicity and propensity to revert back to their wild-type forms have limited their applications in gene delivery. As a safer alternative, nanoparticles have emerged as promising, non-viral based carriers that can form complexes with various genetic materials, protect them from nuclease digestion, and deliver them to target cells/ organelles with reasonable efficiency (albeit less than that of engineered viruses).

In recent years, genetic engineering and genome editing in plants have emerged as exciting concepts in agricultural science owing to the promise of producing resilient crops, such as transgenic cotton and maize plants, leading to higher production efficiency for meeting the ever-expanding global food requirement. The use of clustered regularly interspaced short palindromic repeats (CRISPR), representing a family of gene sequences found in the genome of bacteria and some other prokaryotic organisms, has emerged as a promising gene editing tool for plant genetic engineering. This RNAi guided system facilitates precise modification of target DNA with high efficiency [69].

Traditional gene delivery concepts, such as using viruses, agrobacterium infection, biolistic delivery, etc., suffers from several limitations such as poor delivery, transgene integration in host genome, and genotype dependence. The ability of various nanoparticles to efficiently interact with plants makes them exciting alternative carriers for gene therapy in plants. The use of nanoparticulate systems for the delivery of CRISPR genes/proteins to target plant cells for the editing of plant genomes is another exciting approach. This field of non-viral gene delivery in plants has witnessed a burst of research activity ever since the first report on the use of microscopic silicon carbide fibers to deliver GUS and Bar genes in tobacco suspension cells [70]. Below, we shall underline some selected examples of nanoparticle-mediated genetic engineering applications.

Nanoparticle-mediated delivery of plasmid DNA encoding functional proteins can efficiently augment critical genes in plants responsible for their growth and development. However, permanent integration of such genes at non-target sequences within the host genome can cause unnecessary distress to the plants. Demirer *et al.* have demonstrated the use of unmodified and chemically modified high aspect ratio carbon nanotubes for delivering plasmid DNA into several species of plants (arugula, wheat, and cotton) without transgene integration. High levels of gene delivery and protection from enzymatic digestion were evident based on observation of strong expression of encoded proteins is the target plants [71]. In a recent example of cell-specific gene delivery, Kwak *et al.* have shown the feasibility of chloroplast-selective delivery and co-expression of green fluorescent protein (GFP) and yellow fluorescent protein (YFP) in planta using chitosan-modified single walled carbon nanotubes (SWCNTs) [72].

Nanoparticle-mediated plasmid DNA delivery efficiency in plant cells can further be improved using mechanical stimuli, among which ultrasonic treatment (sonoporation) has emerged as a promising technique. Cationic polyamidoamine (PAMAM) dendrimers, which are hyperbranched polymers with defined three-dimensional nanostructure, can form stable complexes with anionic genes and deliver them to target cells with high efficiency. Amani *et al.* formed complexes with a DNA (encoding the gus A gene) and hyperbranched PAMAM (hPAMAM)-G2 dendrimers, and investigated their delivery in alfalfa cells, without and with ultrasonic treatment. It was found that the most efficient complexes were formed at a N/P ratio of 3 and more, where N represents the number of nitrogen atoms in the dendrimer, and P the number of phosphorus atoms in the DNA. Nanocomplexation was found to protect the immobilized DNA from ultrasonic-induced degradation. The gene internalization efficiency (observed using fluorescently tagged gene) and subsequent gene expression in alfalfa cells treated with the nanocomplexes were found to significantly increase with sonoporation assistance [73]. Recently, Zolghadrnasab *et al.* have shown that sonoporation can enhance the efficiency of delivery of GUS-encoding plasmid DNA complexed with polyethyleneimine (PEI)-functionalized mesoporous silica nanoparticles (MSNs), into suspended tobacco cells. However, the authors have cautioned that excessive ultrasound power and/or time of exposure may negatively effect the health of the target cells, and have outlined the guidelines for safe and efficient use of his technique [74].

Iron-oxide nanoparticles surface modified with suitable cationic polymers can not only form complexes with genes, but also deliver them to target cells aided by the application of external magnetic field (magnetofection). Zhao *et al.* have shown that branched polyethyleneimine-decorated iron oxide nanoparticles could be introduced within pollen cells. This system was used to stably integrate the GUS gene, as well as Bt-resistance genes into the genome of cotton, thus forming transgenic cotton with high pest resistance [75]. Recently, Wang *et al.* have used similar technique to cause genotype independent maize transformation following stable gene transfection in maize pollens using DNA-complexed iron oxide nanoparticles [76].

Short interfering RNA (siRNA) represents double stranded RNA with 22 base pairs that impedes the expression of undesirable/malfunctioning target genes post-transcription in host cells. Therefore, gene silencing using siRNA has promising applications in plant functional genomics for protection and improvement in the quality and production of crops. However, delivering these fragile gene across the plant cell wall poses a formidable challenge, which can be mitigated with the help of nanoparticles. Zhang et al. systematically assessed the ability of different DNA nanostructures incorporating siRNA for internalization into cells and delivery of siRNA to target a specific gene expressed in Nicotiana benthamiana leaves. The size, shape, nanostructure geometry, and siRNA attachment locus were found to be dominant factors that controlled the cellular internalization and subsequent gene silencing ability [77]. Schwartz et al. have used ultrasmall carbon dots for delivering siRNA against the GFP gene in *Nicotiana benthamiana* and tomato (*Solanum lycopersicum*) plants. Surfactant assisted low-pressure spray application led to significant silencing of GFP transgenes in both target plants (Figure 7.7). Furthermore, this technique was extended for the silencing of endogenous genes encoding the enzyme magnesium chelatase, which plays a critical role in chlorophyll synthesis. Gene silencing was confirmed by measurement in both target gene transcript (using quantitative PCR) and protein levels (using Western blot analysis) [78]. The shape of nanoparticles is often found to be a critical determinant of gene silencing ability. For example, in a recent report, Zhang et al. showed that gold nanorods have better ability to cause siRNA-mediated suppression of GFP gene in *N. benthamiana* leaves in comparison to gold nanospheres [79].

Cell penetrating peptides (CPP) have been utilized as efficient mediators for enhanced intracellular delivery of various drugs and genes. Recently, Thagun et al. have developed a CPP-based nanocarrier that complexes with siRNA and efficiently delivers them to plants via foliar spray. The physicochemical parameters of the CPPs dictated their translocation and stomata-dependent uptake in plant cells. Efficient translocation and gene silencing inside the nuclei as well as chloroplasts of target cells could be achieved by modulating the targeting moiety of these nanocarriers [80].

Overall, it can be safely concluded that the potential of nanoparticles to protect and deliver genetic cargos within target cells and intracellular compartments with high efficiency has made them better gene carriers as compared to traditional methods such as viruses. Nanoparticle-mediated gene delivery and gene editing in plants can confer the hosts with a number of

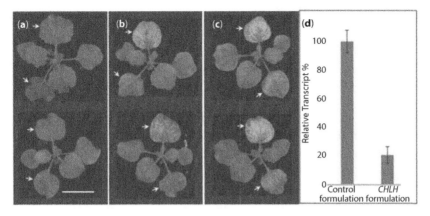

Figure 7.7 Silencing of the magnesium chelatase H or I subunits (CHLH or CHLI) in *N. benthamiana*. Leaves 3 and 4 from 17-d-old *N. benthamiana* plants were sprayed with siRNAs formulated with the purified CD-5K fraction 5. The final concentration of siRNAs in the formulations was 12 ng mL21. Representative plants for each treatment are shown at 4 d after application. The application sites, leaves 3 and 4, are indicated by white arrows. A, Leaves that were sprayed with formulations containing a nontargeting siRNA. Scale bar 5 5 cm. B, Leaves sprayed with a 22-mer siRNA targeting CHLI. C, Leaves sprayed with a 22-mer siRNA targeting CHLH. D, Transcript analysis of CHLH. At 5 d after treatment, leaf 4 from four plants was sampled for RT-qPCR analysis. Error bars represent the mean 6 SE. A Student's t test (two-tailed) was used to compare transcript levels. The reduction in CHLH transcript was statistically significant at P = 3.48E-06. Reprinted with permission from Reference [78].

advantages, such as increased growth and resilience, improved productivity, etc., which are important challenges in this age of agri-tech revolution.

7.7 Conclusions

The effect of nanoparticles on the environment, ecology and agriculture have long been viewed with much scepticism. However, emerging literature from the past few years has indicated that careful use of nanoparticles and nanodevices on plants and other botanical specimens can bring about a positive effect. As it is evident from the examples covered in this chapter, various nanoparticles, with and without incorporated agriculturally relevant cargoes such as fertilizers, pesticides, functional genes, etc., are efficiently uptaken and distributed within plants, with little ill effects. Conversely, several beneficial effects, such as nutrient enrichment, ROS scavenging, biodefense, stress resistance, gene modulation, etc., have been observed as a result of nano-plant interactions. In addition, the controlled and site-specific release of active agents from the nanocarriers enable optimization in their dosages, which provides both economic and ecological benefits by preventing overuse and environmental leakage of these toxic agrochemicals. Overall, nanotechnology promises to significantly improve the resilience, growth, and productivity of crops, with minimum harm to the ecology and the environment.

The multimodality offered by several nanoparticles can be exploited to several smart applications. For example, a cocktail of active components, such as fertilizers, pesticides, functional genes, can be co-incorporated within a single nanoformulation for their combined effects. Emerging genetic engineering probes, such as CRISPR/Cas9 gene editing tools, are being introduced in plants using nanoparticles for better uptake and functional efficiency. On the other hand, nanotechnology-enabled sensing platforms can provide wireless-transmitted information about the state of various botanical species in real-time, which in turn can provide advanced warning signals in response to extreme weather, drought, nutrient depletion, pathogenic invasion, illegal deforestation, and other stress factors. Integration of sensing platforms with smart nanodelivery agents can facilitate on-demand activation and/or release of active molecules in response to specific signal output. The Agri-Tech revolution, ushered by the application of technology-driven solutions to enable enhanced agricultural practices, promises to meet the growing food demand around the world; nanoparticles have already emerged as key players for facilitating this revolution.

With many of the pros and cons already in sight, the future path regarding the applications of nanoparticles in the flora must be treaded carefully. Several parameters, such as composition, mode of synthesis, size, shape, surface charge, crystallinity, porosity, drug-loading efficiency, stimuli-responsiveness, surface coating, dosage, route of delivery, etc., of the nanoparticles must be evaluated carefully for their optimal use in agriculture. Bioinspired, 'green' nanoparticles must be used more extensively as they have shown the maximum safety profile. Extensive guidelines and regulations must be framed based on the knowledge already available in this area, and implemented for the safe and efficient use of nanoparticles in farming, agriculture, and other botanical practices, aimed at the restoration of the delicate balance between living organisms and the environment.

Acknowledgments

The author acknowledges funding support from the Institution of Eminence (IoE), University of Delhi.

References

1. Hussein, A.K., Applications of nanotechnology in renewable energies—A comprehensive overview and understanding. *Renewable Sustainable Energy Rev.*, 42, 460, 2015.
2. Ronteltap, A., Fischer, A.R.H., Tobi, H., Societal response to nanotechnology: Converging technologies–converging societal response research? *J. Nanopart. Res.*, 13, 4399, 2011.
3. Shi, J., Kantoff, P.W., Wooster, R., Farokhzad, O.C., Cancer nanomedicine: Progress, challenges and opportunities. *Nat. Rev. Cancer*, 17, 20, 2017.
4. Ng, T.S.C., Garlin, M.A., Weissleder, R., Miller, M.A., Improving nanotherapy delivery and action through image-guided systems pharmacology. *Theranostics*, 10, 968, 2020.
5. Contera, S., de la Serna, B.J., Tetley, T.D., Biotechnology, nanotechnology and medicine. *Emerg. Top. Life Sci.*, 4, 551, 2020.
6. Bundschuh, M., Filser, J., Lüderwald, S., McKee, M.S., Metreveli, G., Schaumann, G.E., Schulz, R., Wagner, S., Nanoparticles in the environment: Where do we come from, where do we go to? *Environ. Sci. Eur.*, 30, 6, 2018.
7. Pratim Biswas, P. and Wu, C.Y., Nanoparticles and the environment. *J. Air Waste Manage. Assoc.*, 55, 6, 2005.
8. Dinesh, R., Anandaraj, M., Srinivasan, V., Hamza, S., Engineered nanoparticles in the soil and their potential implications to microbial activity. *Geoderma*, 173–174, 19, 2012.
9. Das, S., Carnicer-Lombarte, A., Fawcett, J.W., Bora, U., Bio-inspired nano tools for neuroscience. *Prog. Neurobiol.*, 142, 1, 2016.
10. Sen Gupta, A., Bio-inspired nanomedicine strategies for artificial blood components. *Wiley Interdiscip. Rev. Nanomedicine*, 9, e1464, 2017.
11. Verma, S.K., Das, A.K., Patel., M.K., Shah., A., Kumar, V., Gantait, S., Engineered nanomaterials for plant growth and development: A perspective analysis. *Sci. Total Environ.*, 630, 1413, 2018.
12. Mali, S.C., Raj, S., Trivedi, R., Nanotechnology a novel approach to enhance crop productivity. *Biochem. Biophys. Rep.*, 24, 100821, 2020.
13. Buriak, J.M., Liz-Marzán, L.M., Parak, W.J., Chen, X., Nano and plants. *ACS Nano*, 16, 1681, 2022.
14. Wang, J.W., Cunningham, F.J., Goh, N.S., Boozarpour, N.N., Pham, M., Landry, M.P., Nanoparticles for protein delivery in planta. *Curr. Opin. Plant Biol.*, 60, 102052, 2021.
15. Lowry, G.V., Avellan, A., Gilbertson, L.M., Opportunities and challenges for nanotechnology in the agri-tech revolution. *Nat. Nanotechnol.*, 14, 517, 2019.
16. Zhi, H., Zhou, S., Pan, W., Shang, Y., Zeng, Z., Zhang, H., The promising nanovectors for gene delivery in plant genome engineering. *Int. J. Mol. Sci.*, 23, 8501, 2022.

17. Jat, S.K., Bhattacharya, J., Sharma, M.K., Nanomaterial based gene delivery: A promising method for plant genome engineering. *J. Mater. Chem. B*, 8, 4165, 2020.
18. Kulabhusan, P.K., Tripathi, A., Kant, K., Gold nanoparticles and plant pathogens: An overview and prospective for biosensing in forestry. *Sensors (Basel)*, 22, 1259, 2022.
19. Khater, M., de la Escosura-Muñiz, A., Merkoçi, A., Biosensors for plant pathogen detection. *Biosens. Bioelectron.*, 15, 72, 2017.
20. Sanzari, I., Leone, A., Ambrosone, A., Nanotechnology in plant science: To make a long story short. *Front. Bioeng. Biotechnol.*, 7, 120, 2019.
21. Le, V.N., Rui, Y., Gui, X., Li, X., Liu, S., Han, Y., Uptake, transport, distribution and Bio-effects of SiO2 nanoparticles in Bt-transgenic cotton. *J. Nanobiotechnol.*, 12, 1, 1, 2014.
22. Prasad, A., Astete, C.E., Bodoki, A.E., Windham, M., Bodoki, E., Sabliov, C.M., Zein nanoparticles uptake and translocation in hydroponically grown sugar cane plants. *J. Agric. Food Chem.*, 66, 26, 6544, 2018.
23. Tombuloglu, H., Slimani, Y., AlShammari, T.M., Tombuloglu, G., Almessiere, M.A., Sozeri, H., Baykal, A., Ercan, I., Delivery, fate and physiological effect of engineered cobalt ferrite nanoparticles in barley (Hordeum vulgare L.). *Chemosphere*, 265, 129138, 2021.
24. Wang, W.N., Tarafdar, J.C., Biswas, P., Nanoparticle synthesis and delivery by an aerosol route for watermelon plant foliar uptake. *J. Nanoparticle Res.*, 15, 1, 1, 2013.
25. Hu, P., An, J., Faulkner, M.M., Wu, H., Li, Z., Tian, X., Giraldo, J.P., Nanoparticle charge and size control foliar delivery efficiency to plant cells and organelles. *ACS Nano*, 14, 7, 7970, 2020.
26. Su, Y., Ashworth, V.E.T.M., Geitner, N.K., Wiesner, M.R., Ginnan, N., Rolshausen, P., Roper, C., Jassby, D., Delivery, fate, and mobility of silver nanoparticles in citrus trees. *ACS Nano*, 14, 3, 2966, 2020.
27. Lau, E.C.H.T., Carvalho, L.B., Pereira, A.E.S., Montanha, G.S., Corrêa, C.G., Carvalho, H.W.P., Ganin, A.Y., Fraceto, L.F., Yiu, H.H.P., Localization of coated iron oxide (Fe3O4) nanoparticles on tomato seeds and their effects on growth. *ACS Appl. Bio Mater.*, 3, 7, 4109, 2020.
28. Afzal, S., Sharma, D., Singh, N.K., Eco-friendly synthesis of phytochemical-capped iron oxide nanoparticles as nano-priming agent for boosting seed germination in rice (Oryza sativa L.). *Environ. Sci. Pollut. Res. Int.*, 28, 30, 40275, 2021.
29. Mahakham, W., Sarmah, A.K., Maensiri, S., Theerakulpisut, P., Nanopriming technology for enhancing germination and starch metabolism of aged rice seeds using phytosynthesized silver nanoparticles. *Sci. Rep.*, 7, 1, 1, 2017.
30. Rani, A., Donovan, N., Mantri, N., Review: The future of plant pathogen diagnostics in a nursery production system. *Biosens. Bioelectron.*, 145, 111631, 2019.
31. Kumari, A. and Yadav, S.K., Nanotechnology in agri-food sector. *Crit. Rev. Food Sci. Nutr.*, 54, 975, 2014.

32. Zhang, H., Zhu, J., Gong., Z., Zhu, J.K., Abiotic stress responses in plants. *Nat. Rev. Genet.*, 23, 104, 2022.
33. Chen, G., Roy, I., Yang, C., Prasad, P.N., Nanochemistry and nanomedicine for nanoparticle-based diagnostics and therapy. *Chem. Rev.*, 116, 2826, 2016.
34. Li, J., Wu, H., Santana, I., Fahlgren, M., Giraldo, J.P., Standoff optical glucose sensing in photosynthetic organisms by a quantum dot fluorescent probe. *ACS Appl. Mater. Interfaces*, 10, 34, 28279, 2018.
35. Mou, Q., Xue, X., Ma, Y., Banik, M., Garcia, V., Guo, W., Wang, J., Song, T., Chen, L.Q., Lu, Y., Efficient delivery of a DNA aptamer-based biosensor into plant cells for glucose sensing through thiol-mediated uptake. *Sci. Adv.*, 8, 26, 902, 2022.
36. Ruiz-gonzalez, A., Kempson, H., Haseloff, J., *In vivo* sensing of PH in tomato plants using a low-cost and open-source device for precision agriculture. *Biosensors*, 12, 7, 447, 2022.
37. Wang, Z., Xue, L., Li, M., Li, C., Li, P., Li, H., Au@SnO 2-vertical graphene-based microneedle sensor for *in-situ* determination of abscisic acid in plants. *Mater. Sci. Eng. C. Mater. Biol. Appl.*, 127, 112237, 2021.
38. Giust, D., Lucío, M.I., El-Sagheer, A.H., Brown, T., Williams, L.E., Muskens, O.L., Kanaras, A.G., Graphene oxide-upconversion nanoparticle based portable sensors for assessing nutritional deficiencies in crops. *ACS Nano*, 12, 6, 6273, 2018.
39. Hao, L., Li, M., Peng, K., Ye, T., Wu, X., Yuan, M., Cao, H., Yin, F., Gu, H., Xu, F., Fluorescence resonance energy transfer aptasensor of ochratoxin a constructed based on gold nanorods and DNA tetrahedrons. *J. Agric. Food Chem.*, 70, 34, 10662, 2022.
40. Cortijo, S. and Locke, J.C.W., Does gene expression noise play a functional role in plants? *Trends Plant Sci.*, 25, 1041, 2020.
41. Crawford, B.M., Strobbia, P., Wang, H.N., Zentella, R., Boyanov, M.I., Pei, Z.M., Sun, T.P., Kemner, K.M., Vo-Dinh, T., Plasmonic nanoprobes for *in vivo* multimodal sensing and bioimaging of MicroRNA within plants. *ACS Appl. Mater. Interfaces*, 11, 8, 7743, 2019.
42. Ghazi, Y., Haddadi, F., Kamaladini, H., Gold nanoparticle biosensors, a novel application in gene transformation and expression. *Mol. Cell. Probes*, 41, 1, 2018.
43. Tarafdar, J.C., Raliya, R., Mahawar, H., Rathore, I., Development of zinc nanofertilizer to enhance crop production in pearl millet (Pennisetum americanum). *Agric. Res.*, 3, 3, 257, 2014.
44. Quiterio-Gutiérrez, T., Ortega-Ortiz, H., Cadenas-Pliego, G., Hernández-Fuentes, A.D., Sandoval-Rangel, A., Benavides-Mendoza, A., Cabrera-De La Fuente, M., Juárez-Maldonado, A., The application of selenium and copper nanoparticles modifies the biochemical responses of tomato plants under stress by alternaria solani. *Int. J. Mol. Sci.*, 20, 8, 1950, 2019.
45. Ma, C., Borgatta, J., De La Torre-Roche, R., Zuverza-Mena, N., White, J.C., Hamers, R.J., Elmer, W.H., Time-dependent transcriptional response of

tomato (Solanum lycopersicum L.) to Cu nanoparticle exposure upon infection with Fusarium oxysporum f. sp. lycopersici. *ACS Sustain. Chem. Eng.*, 7, 11, 10064, 2019.

46. Sarkar, R.D. and Kalita, M.C., Se nanoparticles stabilized with Allamanda cathartica L. flower extract inhibited phytopathogens and promoted mustard growth under salt stress. *Heliyon*, 8, 3, e09076, 2022.
47. Chaudhry, C. and Sidhu, G.P.S., Climate change regulated abiotic stress mechanisms in plants: A comprehensive review. *Plant Cell Rep.*, 41, 1, 2022.
48. Wu, H., Tito, N., Giraldo, J.P., Anionic cerium oxide nanoparticles protect plant photosynthesis from abiotic stress by scavenging reactive oxygen species. *ACS Nano*, 11, 11, 11283, 2017.
49. Liu, J., Li, G., Chen, L., Gu, J., Wu, H., Li, Z., Cerium oxide nanoparticles improve cotton salt tolerance by enabling better ability to maintain cytosolic K+/Na+ ratio. *J. Nanobiotechnol.*, 19, 1, 1, 2021.
50. Ashraf, H., Anjum, T., Riaz, S., Naseem, S., Microwave-assisted green synthesis and characterization of silver nanoparticles using melia azedarach for the management of fusarium wilt in tomato. *Front. Microbiol.*, 11, 238, 2020.
51. Zou, C., Lu, T., Wang, R., Xu, P., Jing, Y., Wang, R., Xu, J., Wan, J., Comparative physiological and metabolomic analyses reveal that Fe3O4 and ZnO nanoparticles alleviate Cd toxicity in tobacco. *J. Nanobiotechnol.*, 20, 1, 302, 2022.
52. Velikova, V., Petrova, N., Kovács, L., Petrova, A., Koleva, D., Tsonev, T., Taneva, S., Petrov, P., Krumova, S., Single-walled carbon nanotubes modify leaf micromorphology, chloroplast ultrastructure and photosynthetic activity of pea plants. *Int. J. Mol. Sci.*, 22, 9, 4878, 2021.
53. Santana, I., Jeon, S.J., Kim, H.I., Islam, M.R., Castillo, C., Garcia, G.F.H., Newkirk, G.M., Giraldo, J.P., Targeted carbon nanostructures for chemical and gene delivery to plant chloroplasts. *ACS Nano*, 16, 8, 12156, 2022.
54. Salinas, F., Astete, C.E., Waldvogel, J.H., Navarro, S., White, J.C., Elmer, W., Tamez, C., Davis, J.A., Sabliov, C.M., Effects of engineered lignin-graft-PLGA and zein-based nanoparticles on soybean health. *NanoImpact*, 23, 100329, 2021.
55. Tripathi, D.K., Shweta, Singh, S., Singh, S., Pandey, R., Singh, V.P., Sharma, N.C., Prasad, S.M., Dubey, N.K., Chauhan, D.K., An overview on manufactured nanoparticles in plants: Uptake, translocation, accumulation and phytotoxicity. *Plant Physiol. Biochem.*, 110, 2, 2017.
56. Kumar, A., Choudhary, A., Kaur, H., Mehta, S., Husen, A., Smart nanomaterial and nanocomposite with advanced agrochemical activities. *Nanoscale Res. Lett.*, 16, 156, 2021.
57. Abdel-Aziz, H.M.M., Hasaneen, M.N.A., Ome, A.M., Nano chitosan-NPK fertilizer enhances the growth and productivity of wheat plants grown in sandy soil. *Span. J. Agric. Res.*, 14, 1, e0902, 2016.
58. Karny, A., Zinger, A., Kajal, A., Shainsky-Roitman, J., Schroeder, A., Therapeutic nanoparticles penetrate leaves and deliver nutrients to agricultural crops. *Sci. Rep.*, 8, 1, 1, 2018.

59. Ramírez-Rodríguez, G.B., Miguel-Rojas, C., Montanha, G.S., Carmona, F.J., Dal Sasso, G., Sillero, J.C., Pedersen, J.S., Masciocchi, N., Guagliardi, A., Pérez-De-luque, A., Delgado-López, J.M., Reducing nitrogen dosage in triticum durum plants with urea-doped nanofertilizers. *Nanomater. (Basel, Switzerland)*, 10, 6, 1043, 2020.
60. Oliveira, H.C., Stolf-Moreira, R., Martinez, C.B.R., Grillo, R., De Jesus, M.B., Fraceto, L.F., Nanoencapsulation enhances the post-emergence herbicidal activity of atrazine against mustard plants. *PloS One*, 10, 7, e0132971, 2015.
61. Rashidipour, M., Maleki, A., Kordi, S., Birjandi, M., Pajouhi, N., Mohammadi, E., Heydari, R., Rezaee, R., Rasoulian, B., Davari, B., Pectin/chitosan/tripolyphosphate nanoparticles: Efficient carriers for reducing soil sorption, cytotoxicity, and mutagenicity of paraquat and enhancing its herbicide activity. *J. Agric. Food Chem.*, 67, 20, 5736, 2019.
62. De La Torre-Roche, R., Hawthorne, J., Deng, Y., Xing, B., Cai, W., Newman, L.A., Wang, C., Ma, X., White, J.C., Fullerene-enhanced accumulation of p, p'-DDE in agricultural crop species. *Environ. Sci. Technol.*, 46, 17, 9315, 2012.
63. Tong, Y., Wu, Y., Zhao, C., Xu, Y., Lu, J., Xiang, S., Zong, F., Wu, X., Polymeric nanoparticles as a metolachlor carrier: Water-based formulation for hydrophobic pesticides and absorption by plants. *J. Agric. Food Chem.*, 65, 7371, 2017.
64. Wang, X., Yan, M., Zhou, J., Song, W., Xiao, Y., Cui, C., Gao, W., Ke, F., Zhu, J., Gu, Z., Hou, R., Delivery of acetamiprid to tea leaves enabled by porous silica nanoparticles: Efficiency, distribution and metabolism of acetamiprid in tea plants. *BMC Plant Biol.*, 21, 337, 2021.
65. Luo, J., Gao, Y., Liu, Y., Huang, X., Zhang, D., Cao, H., Jing, T., Liu, F., Li, B., Self-assembled degradable nanogels provide foliar affinity and pinning for pesticide delivery by flexibility and adhesiveness adjustment. *ACS Nano*, 15, 14598, 2021.
66. Bapat, G., Zinjarde, S., Tamhane, V., Evaluation of silica nanoparticle mediated delivery of protease inhibitor in tomato plants and its effect on insect pest Helicoverpa armigera. *Colloids Surf. B Biointerfaces*, 193, 111079, 2020.
67. Liang, Y., Song, J., Dong, H., Huo, Z., Gao, Y., Zhou, Z., Tian, Y., Li, Y., Cao, Y., Fabrication of pH-responsive nanoparticles for high efficiency pyraclostrobin delivery and reducing environmental impact. *Sci. Total Environ.*, 787, 147422, 2021.
68. Wu, H., Hu, P., Xu, Y., Xiao, C., Chen, Z., Liu, X., Jia, J., Xu, H., Phloem delivery of fludioxonil by plant amino acid transporter-mediated polysuccinimide nanocarriers for controlling fusarium wilt in banana. *J. Agric. Food Chem.*, 69, 2668, 2021.
69. Alghuthaymi, M.A., Ahmad, A., Khan, Z., Khan, S.H., Ahmed, F.K., Faiz, S., Nepovimova, E., Kuća, K., Abd-Elsalam, K.A., Exosome/liposome-like nanoparticles: New carriers for CRISPR genome editing in plants. *Int. J. Mol. Sci.*, 22, 7456, 2021.

70. Kaeppler, H.F., Somers, D.A., Rines, H.W., Cockburn, A.F., Silicon carbide fiber-mediated stable transformation of plant cells. *Theor. Appl. Genet.*, 84, 5, 560, 1992.
71. Demirer, G.S., Zhang, H., Matos, J.L., Goh, N.S., Cunningham, F.J., Sung, Y., Chang, R., Aditham, A.J., Chio, L., Cho, M.J., Staskawicz, B., Landry, M.P., High aspect ratio nanomaterials enable delivery of functional genetic material without DNA integration in mature plants. *Nat. Nanotechnol.*, 14, 5, 456, 2019.
72. Kwak, S.Y., Lew, T.T.S., Sweeney, C.J., Koman, V.B., Wong, M.H., Bohmert-Tatarev, K., Snell, K.D., Seo, J.S., Chua, N.H., Strano, M.S., Chloroplast-selective gene delivery and expression in planta using chitosan-complexed single-walled carbon nanotube carriers. *Nat. Nanotechnol.*, 14, 5, 447, 2019.
73. Amani, A., Zare, N., Asadi, A., Asghari-Zakaria, R., Ultrasound-enhanced gene delivery to alfalfa cells by HPAMAM dendrimer nanoparticles. *Turk. J. Biol.*, 42, 1, 63, 2018.
74. Zolghadrnasab, M., Mousavi, A., Farmany, A., Arpanaei, A., Ultrasound-mediated gene delivery into suspended plant cells using polyethyleneimine-coated mesoporous silica nanoparticles. *Ultrason. Sonochem.*, 73, 105507, 2021.
75. Zhao, X., Meng, Z., Wang, Y., Chen, W., Sun, C., Cui, B., Cui, J., Yu, M., Zeng, Z., Guo, S., Luo, D., Cheng, J.Q., Zhang, R., Cui, H., Pollen magnetofection for genetic modification with magnetic nanoparticles as gene carriers. *Nat. Plants*, 3, 12, 956, 2017.
76. Wang, Z.P., Zhang, Z.B., Zheng, D.Y., Zhang, T.T., Li, X.L., Zhang, C., Yu, R., Wei, J.H., Wu, Z.Y., Efficient and genotype independent maize transformation using pollen transfected by DNA-coated magnetic nanoparticles. *J. Integr. Plant Biol.*, 64, 6, 1145, 2022.
77. Zhang, H., Demirer, G.S., Zhang, H., Ye, T., Goh, N.S., Aditham, A.J., Cunningham, F.J., Fan, C., Landry, M.P., DNA nanostructures coordinate gene silencing in mature plants. *Proc. Natl. Acad. Sci. U.S.A.*, 116, 15, 7543, 2019.
78. Schwartz, S.H., Hendrix, B., Hoffer, P., Sanders, R.A., Zheng, W., Carbon dots for efficient small interfering RNA delivery and gene silencing in plants. *Plant Physiol.*, 184, 2, 647, 2020.
79. Zhang, H., Goh, N.S., Wang, J.W., Pinals, R.L., González-Grandío, E., Demirer, G.S., Butrus, S., Fakra, S.C., Del Rio Flores, A., Zhai, R., Zhao, B., Park, S.J., Landry, M.P., Nanoparticle cellular internalization is not required for RNA delivery to mature plant leaves. *Nat. Nanotechnol.*, 17, 2, 197, 2022.
80. Thagun, C., Horii, Y., Mori, M., Fujita, S., Ohtani, M., Tsuchiya, K., Kodama, Y., Odahara, M., Numata, K., Non-transgenic gene modulation via spray delivery of nucleic acid/peptide complexes into plant nuclei and chloroplasts. *ACS Nano*, 16, 3, 3506, 2022.

8

A Clean Nano-Era: Green Synthesis and Its Progressive Applications

Susmita Das[1]* and Kajari Dutta[2]†

[1]*Department of Chemistry, Amity Institute of Applied Sciences, Amity University Kolkata, Kadampukur Village, Rajarhat, Newtown, Kolkata, West Bengal, India*
[2]*Department of Physics, Amity Institute of Applied Sciences, Amity University Kolkata, Kadampukur Village, Rajarhat, Newtown, Kolkata, West Bengal, India*

Abstract

Population explosion, automation and urbanization has resulted in numerous harsh environmental effects and ultimately to climate change. The future of this globe thus seeks immediate attention and investment of thoughts to restrict the use of hazardous chemicals and thus arrest the further deterioration of the environment. Nanotechnology has been an indispensable arena, which has extended its wings into every aspect of modernization and thus green synthetic protocols are extensively researched to inhibit the harmful effects of chemical remains and reduce chemical wastes. Green synthesis of nanoparticles generally involve use of plant extracts or microbial action to replace the chemical reducing agents. Biogenic reduction of metal salts generally result in nanoparticles possessing unique properties than those produced using physico-chemical techniques. Thus, green synthetic techniques are ecofriendly, economic, and appropriate for mass production. Herein, in this chapter we have provided a detailed review and analysis of the various green synthetic approaches for developing nanoparticles, their distinctive characteristics and their applications. The chapter also emphasizes on the applications and improved properties of the nanomaterials obtained using green synthesis.

Keywords: Green synthesis, nanoparticles, biosynthesis, plant extracts, applications

---

*Corresponding author: sdas@kol.amity.edu; ssmtdas@gmail.com
†Corresponding author: kdutta@kol.amity.edu

Mousumi Sen and Monalisa Mukherjee (eds.) Bioinspired and Green Synthesis of Nanostructures: A Sustainable Approach, (189–206) © 2023 Scrivener Publishing LLC

8.1 Introduction

Green synthesis is a non-toxic, low temperature and one-pot synthetics method that uses natural resources and environmentally friendly reagents and solvents. Nanoparticles, especially metal or metal oxide nanoparticles have gained tremendous attention over past two decades for their versatile applications in the advancement of every possible arena ranging from mechanics to medicine [1].

Nanoscience research continuously strives for the development of new materials and on fabrication of nanodevices with great array of properties and applications. Nanomaterial synthesis is generally carried out with either top down or bottom up approaches. Top down approaches involved etching, sputtering, milling, grinding etc. while the bottom up approaches involve formation of nanoscale materials from simple precursor molecules in presence of oxidizing/reducing agents or reactions like displacement, double-displacement and hydrolysis reactions. Additionally, desired properties are incorporated into the nanoparticles using target precursors. Thus nanoparticle synthesis involves the use of multiple reagents, a part of which is discarded into the environment. Hence, development of eco-friendly and sustainable approaches for nanoparticle synthesis is essential for wide range applications and for effective use in biomedical devices [2].

Green synthesis of nanoparticles are generally based on bottom up approaches in the presence of microorganisms, plant extracts, carbohydrates, vitamins, enzymes, biodegradable polymers etc. In this chapter, we have elaborated the various green synthetic approaches and have presented a review of various nanomaterials that are synthesized using the reported green synthetic techniques.

8.2 Green Synthetic Approaches

8.2.1 Microorganism-Induced Synthesis of Nanoparticles

Green synthesis of nanoparticles by microbial actions can be achieved from simple prokaryotic cells like bacteria to eukaryotic organisms like fungi or plants. Such an approach require in general the optimization of the most effective organism, optimal cell growth conditions including the amount of light, nutrients, inoculum size, pH, agitation speed, temperature and so on as well as optimal reaction conditions [2].

8.2.2 Biosynthesis of Nanoparticles Using Bacteria

Green nanotechnology has used a wide range of bacterial species. Biosynthesis of nanoparticles using bacteria has successfully yielded metal nanoparticles like gold, silver, platinum, palladium, titanium, titanium dioxide etc. Toxicity of heavy metals and heavy metal ions induce a defensive mechanism that enables the bacteria to reduce metal ions to metal nanoparticles [3]. However, some bacteria survive and grow even in high metal ion concentration e.g. *Pseudomonas aeruginosa*. It is reported by Brock and Gustafson that certain strains like *Thiobacillus ferooxidans, T. thiooxidans* etc. reduce Fe^{3+} to Fe^{2+} using elemental sulphur as energy source either aerobically or anaerobically [4]. *Enterobacter cloacae, Desulfovibrio desulfuricans, Rhodospirillium rubrum* reduce selenite to selenium. A few other bacteria such as *E. coli, P. aerugenosa, Bacillus subtilis* are used for removal of metal ions like Ag^+, Cu^{2+}, Cd^{2+}, and La^{3+} from solution [5].

8.2.3 Biosynthesis of Nanoparticles Using Fungi

Fungi are used for the synthesis of metallic nanoparticles as they can serve as reducing and stabilizing agents. Fungi exhibit high binding capacity and metal accumulation ability unlike bacteria resulting in high nanoparticle yields. Use of fungi for nanoparticle generation is more beneficial as compared to other organisms as they grow faster. Fungi exhibit a different mechanism compared to bacteria towards nanoparticle synthesis. They secret a large amount enzymes that reduce metal ions to metal nanoparticles. Naphthoquinones and anthraquinones are extracellular enzymes that facilitate reduction [6]. Fungus-mediated Ag and Au nanoparticle synthesis have been reported. *Verticillium luteoalbum* was used for intracellular gold nanoparticle generation that demonstrated a pH dependent size variation. Au, Ag, and Ag–Au nanoparticles were achieved using saprophytic straw mushroom fungus extract, *Volvariella volvacea*. The fungus *Aperigillus flavus* yield Ag nanoparticles [7] that absorb at 420 nm and emits at 553 nm. Fusarium oxysporum was used to reduce Ag^+ by the nitrate reductase enzymes and anthraquinones [8]. Besides the enzymatic action, certain biomolecules like NaDPH/NaDH can react with Ag ions to give Ag nanoparticle. Thus presence of nitrate reductase enzymes were not mandatory for nanoparticle synthesis as reported by Hietzshold *et al*. [9].

8.2.4 Biosynthesis of Nanoparticles Using Actinomycetes

Actinomycetes are unicellular filamentous bacteria that form branching network of filaments and produce spores. They are mostly aerobic, while some anaerobic strains are also reported. Actinomycetes are reported for the production of size controlled and stable nanoparticle dispersion which may be produced either intracellularly or extracellularly. Ahmed *et al.* reported that nanoparticles are formed by intracellular reduction of metal ions on the surface of mycelial cytoplasmic membrane [10]. Sunitha and co-workers ascribed the intracellular reduction of metal ions as a result of electrostatic attraction of M^+ (metal ion) and negatively charged COO^- group of mycelia cell wall enzymes [11]. Extracellular synthesis of Au and Ag nanoparticles are achieved by NADH dependent nitrate reductase enzymes [8]. Electron shuttle enzymatic metal reduction is considered as a possible mechanism for reduction of Ag^+ ions. Biogenic reduction of Au is carried out by oxidation of NADH by NADH dependent reductase that serve as electron carriers [12, 13].

Although biogenic synthesis of nanomaterials is a slow process, culture supernatants of *Klebsiella pneumoniae, Escherichia coli, Enterobacter cloaceae,* or *Staphylococcus aureus* demonstrated synthesis of Ag^+ nanoparticles within 5 minutes of contact time [14, 15]. However, it is observed that actinomycetes demonstrate longer duration ranging from 24 to 120 hours for the synthesis of nanoparticle unlike other bacterial strains [16–21]. Although biogenic synthesis of nanoparticles is not considered for large-scale application, it is observed that these nanoparticles are generally protein capped and hence more stable and less polydisperse in nature [22–27]. Such nanoparticles also demonstrate higher antimicrobial activities due to protein capping [28].

8.2.5 Biosynthesis of Nanoparticles Using Algae

Algae are wide group of photosynthetic, autotrophic organisms ranging from unicellular to multicellular forms. Algae are also reported for biogenic synthesis of metal nanoparticles through bioreduction of metal ions although the reports are limited in comparison to other organisms. *Spirulina platensis*, an edible blue green algae, is used for extracellular synthesis of Ag and Au nanoparticles of variable sizes ranging from 7-16 nm and 6-10 nm respectively with absorption around 424 and 550 nm [29]. *Sargassum wightii and Kappaphycus alvarezii* reported the synthesis of intracellular metal nanoparticles. Bioreduction of Au(III) was also achieved by using biomass of the brown alga *Fucus vesiculosus* [30].

8.2.6 Plant Extracts for Biosynthesis of Nanoparticles

Use of plant extracts for biosynthesis of nanoparticles is advantageous over synthesis using micro-organisms because this excludes the need for harvesting bacteria and fungi. Plants and plant extracts can extracellularly or intracellularly produce metal nanoparticles [31]. Live lucerne plants were demonstrated for the first time for the synthesis of Au and Ag nanoparticles. Crushed lucerne seeds placed in nutrient media lead to their germination and metal salts placed in contact with the germinated seeds result in reduction to metal nanoparticles [32]. *Geranium, Azadirachta indica (neem), Magnolia kobus, Aloe vera, Cycas* and *Pelargonium graveolens* leaves as well as *Stevia rebaudiana* leaves from the *sunflower Asteraceae* family were also used for extracellular synthesis of Ag nanoparticles. The protocol for leaf extract preparation varied from plant to plant, however the prime methodology is reduction of metal salts to the metal. Generally, leaf extracts are prepared by boiling fresh cut leaves in distilled water for a certain period of time [1]. Extracellular synthesis is more preferred over intracellular methods due to the ease of extraction of the resulting nanoparticles. *Aloe vera* extracts are used to obtain Au nanotriangles and silver nanoparticles [1]. Fruit extract from *Dillenia indica* is known to contain high phenolic content and has been demonstrated for production of polydispersed Au nanomaterials of various shapes [33, 34].

8.3 Nanoparticles Obtained Using Green Synthetic Approaches and Their Applications

8.3.1 Synthesis of Silver (Ag) and Gold (Au)

Various microorganisms like bacteria, fungus, and algae and different plant extracts were used to synthesize Ag and Au nanoparticles. Behravan *et al.* reported green synthesis of monodispersed Ag nano particles of size ~5 to 20 nm using the aqueous plant extract of *Berberis vulgaris* leaf, root and different concentration of silver nitrate. The antibacterial efficacy of Ag nanoparticles on the bacteria *S. aureus, E. coli, Bacillus subtilis* and *Proteus mirabilis* were also studied [35]. H.M. Ibrahim *et al.* used banana peel extract as a reducing and capping agent for green synthesis of spherical shaped and monodispersed Ag nanoparticles. They found that the optimum reduction conditions for Ag ions to Ag nanoparticles was silver nitrate 1.75 mM, dried banana peel powder 20.4 mg, pH 4.5 and incubation time 72 h. Rapid reduction was observed by banana peel extract within 5 mins of heating at the temperature 400 °C. In this study, the antibacterial activity of Ag nanoparticles against representative pathogens of bacteria and yeast was reported. They also investigated the synergistic antimicrobial activity of Ag nanoparticles and the standard antibiotic levofloxacin against Gram-positive and Gram-negative bacteria [36]. Banerjee et al. reduced the Ag ions in 1 mM aqueous $AgNO_3$ to Ag metal nanoparticles by using leaf extracts of three plants, Musa balbisiana (banana), Azadirachta indica (neem) and Ocimum tenuiflorum (black tulsi). These Ag nanoparticles exhibited good antimicrobial against the bacteria Escherichia coli (E. coli) and Bacillus sp. Moreover, they carried out a study of toxicity evaluation on Moong Bean (Vigna radiata) and Chickpea (Cicer arietinum) seeds and observed the rates of germination and oxidative stress enzyme activity, which revealed that the green synthesized Ag nanoparticles are safe to be released in environment [37]. Kalishwaralal et al. reported the green synthesis of Ag nanoparticles of size ~40 nm using a non-pathogenic bacterium Bacillus licheniformis as an extracellular reducing agent of aqueous Ag^+ ions [38]. Ahmad et al. observed the size variation of green synthesized Ag nanoparticles at room temperature for two different plant extracts prepared from dried stem and root of Ocimum sanctum. The phytochemicals present in the stem and root of the Ocimum plant are responsible for the reduction of Ag^+ ions to Ag metal. The chemicals also effectively wrapped around the metal and prevented nanoparticle agglomeration. The Ag nanoparticles synthesized using stem and roots were of sizes 5 ± 1.5 nm and 10 ± 2 nm, respectively [39]. Zayed et al. investigated the effectiveness of different plant leaf extracts (Malva parviflora, Beta vulgaris subsp. Vulgaris, Anethum graveolens, Allium kurrat, and Capsicum frutescens) as the bio reducing agents of Ag^+ ions for synthesis of Ag nanoparticles. The study revealed that the plant M. parviflora (Malvaceae) shows the best reducing and capping action in terms of synthesis rate and mono-dispersity of Ag nanoparticles of diameters 19–25 nm. The presence of the capping agent like proteins secreted by the biomass on nanoparticle surface was confirmed by the FTIR analysis [40].

There are many reports on green synthesis of Au nanoparticles. In one study, Ananas comosus were used as reducing agent of Au^+ ions to Au metal nanoparticles of size ~16 nm, which exhibit the surface plasmon resonance peak at 546 nm. The antibacterial activity against gram positive and negative pathogens found in water were studied, which showed good efficacy. Hence, synthesized Au nanoparticles were applied in water purification system [41]. Yang et al. synthesized Au nanoparticles of size ~ 6–18 nm using mango peel extract. The biological cytotoxicity of synthesized Au nanoparticles at a concentration of 160 μg/ml was tested on African green monkey's kidney normal cells and Normal human fatal lung fibroblast cells, which confirmed the less toxicity of Au in mammals, hence these Au nanoparticles may be used in a drug delivery system for therapy of different cancer cells. Tahir et al. prepared Au nanoparticles using aqueous leaf extracts of *N. oleander* playing dual role as reducing and stabilizing agent. They also studied the antioxidant activity of synthesized Au nanoparticles [42]. In one study, a green synthesis of both Au and Ag nanoparticles by using *B. monosperma* leaf extract were reported. These Au and Ag nanoparticles showed high stability in different biological buffers for several weeks and biocompatible towards normal endothelial cells (HUVEC, ECV-304). The noble metals Au and Ag nanoparticle-based drug delivery systems using FDA approved anticancer drug doxorubicin were developed, which exhibited better therapeutic efficacy compared to free drug [43]. Raju et al. synthesized cationic Au nanoparticles (C-GNPs) by using peanut leaf extract in the presence of cysteamine for biological applications. They determined the binding of C-GNPs with plasmid DNA by agarose gel electrophoresis [44]. In another study, microorganisms culture filtrate (FCF) of *Alternaria alternata* was used to synthesis of Au nanoparticles of size ~12 ± 5 nm. They proved the stabilization of nanoparticles by confirming the presence of the protein shell outside the nanoparticles [45]. A synthesis of 25 nm sized Au nanotriangles and 30 nm sized spherical shaped Au nanoparticles were synthesized using photosynthetic microorganisms like eukaryotic and prokaryotic. Here potent antioxidant property of Biogenic gold nanoparticles was confirmed by the interaction of gold nanoparticles with DNA [46].

8.3.2 Synthesis of Palladium (Pd) Nanoparticles

Palladium (Pd) nanoparticles are important as a photocatalyst. Arsiya et al. reported the green synthesis of Pd nanoparticles of average size ~15 nm by using the aqueous extract *Chlorella vulgaris*. The functional groups polyol and amide present in *Chlorella vulgaris* play the major role to reduce the Pd^{2+} ions [47]. A green synthesis route of Pd nanoparticles using natural black tea leave (*Camellia sinensis*) extracts was reported by S. Lebaschi et al. The nanoparticles exhibited good catalytic efficiency as a heterogeneous catalyst for Suzuki coupling reactions of phenylboronic acid and a range of aryl halides (X = I, Br, Cl). The Pd nanoparticles also act as catalyst to reduce 4-nitrophenol using sodium borohydride in an eco-friendly medium [48]. In another study, several biological applications like anticancer, antioxidant, antimicrobial and DNA cleavage activities of green synthesized Pd nanoparticles of size ~7.44 ± 1.94 nm were described.

The therapeutic effects of Pd nanoparticles were enhanced tremendously for non-toxic green synthesis of the nanoparticles. The antioxidant activity of Pd nanoparticles was screened by DPPH free radical scavenging, which was determined as 79.6% at 500 mg/L. The nanoparticles showed high antimicrobial efficacy to kill the gram-negative bacteria and double strain DNA cleavage activity. For investigating the cytotoxic effects of Pd nanoparticles, various cancer cell lines (MDA-MB-231 breast cancer cell line, HT-29 colon cancer cell line, Mia Paca-2 human pancreatic cancer cell line, and healthy cell line L929-Murine fibroblast cell line) were used, which exhibited IC_{50} values of Pd nanoparticles against MDA-MB-231, HT-29, and MIA PaCa-2 cancer cell lines as 31.175, 20.383, and 29.335 µg/ml, respectively. However, no significant cytotoxic effect of the nanoparticles against healthy lines L929 was observed [49].

8.3.3 Synthesis of Copper (Cu) Nanoparticles

Ghosh *et al.* stated the green synthesis of Cu nanoparticles of size ~10 nm using the *Jatropha curcas* leaf extract. The presence of flavonoids, tannins, glycosides, and alkaloids in the leaf extract were responsible for reducing Cu^+ or Cu^{2+} ions to Cu nanoparticles. They also act as strong stabilizing and capping agents of the nanoparticles, as the green synthesized Cu nanoparticles were not coagulated even after six months. The Cu nanoparticles showed the surface plasmon resonance (SPR) peaks at 266 and 337 nm. In addition, they delivered good photocatalytic activity (rate constant (k) ~ 2.30×10^{-4} s^{-1}) in degradation of methylene blue (MB) dye under sunlight illumination. They also exhibited a good binding property (binding constant (K_b) is 1.024×10^2 M^{-1}) with CT-DNA through an intercalation mode [50]. In another report, the environmental-friendly Cu nanoparticles of average diameter ~5 nm were synthesized via green synthesis method using *Celastrus paniculatus* leaf extract. The biosynthesized nanoparticles displayed an absorption peak at 269 nm. The functional groups OH, C=C and C-H of leaf extract trigger the reduction of Cu ions and stabilization of Cu nanoparticles. The nanoparticles also displayed a good photocatalytic property (degradation rate constant k~ 0.0172 min^{-1}) in degradation of methylene blue dye. Additionally, the Cu nanoparticles showed a good antifungal property against plant pathogenic fungi Fusarium oxysporum, where maximum mycelial inhibition was ~76.29 ± 1.52 [51].

The green synthesis of metal-oxides has paved its way for the applications in the field of photocatalyst and antibiotic owing to a stable homogeneous particle size distribution. Several research groups have reported the green synthetic methods of different oxide nanoparticles from various plant extracts as well as microorganisms and their effectiveness as photocatalysts and antibiotics were also examined. In general, crystalline oxides were obtained via annealing of green synthesized oxide nanoparticles.

8.3.4 Synthesis of Silver Oxide (Ag_2O) Nanoparticles

Rashmi et al. synthesized Ag_2O nanoparticles by a green combustion method using *Centella Asiatica* and *Tridax* plant powder. These two plants contain the inimitable ingredients: pentacyclic and triterpene, which contributed various photocatalytic, electrochemical and biological activities. The synthesized Ag_2O nanoparticles exhibited a good efficiency to degrade the dye acid orange 8 (AO8). The nanoparticles also displayed a hysteresis in the range of 0.3 V to −0.1 V observed in the cyclic voltametric measurement. In addition, the nanoparticles behaved as an anti-bacterial and anti-fungal agent against many disease-causing pathogens like bacteria: *S. epidermidis, A. aureus, S. aureus* and fungus: *A. fumigates* [52]. In another study, *Cantaloupe (Cucumis melo)* seeds were used to prepare Ag_2O nanoparticles using a green synthesis method. The synthesized nanoparticles were crystalline in cubic crystal structure and spherical in morphology wise. Here the antimicrobial property of the nanoparticles was examined by the disc diffusion method against Gram-positive and Gram-negative microbial strains [53]. Rokade et al. described the preparation of disc and faceted shaped Ag_2O nanoparticles using starch as stabilizing agent. The synthesized nanoparticles showed good antimicrobial property against the pathogenic bacteria causing food poisoning [54].

8.3.5 Synthesis of Titanium Dioxide (TiO_2) Nanoparticles

Kaur et al. introduced *Carica papaya* leaf extract to synthesis spherical porous TiO_2 nanoparticles of average size ~15.6 nm crystalline in anatase phase. The mesoporous nature of TiO_2 particles was determined by Brunauer-Emmett-Teller (BET) surface area (BET) analysis, which provided the surface area and mean pore size of 81.653 m^2g^{-1} and 8.0615 nm, respectively. The synthesized TiO_2 photocatalyst having absorptivity in UV region demonstrates notable photocatalytic efficiency (91.19%) in degradation of RO-4 dye under 180 min of UV exposure [55]. In another work, Amanull illustrated the green synthesis method of TiO_2 nanoparticles by reducing titanium tetrachloride from *orange peel* extract. The purity of the prepared nanoparticles was examined using characterization techniques like X-ray diffraction, FESEM, TEM and FTIR. The nanoparticles were employed to fabricate humidity sensor and studied efficiency at different humid atmosphere (RH 5% to RH 98%). In addition, the antimicrobial studies of the nanoparticles illustrated the good activity against Gram-positive *Staphylococcus aureus* and Gram-negative *Escherichia coli* and *Pseudomonuas aeruginosa* bacteria respectively. The cytotoxicity study of anatase TiO_2 NPs in A549 cell lines were measured using MTT assay, which revealed that the nanoparticles synthesized by green synthesis method exhibit better antimicrobial property in comparison to that prepared by harsh chemical process [56]. In another green method, TiO_2 nanoparticles were synthesized from $TiCl_4$ by using remnant water collected from soaked Bengal gram beans (*Cicer arietinum L.*). Here *Cicer arietinum L.* extract also played a major role as stabilizing agent of TiO_2 nanoparticles resulting uniform size distribution even after calcination. Here the nanoparticles were used to form a Li-ion battery by insertion of Li ions into bio-synthesized TiO_2, whose properties were assessed as anodes in the half-cell configuration. The prepared half-cell Li ion battery demonstrated a good cyclability. 98% of its initial reversible capacity was retained after completion of 60 galvanostatic cycles [57]. Goutam *et al.* reported the synthesis of green TiO_2 nanoparticles using leaf extract of the biodiesel plant, *Jatropha curcas L.* and studied the photocatalytic performance of synthesized TiO_2 for the treatment of Tannery Wastewater (TWW). Various characterization techniques confirmed the anatase phase of the spherical TiO_2 nanoparticles. FTIR measurement also disclosed the existence of phytochemicals in leaf extract, which might play key roll to stabilize the nanoparticles and prevent the random growth by capping their surfaces. Moreover, the present work stated that the green synthesized TiO_2 nanoparticles were used to affirm its efficiency for instantaneous elimination of 82.26% chemical oxygen demand (COD) and 76.48% chromium (Cr) from secondary treated TWW in a designed and fabricated Parabolic Trough Reactor (PTR). This result demonstrates the huge application of the green synthesized nanoparticles in industrial wastewater treatment [58].

8.3.6 Synthesis of Zinc Oxide (ZnO) Nanoparticles

Pillai *et al.* reported a green synthesis method of ZnO using different plant extracts: *Beta vulgaris, Cinnamomum tamala, Cinnamomum verum, Brassica oleracea var. Italica*. The synthesized ZnO nanoparticles using different plant extracts displayed good antimicrobial efficacy against both gram-negative bacteria *Escherichia coli* and gram-positive bacteria *Staphylococcus aureus*. However, the nanoparticles obtained from *Beta vulgaris* were found to be inactive towards *S. aureus*. The ZnO nanoparticles prepared using *Beta vulgaris* displayed antifungal action against the fungal strain *Aspergillus niger*, while the ZnO prepared from *Cinnamomum tamala* actively treated the fungal infection by *Candida albicans*. Moreover, ZnO nanoparticles prepared from the extract of *Brassica oleracea var. italica* showed the antifungal activity against both the fungal stains [59]. In another work, different shaped ZnO nanoparticles were synthesized using microorganisms. Hxagonal nanoparticles were synthesized from fungal strain *Fusarium keratoplasticum* and nanorods were obtained from the strain *Aspergillus niger*. Here the shape-dependency of biocidal activity of the synthesized ZnO was studied. ZnO nano-rods displayed improved antimicrobial properties against pathogenic bacteria and better UV-protection index in comparison to that of the hexagonal ZnO [60]. Agarwal *et al.* illustrated the potential of green synthesized ZnO nanoparticles as an anti-inflammatory drug or a vector for drug delivery. Inflammation plays a significant role in the pathogenesis of several diseases like rheumatoid arthritis, asthma, atherosclerosis, and cancer. The pharmacokinetic activity of ZnO nanoparticles inside the cells was highlighted here. They also established the inflammatory model for illustrating the anti-inflammatory behaviour of ZnO [61].

8.3.7 Synthesis of Iron Oxide Nanoparticles

Demirezen *et al.* reported green synthesized iron oxide nanoparticles (size~ 7 nm) using Carob pod (*Ceratonia siliqua*). The synthesized iron oxide nanoparticles removed beta-lactam antibiotic amoxicillin (AMX) in aqueous solution due to their high surface area (~7.67 m^2/g). The amoxicillin removal efficiency of iron oxide depends on several factors like initial pH, concentration of AMX and iron oxide and temperature. 99% removal efficiency was found at a molar ratio of AMX: iron oxide ~ 1:50 in 200 min at pH 2. Kinetic study was used to analyse the removal process of amoxicillin by iron oxide nanoparticles in aqueous solution. The removal process was established by adsorption curve followed by the pseudo-first-order fitting, which showed a chemically surface-controlled reaction due to the high activation energy of 87 $kJ.mol^{-1}$ [62]. In another study, a magnetic hybrid material, encapsulation of iron oxide nanoparticles into a chitosan matrix were prepared. The green synthesis method was employed to synthesis metallic iron nanoparticles using eucalyptus extract as reducing agent. Iron oxide nanoparticles were formed after annealing. Then the magnetic hybrid organic/inorganic material was developed by encapsulation of the synthesized iron oxide nanoparticles in chitosan beads. The hybrid materials were demonstrated as good sorbent of arsenic from water [63].

8.4 Conclusion

Summarizing, chemical methods of nanomaterial synthesis has posed severe environmental threat due to large-scale production and increased applications of nanomaterials. Furthermore, the nanoparticles obtained via chemical methods require additional stabilizing agent to prevent nanoparticle agglomeration. Green synthetic approaches of nanomaterials has excluded the use of hazardous chemicals and have demonstrated the capacity of naturally occurring microbes and plants in the production of nanoparticles. Various bacteria, fungi, algae and plant extracts have reported the capability of reducing the metal salts to metal nanoparticles. The produced metal nanoparticles are generally more stable due to protein or other biomolecule capping and less polydisperse. Nanoparticle properties are highly size and shape dependent and polydispersity in a preparation negatively affect the nanoparticle activity. The nanoparticles exhibit a microbial strain dependent size, shape, properties and hence the resulting applications. Antibacterial activity of the synthesized nanoparticles are more compared to those obtained by chemical methods due to the protein capping on the nanoparticles. Additionally, nanoparticles also demonstrated improved photocatalytic degradation and good performance in batteries. Although, several green synthetic approaches have been developed, still large-scale synthesis of nanoparticles require the involvement of chemical methods. Thus, further studies must aim at developing green synthetic approaches for large-scale nanoparticle synthesis to minimize the use of harmful chemical and thus reduce eco-toxicity.

References

1. Lee, H.J., Lee, G., Jang, N.R., Yun, J.H., Song, J.Y., Kim, B.S., Biological synthesis of copper nanoparticles using plant extract. *Nanotechnology*, 1, 1, 371–374, 2011.
2. Virkutyte, J. and Varma, R.S., Green synthesis of nanomaterials: Environmental aspects, in: *Sustainable Nanotechnology and the Environment: Advances and Achievements*, ch2, pp. 11–39, ACS Syposium Series, 2013.
3. Iravani, S., Green synthesis of metal nanoparticles using plants. *Green Chem.*, 13, 10, 2638–2650, 2011.
4. Brock, T.D. and Gustafson, J.O.H.N., Ferric iron reduction by sulfur- and iron-oxidizing bacteria. *Appl. Environ. Microbiol.*, 32, 4, 567–571, 1976.
5. Lloyd, J.R., Ridley, J., Khizniak, T., Lyalikova, N.N., Macaskie, L.E., Reduction of technetium by Desulfovibrio desulfuricans: Biocatalyst characterization

and use in a flowthrough bioreactor. *Appl. Environ. Microbiol.*, 65, 6, 2691–2696, 1999.
6. Pal, G., Rai, P., Pandey, A., Green synthesis of nanoparticles: A greener approach for a cleaner future, in: *Green Synthesis, Characterization and Applications of Nanoparticles*, Micro and Nano Technologies, Ch-1, pp. 1–26, Elsevier, 2019.
7. Vigneshwaran, N., Ashtaputre, N.M., Varadarajan, P.V., Nachane, R.P., Paralikar, K.M., Balasubramanya, R.H., Biological synthesis of silver nanoparticles using the fungus Aspergillus flavus. *Mater. Lett.*, 61, 6, 1413–1418, 2007.
8. Durán, N., Marcato, P.D., Alves, O.L., De Souza, G.I., Esposito, E., Mechanistic aspects of biosynthesis of silver nanoparticles by several Fusarium oxysporum strains. *J. Nanobiotechnol.*, 3, 1, 1–7, 2005.
9. Hietzschold, S., Walter, A., Davis, C., Taylor, A.A., Sepunaru, L., Does nitrate reductase play a role in silver nanoparticle synthesis? Evidence for NADPH as the sole reducing agent. *ACS Sustain. Chem. Eng.*, 7, 9, 8070–8076, 2019.
10. Sastry, M., Ahmed, A., Khan, M.I., Kumar, R., Biosynthesis of metal nanoparticles using fungi and actinomycetes. *Curr. Nanosci.*, 85, 2, 162–170, 2003.
11. Sunitha, A., Geo, S., Sukanya, S., Praseetha, P.K., Dhanya, R.P., Biosynthesis of silver nanoparticles from actinomycetes for therapeutic applications. *Int. J. Nano Dimens.*, 5, 155–162, 2014.
12. Shah, R., Oza, G., Pandey, S., Sharon, M., Biogenic fabrication of gold nanoparticles using Halomonas salina. *J. Microbiol. Biotechnol. Res.*, 2, 4, 485–492, 2012.
13. Waghmare, S.S., Deshmukh, A.M., Sadowski, Z., Biosynthesis, optimization, purification and characterization of gold nanoparticles. *Afr. J. Microbiol. Res.*, 8, 138–146, 2014.
14. Shahverdi, A.R., Minaian, S., Shahverdi, H.R., Jamalifar, H., Nohi, A.A., Rapid synthesis of silver nanoparticles using culture supernatants of Enterobacteria: A novel biological approach. *Process Biochem.*, 42, 919–923, 2007.
15. Nanda, A. and Saravanan, M., Biosynthesis of silver nanoparticles from S. aureus and its antimicrobial activity against MRSA and MRSE. *Nanomedicine*, 5, 452–456, 2009.
16. Ahmad, A., Senapati, S., Khan, M.I., Kumar, R., Ramani, R., Srinivas, V., Sastry, M., Extracellular biosynthesis of monodisperse gold nanoparticles by a novel extremophilic actinomycete Thermomonospora sp. *Langmuir*, 19, 3550–3553, 2003b.
17. Ahmad, A., Senapati, S., Khan, M.I., Kumar, R., Ramani, R., Srinivas, V., Sastry, M., Intracellular synthesis of gold nanoparticles by a novel alkalotolerant actinomycete, Rhodococcus species. *Nanotechnology*, 14, 824, 2003c.
18. Alani, F., Moo-Young, M., Anderson, W., Biosynthesis of silver nanoparticles by a new strain of Streptomyces sp. compared with Aspergillus fumigatus. *World J. Microbiol. Biotechnol.*, 28, 3, 1081–1086, 2012.

19. Manivasagan, P., Venkatesan, J., Senthilkumar, K., Sivakumar, K., Kim, S., Biosynthesis, antimicrobial and cytotoxic effect of silver nanoparticles using a novel Nocardiopsis sp. MBRC-1. *BioMed. Res. Int.*, 2013, 9, 2013. Article ID 287638.
20. Sastry, M., Ahmed, A., Khan, M.I., Kumar, R., Biosynthesis of metal nanoparticles using fungi and actinomycetes. *Curr. Nanosci.*, 85, 2, 162–170, 2003.
21. Selvakumar, P., Viveka, S., Prakash, S., Jasminebeaula, S., Uloganathan, R., Antimicrobial activity of extracellularly synthesized silver nanoparticles from marine derived Streptomyces rochei. *Int. J. Pharm. Biol. Sci.*, 3, 188–197, 2012.
22. Bawaskar, M., Gaikwad, S., Ingle, A., Rathod, D., Gade, A., Duran, N., Marcato, P., Rai, M., A new report on mycosynthesis of silver nanoparticles by Fusarium culmorum. *Curr. Nanosci.*, 6, 4, 376–380, 2010.
23. Ingle, A., Rai, M., Gade, A., Bawaskar, M., Fusarium solani: A novel biological agent for the extracellular synthesis of silver nanoparticles. *J. Nanopart. Res.*, 11, 8, 2079–2085, 2009.
24. Prakasham, R.S., Buddana, S.K., Yannam, S.K., Guntuku, G.S., Characterization of silver nanoparticles synthesized by using marine isolate Streptomyces albidoflavus. *J. Microbiol. Biotechnol.*, 22, 614–621, 2012.
25. Raheman, F., Deshmukh, S., Ingle, A., Gade, A., Rai, M., Silver nanoparticles: Novel antimicrobial agent synthesized from a endophytic fungus Pestalotia sp. isolated from leaves of Syzygium cumini (L.). *Nano Biomed. Eng.*, 3, 3, 174–178, 2011.
26. Sanghi, R. and Verma, P., Biomimetric synthesis and characterization of protein capped silver nanoparticles. *Bioresour. Technol.*, 100, 501–504, 2009.
27. Sanjenbam, P., Gopal, J.V., Kannabiran, K., Anticandidal activity ofsilver nanoparticles synthesized using Streptomyces sp. VITPK1. *J. Mycol. Méd.*, 24, 3, 211–219, 2014, Available from: http://dx.doi.org/10.1016/j.mycmed.2014.03.004.
28. Kora, A.J. and Rastogi, L., Enhancement of antibacterial activity of capped silver nanoparticles in combination with antibiotics, on model gram-negative and gram-positive bacteria. *Bioinorg. Chem. Appl.*, 2013, 7, 2013, Article ID 871097.
29. Shakibaie, M., Forootanfar, H., Mollazadeh-Moghaddam, K., Bagherzadeh, Z., Nafissi-Varcheh, N., Shahverdi, A.R., Faramarzi, M.A., Green synthesis of gold nanoparticles by the marine microalga Tetraselmis suecica. *Biotechnol. Appl. Biochem.*, 57, 2, 71–75, 2010.
30. Mata, Y.N., Torres, E., Blázquez, M.L., Ballester, A., González, F., Muñoz, J.A., Gold(III) biosorption and bioreduction with the brown alga Fucus vesiculosus. *J. Hazard. Mater.*, 166, 2-3, 612–8, 2009.
31. Kumar, A., Kaur, K., Sharma, S., Synthesis, characterization and antibacterial potential of silver nanoparticles by Morus nigra leaf extract. *Indian J. Pharm. Biol. Res.*, 1, 4, 16–24, 2013.
32. Gardea-Torresdey, J.L., Peralta-Videa, J.R., Parsons, J.G., Mokgalaka, N.S., de la Rosa, G., Production of metal nanoparticles by plants and plant-derived

materials, in: *Metal Nanoclusters in Catalysis and Materials Science: The Issue of Size Control*, B. Corain, G. Schmid, N. Toshima, (Eds.), pp. 401–411, Elsevier, New York, Boston, 2008.
33. Sett, A., Gadewar, M., Sharma, P., Deka, M., Bora, U., Green synthesis of gold nanoparticles using aqueous extract of Dillenia indica. *Adv. Nat. Sci. Nanosci. Nanotechnol.*, 7, pp-025005, 2016.
34. Sadowski, Z., Green synthesis of silver and gold nanoparticles using plant extracts, in: *Green biosynthesis of nanoparticles: mechanisms and applications* M. Rai and C. Posten (Eds.), CAB International, Books, CABI International. doi: 10.1079/9781780642239.0079. 2013.
35. Behravan, M., Panahi, A.H., Naghizadeh, A., Ziaee, M., Mahdavi, R., Mirzapour, A., Facile green synthesis of silver nanoparticles using Berberis vulgaris leaf and root aqueous extract and its antibacterial activity. *Int. J. Biol. Macromol.*, 124, 148–154, 2019.
36. Ibrahim, H.M., Green synthesis and characterization of silver nanoparticles using banana peel extract and their antimicrobial activity against representative microorganisms. *J. Radiat. Res. Appl. Sci.*, 8, 3, 265–275, 2015.
37. Banerjee, P., Satapathy, M., Mukhopahayay, A., Das, P., Leaf extract mediated green synthesis of silver nanoparticles from widely available Indian plants: Synthesis, characterization, antimicrobial property and toxicityanalysis. *Bioresour. Bioprocess.*, 1, 1–10, 2014, https ://doi.org/10.1186/s4064 3-014-0003-y.
38. Kalishwaralal, K., Deepak, V., Ramkumarpandian, S., Nellaiah, H., Sangiliyandi, G., Extracellular biosynthesis of silver nanoparticles by the culture supernatant of Bacillus licheniformis. *Mater. Lett.*, 62, 29, 4411–4413, 2008.
39. Ahmad, N., Sharma, S., Alam, M.K., Singh, V.N., Shamsi, S.F., Mehta, B.R., Fatma, A., Rapid synthesis of silver nanoparticles using dried medicinal plant of basil. *Colloids Surf. B: Biointerfaces*, 81, 1, 81–86, 2010.
40. Zayed, M.F., Eisa, W.H., Shabaka, A.A., Malva parviflora extract assisted green synthesis of silver nanoparticles. *Spectrochim. Acta Part A: Mol. Biomol. Spectrosc.*, 98, 423–428, 2012.
41. Bindhu, M.R. and Umadevi, M., Antibacterial activities of green synthesized gold nanoparticles. *Mater. Lett.*, 120, 122–125, 2014.
42. Yang, N., WeiHong, L., Hao, L., Biosynthesis of Au nanoparicles using agricultural waste mango peel extract and its *in vitro* cytotoxic efect on two normal cells. *Mater. Lett.*, 134, 67–70, 2014, https://doi.org/10.1016/j.matlet.2014.07.025.
43. Patra, S., Mukherjee, S., Barui, A.K., Ganguly, A., Sreedhar, B. et al., Green synthesis, characterizaion of gold and silver nanoparicles and their potenial applicaion for cancer therapeuics. *Mater. Sci. Eng. C*, 53, 298–309, 2015, https://doi.org/10.1016/j.msec.2015.04.048.
44. Raju, D., Vishwakarma, R.K., Khan, B.M., Mehta, U.J., Ahmad, A., Biological synthesis of caionic gold nanoparicles and binding of plasmid DNA. *Mater. Lett.*, 129, 159–161, 2014, https://doi.org/10.1016/j.matlet.2014.05.021.

45. Sarkar, J., Ray, S., Chattopadhyay, D., Laskar, A., Acharya, K., Mycogenesis of gold nanoparticles using a phytopathogen Alternaria alternata. *Bioprocess Biosyst. Eng.*, 35, 4, 637–43, 2012.
46. MubarakAli, D., Arunkumar, J., Nag, K.H., SheikSyedIshack., K.A., Baldev, E., Pandiaraj, D., Thajuddin, N., Gold nanoparticles from Pro and eukaryotic photosynthetic microorganisms Comparative studies on synthesis and its application on biolabelling. *Colloids Surf. B: Biointerfaces*, 103, 166–173, 2013, https://doi.org/10.1016/j.colsurfb.2012.10.014.
47. Arsiya, F., Sayadi, H.M., Sobhani, S., Green synthesis of palladium nanoparticles using Chlorella vulgaris. *Mater. Lett.*, 186, 113–115, 2017, https://doi.org/10.1016/j.matlet.2016.09.101.
48. Lebaschi, S., Hekmati, M., Veisi, H., Green synthesis of palladium nanoparticles mediated by black tea leaves (Camellia sinensis) extract: Catalytic activity in the reduction of 4-nitrophenol and Suzuki-Miyaura coupling reaction under ligand-free conditions. *J. Colloid Interface Sci.*, 485, 223–231, 2017, https://doi.org/10.1016/j.jcis.2016.09.027.
49. Gulbagca, F., Aygün, A., Gülcan, M., Ozdemir, S., Gonca, S., Şen, F., Green synthesis of palladium nanoparticles: Preparation, characterization, and investigation of antioxidant, antimicrobial, anticancer, and DNA cleavage activities. *Appl. Organomet. Chem.*, 35, e6272, 2021, https://doi.org/10.1002/aoc.6272.
50. Ghosh, M.K., Sahu, S., Gupta, I., Ghorai., T.K., Green synthesis of copper nanoparticles from an extract of *Jatropha curcas* leaves: Characterization, optical properties, CT-DNA binding and photocatalytic activity. *RSC Adv.*, 10, 22027–22035, 2020.
51. Mali, S.C., Dhaka, A., Githala, C.K., Trivedi, R., Green synthesis of copper nanoparticles using Celastrus paniculatus Willd. leaf extract and their photocatalytic and antifungal properties. *Biotechnol. Rep.*, 27, e00518, 2020.
52. Rashmi, B.N., Harlapur, S.F., Avinash, B., Ravikumar, C.R., Nagaswarupa, H.P., Kumar, M.A., Santosh, M.S., Facile green synthesis of silver oxide nanoparticles and their electrochemical, photocatalytic and biological studies. *Inorg. Chem. Commun.*, 111, 107580, 2020.
53. Vinay, S.P., Sumedha, H.N., Nagaraju, G., Harishkumar, S., Chandrasekhar, N., Facile combustion synthesis of Ag2O nanoparticles using cantaloupe seeds and their multidisciplinary applications. *Appl. Organomet. Chem.*, 34, 10, e5830, 2020.
54. Rokade, A.A., Patil, M.P., Yoo, S.I., Lee, W.K., Park, S.S., Pure green chemical approach for synthesis of Ag2O nanoparticles. *Green Chem. Lett. Rev.*, 9, 4, 216–222, 2016.
55. Kaur, H., Kaur, S., Singh, J., Rawat, M., Kumar, S., Expanding horizon: Green synthesis of TiO2 nanoparticles using Carica papaya leaves for photocatalysis application. *Mater. Res. Express*, 6, 9, 095034, 2019.

56. Amanulla, A.M. and Sundaram, R. J. M. T. P., Green synthesis of TiO2 nanoparticles using orange peel extract for antibacterial, cytotoxicity and humidity sensor applications. *Mater. Today: Proc.*, 8, 323–331, 2019.
57. Kashale, A.A., Gattu, K.P., Ghule, K., Ingole, V.H., Dhanayat, S., Sharma, R., Ghule, A.V., Biomediated green synthesis of TiO2 nanoparticles for lithium ion battery application. *Compos. Part B: Eng.*, 99, 297–304, 2016.
58. Goutam, S.P., Saxena, G., Singh, V., Yadav, A.K., Bharagava, R.N., Thapa, K.B., Green synthesis of TiO2 nanoparticles using leaf extract of Jatropha curcas L. for photocatalytic degradation of tannery wastewater. *Chem. Eng. J.*, 336, 386–396, 2018.
59. Pillai, A.M., Sivasankarapillai, V.S., Rahdar, A., Joseph, J., Sadeghfar, F., Rajesh, K., Kyzas, G.Z., Green synthesis and characterization of zinc oxide nanoparticles with antibacterial and antifungal activity. *J. Mol. Struct.*, 1211, 128107, 2020.
60. Mohamed, A.A., Fouda, A., Abdel-Rahman, M.A., El-Din Hassan, S., El-Gamal, M.S., Salem, S.S., Shaheen, T.I., Fungal strain impacts the shape, bioactivity and multifunctional properties of green synthesized zinc oxide nanoparticles. *Biocatal. Agric. Biotechnol.*, 19, 101103, 2019.
61. Agarwal, H. and Shanmugam, V., A review on anti-inflammatory activity of green synthesized zinc oxide nanoparticle: Mechanism-based approach. *Bioorg. Chem.*, 94, 103423, 2020.
62. Demirezen, D.A., Yıldız, Y. Ş., Yılmaz, D.D., Amoxicillin degradation using green synthesized iron oxide nanoparticles: Kinetics and mechanism analysis. *Environ. Nanotechnol. Monit. Manage.*, 11, 100219, 2019.
63. Martínez-Cabanas, M., López-García, M., Barriada, J.L., Herrero, R., de Vicente, M.E.S., Green synthesis of iron oxide nanoparticles. Development of magnetic hybrid materials for efficient As (V) removal. *Chem. Eng. J.*, 301, 83–91, 2016.

9

A Decade of Biomimetic and Bioinspired Nanostructures: Innovation Upheaval and Implementation

Vishakha Sherawata, Anamika Saini, Priyanka Dalal and Deepika Sharma*

Institute of Nano Science and Technology, Knowledge City, Mohali, Punjab, India

Abstract

The interacting chemistry between biological components and nanostructured materials is increasingly gaining attention due to the prospect of engineering advanced bioinspired and biomimetic materials with specialized biological functions. The biomimetic and bioinspiration approach is based on learning from living nature entities and has seen an unprecedented surge in the last decade, spurred by advances in nanoscience and technology. In recent years, novel green synthesis protocols have emerged as one of the environment-conscious alternative methods for nanoparticle synthesis. In response to the growing demand over the last century for green, affordable, and sustainable solutions, scientists are turning to nature for inspiration to design nanomaterials that have unique characteristics and properties, for example, adaptability, miniaturization, and hierarchical organization. Additionally, the poise of hierarchical structures confers multifunctionality in biological systems, which provides precise control over the synthesis of nanoparticles to enable materials design with specific functionalities. Therefore, environmentally friendly, non-toxic, and biomimetic approaches are gaining prominence over synthetically engineering complex materials. This chapter attempts to explain the advances in biomimetic and bioinspired nanostructures and present them as promising solutions to many unresolved problems in the biomedical field. Biomimetic nanostructures regulate cell behavior, which is reported in *in vitro* studies where they play an important role in cell nuclear alignment, cell spreading, cell differentiation, phagocytosis, and viability. Here, we will present the recent developments in the preparation of bioinspired and biomimetic nanostructures through different routes of synthesis. The different templates used for the synthesis

---

*Corresponding author: deepika@inst.ac.in

Mousumi Sen and Monalisa Mukherjee (eds.) Bioinspired and Green Synthesis of Nanostructures: A Sustainable Approach, (207–230) © 2023 Scrivener Publishing LLC

of nanostructures and binding the template with other useful materials to enhance the therapeutic efficacy have also been discussed. The chapter will also highlight the repair of cardiac and bioanalysis of dietary supplements using biomimetic nanostructures, the signal amplification by bioinspired structure, increased drug efficacy and other biomedical advantages have also been mentioned, thereby conclude the chapter with future outlooks of biomimetic and bioinspired structures in the medical field.

Keywords: Bioinspired, nanostructure, biomimetic, immunogenicity, cancer

9.1 Introduction

Learning from nature has given us different ways to address problems, which can be termed as biologically inspired and biomimetic strategy to develop novel materials. Biologically inspired and biomimetic strategies rely on learning from surrounding entities and have been experiencing an unprecedented surge in the last decade, spurred by advances in nanoscience and technology. The abovementioned terms are defined as the study of structure, function, and formation of biologically produced material or substances and their biological mechanism, especially to synthesize similar products by an artificial mechanism that mimics the natural processes [1]. Bioinspired, biomimetic, and bioengineered drug delivery systems (DDS) offer the potential to provide novel solutions with inherent biocompatibility over natural barriers for non-invasive administration. Generally, bioinspired and biomimetic DDS designs are based on biomaterials with unique properties, special morphologies, and mechanisms [2]. Taking inspiration from nature, scientists are incorporating nanomaterials with unique characteristics, such as adaptability, miniaturization, and hierarchical organization, to meet the growing demand for green, affordable, and sustainable solutions. There are a number of materials that are inspired by the structural properties of the natural organism. Different types of bioinspired materials are used to coat the nanoparticles to make them compatible with biomimicry ability such as cell membranes derived from natural killer cells, neutrophils, erythrocytes, extracellular vesicles, macrophages, and cancer cells. Additionally, there are some nature-inspired biomaterials such as viral capsids, natural proteins, monoclonal antibodies, and synthetic biomaterials, such as targeting peptides and aptamers [3, 4].

This chapter attempts to explain alternative ways of achieving biomimetic synthesis through imitation of biological processes and categorizing them in different groups along with various fabrication strategies.

Researchers worldwide are searching for synthetic materials that mimic the structural, physical, and chemical characteristics of biological organisms in an attempt to mimic the natural processes. As a result of billions of years of technological evolution, nature has offered solutions to many complex technological and material problems. A wide variety of applications are possible for biomimetic materials in a variety of fields, including biology, materials science, electronics, and chemistry. Biomimicking materials are currently fabricated successfully by today's state-of-the-art technology for biomedical applications, especially tumor regression [5]. Several advances have been made in biomimetic nanoparticle fabrication for the purpose of combating cancer, which will be discussed in this chapter. Furthermore, there are a number of applications of biomimetic structures, which are explained as well to elucidate on the fact that it is not about making an exact clone of biological entities but to develop new materials by applying the underlying principles.

9.2 Bioinspired Nanostructures

Bioinspired materials are natural materials that provide an excellent example of a functional system where pieces are put together to provide something useful for the biological system to help it survive. A myriad of new materials, some with extraordinary properties and novelty, can be created by combining science and nature together. As a result of their expected biocompatibility and nontoxic properties, bioinspired nanomaterials are ideal for nanomedicine applications [1]. Nature provides insight into how nanostructured surfaces interact with liquids, which can help us better comprehend the living biological interfaces formed by these nanostructures and further the development of innovative materials contacting water [6]. In case of biological hard materials, nature grows hierarchically structured organic/inorganic composites in which soft materials (e.g., proteins, membranes, and fibers) organized at lengths of 1 to 100 nm are used as frameworks for the growth of specifically oriented and shaped inorganics (e.g., $CaCO_3$, SiO_2, $Fe3O_4$, hydroxyapatite) with small unit cells (~1 nm) [7]. Biominerals are inorganic yet biogenic solid-state materials, i.e., bioceramics. They are used by a variety of organisms as sensors, tools, skeletons, weapons, or protection against predators. Biologically inspired materials or engineered molecules and structures that mimic the structure found in biological structure can be classified into four categories: (i) materials inspired by structural properties of a natural organism, (ii) materials inspired by biological functions, (iii) materials inspired by natural recycling methods, and (iv) biological processes inspired materials [7–9] (Table 9.1).

9.2.1 Materials Inspired by Structural Properties of Natural Organism

As a part of recent efforts, surface modification of nanoparticles by conjugating with molecules, such as carbohydrates and peptides, enable the accumulation of nanoparticles and could selectively bind to a specific receptor or antigen on the targeted site. Notably, due to these kinds of modification, the nanomaterial could not fully simulate the complex innate interfaces, and might even increase the immunogenicity of NPs, which hindering their translation approach from research to clinical application [10]. For novel drug delivery design, green technology-based nanomaterials have the potential to bring about significant improvement in efficient and targeted drug delivery with superior binding capability, biocompatibility, and biorenewability.

For these reasons, cell membrane-veiling approaches attaining large interest as they make nanoparticles compatible with innate advantages and surface diversity of the source cells were going to be extensively explored in current years. Platelets (PLTs) are micron-sized and discoid-shaped cells in the blood, which play an indispensable role in hemostasis, thrombosis, and other physiological/pathological processes. Natural platelets plasma membrane is translocated on the synthetic DOX-PFP-CNs@PLGA (poly-lactic-co-glycolic acid) skeletons were co-encapsulated with perfluoropentane (PFP) and doxorubicin (DOX) NPs and established a PM-coated biomimetic targeted drug delivery system for guided multimodal PTT/chemotherapy of breast cancer. High-density lipoproteins (HDL) are a green concept for manufacturing environmentally friendly biomaterials derived from homologous lipids and endogenous apoA-I. iRGD is anchored to the lipid monolayer, thus enhancing tumor penetration. HDLs have demonstrated synergistic chemo-phototherapy in vitro and in vivo by simultaneously entrapping PTX and ICG, and burst releases of the cargo triggered by NIR irradiation [11]. Many living organisms contain melanin, an abundant biopigment and heterogeneous in nature. On the basis of their source, such as plants, fungal, bacterial, and animal melanin, pigments can be classified into different types based on their physical and chemical properties as pheomelanin, eumelanin, neuromelanin, pyomelanin, and allomelanin [12]. Polymeric nanoparticles are solid colloidal particles ranging in size from 10 to 1000 nm. Because these systems have a very high surface

Table 9.1 Various bioinspired materials and their functions.

Sr. no.	Bioinspired material type	Material used	Function	References
1	Structural material	Bio-inspired hierarchically structured surfaces (e.g., protein, membrane, fibers)	• High surface area affords a higher nominal coverage of anti-EpCAM • Increases the binding sites for cell capture	[6, 16]
		Bioinspired versatile spore coat nanomaterial	• Probiotics protective delivery	[17]
		Morphology-controlled poly(aminophenyl boronic acid) nanostructures	• Enhanced capture and release efficiencies toward CCRF-CEM • Less damage	[18]
		Nanoplatelets (DOX-PFP-CNs@PLGA/PM NPs)	• Antigen–antibody interactions • Active targeting to cancer cells • Phagocytosis escape	[19]
2	Functional material	RBC–(Molybdenum diselenide) MoSe2 nanosheet	• Camouflage 2D MoSe2 nanosheets High photothermal conversion efficiency enhanced hemocompatibility • Enhanced reaction time by preventing macrophage phagocytosis	[20]

(Continued)

Table 9.1 Various bioinspired materials and their functions. (*Continued*)

Sr. no.	Bioinspired material type	Material used	Function	References
		Saccharide-responsive smart copolymer (SRSC)	• Cell–cell and cell–microenvironment interactions	[21]
3	Stimulus-responsive materials	pH dependent soluble CaCO3 capsules	• Bioresorbable • Soluble under acidic conditions	[7]
		Chitosan calcium phosphate flower-like microparticles	• pH-sensitive drug delivery • Immobilization of enzymes	[22]
4	Bioactivity based material	Polymeric nanoparticles	• Colon-specific drug delivery	[23]
		Carbon nanosphere and protein cage magnetic nanoparticle	• Signal amplification in immunoassay	[24]
		Macrophage membrane decorated liposome	• Treatment efficacy	[25]

area, drugs may also be absorbed, entrapped, encapsulated, or attached to a nanoparticle matrix. As a result of their biocompatibility, biodegradability, and low-cost polymers have drawn a great deal of attention in a range of fields, including medicine, bioinspired materials, pharmaceuticals, and agriculture [13]. By virtue of its nanometer size, the polymeric nanostructure is able to permeate through cell membranes and remain stable in the bloodstream. Polymer nanostructures with modulated morphologies and properties can be easily achieved from different fabricated methods [14]. To give the polymer better biological properties, a CH_2COOH function would be added to the carboxymethyl chitin (CMC) and carboxymethyl chitosan (CMCS), which would make their derivatives water soluble, and these function into the polymer, which endow it with better application. The functional group makes CMC/CMCS nanoparticles (NPs) efficient vehicles for the delivery of DNA, proteins, and drugs. CMCS and CMC can be used for a variety of purposes, including adsorbing metal ions, wound healing, antimicrobials, drug delivery systems, tissue engineering, food, cosmetics, and antitumor activities. Since these polymers possess reactive moieties, such as the carboxyl group, which will easily combined to drugs with an amino group by forming an amide bond [15].

9.3 Biomimetic Structures

Biomimetic structures are artificially derived materials, devices, and system that imitate the nature and play significant role in biological applications. However, these conventional nanocarriers with some improved targeting property and bioavailability are a safe and effective approach but still have some limitations in clinical application, such as low cytotoxicity, not easy recognition due to low reaction time between system and carriers and removed by the immune system [26, 27]. There have been a number of biomimetic nanomaterials that have been used as drug carriers, including organic materials like metal-organic frameworks (MOF), liposomes, micelles, nanogel, liposomes, dendrimers, nanoemulsion, and inorganic material like *Mesoporous silica* nanoparticles, carbon nanotube, iron oxide nanoparticle, noble metal-based nanoparticle etc. For various nanoformulation, synthesis routes and specific material provide the nanocarrier inimitable features in different applications. However, great progress has been made by these nanocarriers in cancer therapy, still nanocarriers are having limitation in clinical application, such as high toxicity, difficult to control release ratio, and low encapsulation efficiency in vivo [28].

By mimicking the function and structure of various natural particles, such as endogenous proteins, different cell membrane, pathogens, and so on, biomimetic drug delivery systems have low immunogenicity effects, more reaction time and good targeting. These properties help make them a suitable candidate for treating cancer and other diseases, as well as reduce their adverse effects on the body [29]. Endogenous protein, such as albumin, lipoproteins, ferritin, lactoferrin, and hemoglobin, have a wide variety of functions and properties, which were used to easily modify the structure of biomimetic nanoparticles and applied in drug delivery and cell imagining with the advancement of biomaterial technology and molecular biology, endogenous protein application will be further expanded.

Due to their simple structure pathogen, like virus (VNP, VLP, virosomes, viral vector), bacteria, and fungi are the emerging microbe-based biomimetic drug delivery carriers. Virus and bacteria are most commonly studied carriers for drugs delivery while fungi yet to be explore [30]. Microbes have been demonstrated to be capable of containing drugs efficiently and have also been shown to possess potential toxicity. Furthermore, it is unclear how this will affect the immune system, which needs to be explored further. Nevertheless, they are achieving wide recognition as a key component of nature. A membrane-coated nanoparticle can also play a significant role in easy cellular uptake, the immune response, and the regulation of multiple cell processes. Cell membranes, such as platelets, red blood cells, cancer cells, macrophages, lymphocytes, stem cells (neural stem cells and mesenchymal stem cells), fibroblasts, and hybrid cells, can also be modified to actively target a specific site [31]. These abovementioned biomimetic structures attaining great attention due to their good immunocompatibility.

9.4 Biomimetic Synthesis Processes and Products

Biomimetic synthesis can be achieved by replicating the characteristics of natural compounds or by imitating the biochemical processes that organisms use to synthesize substances. Biomimetic synthesis can be characterized into two groups: process biomimetic synthesis (PBS) and functional biomimetic synthesis (FBS). Functional biomimetic synthesis attempts to mimic certain attributes of natural materials/systems by utilizing multifarious substances and different procedures. For instance, artificial bones have been manufactured by different methods (e.g., nucleic-acid-template biomimetic synthesis process) of various materials (polymers, composites, bioceramics, etc.), which possess biocompatibility, toughness, and high strength, whereas process biomimetic synthesis endeavors to synthesize diverse high-value nanomaterials by imitating the synthesis routes, methods, and processes of natural substances and materials. For example, novel nanostructures like satellite structures, dendrimer-like structures [32], pyramids [33], cubes [34], 2D nanoparticle arrays [35], and 3D AuNP tubes [35] etc., have been assembled in vitro by mimicking the protein synthesis process. As a result of the above discussion, we can conclude that FBS and PBS have some significant differences. While the former mimics "properties," the latter mimics "methods." To make research more targeted, it is critical to differentiate and outline these terms clearly. PBS is an excellent choice for researchers who are looking for mild, efficient, and green synthetic routes. Alternatively, when researchers are interested in a material's properties, they may consider FBS. By using FBS, a variety of synthesis routes can be used to produce nanomaterials/nanostructures with specific properties (i.e., different synthesis processes produce the same product); by using PBS, nanomaterials/nanostructures can be constructed using different biomimetic pathways (i.e., branches can be derived from the same source). They can even interrelate: FBS can use biomimetic synthesis approaches for process synthesis, while PBS can manufacture nanomaterials and structures for functional biomimetics.

With the advent of nanotechnology, scientists have discovered milder, more effective, and environmentally friendly ways to produce nanomaterials when traditional physical and chemical methods fail to meet their challenges. Hence, process-bioinspired and process-biomimetic procedures are emerging and resurface as needed. It is interesting to note that in recent decades most publications mentioning "biomimetic synthesis," or "bioinspired synthesis" refer to functional biomimetic synthesis. It is also important to note that a few papers have a description of biomimetic synthesis, but they are only isolated episodes of biomimetic synthesis and are used under the terms "biomineralization," "template synthesis," etc., [36–38] although not using terms, such as "biomimetic synthesis" or "process biomimetic synthesis" possibly points to the lack of explicit differentiation between FBS and PBS. As a matter of fact, PBS has a wide range of applications and has tremendous potential. PBS can be performed through elementary, living, and high-level/elementary biomimetic systems. Organisms form biological units from a combination of fundamental biological processes to fulfill specific tasks, like biomineralization, which involves fundamental biological mechanisms like supramolecular reorganization, molecular recognition, cellular processing and template regulation. Xu and co-authors [39] synthesized TiO_2-Ag-AgCl photocatalyst through polydopamine-mediated biomimetic mineralization. When irradiated with visible light, the prepared composite degrades Rhodamine B (RhB) and ciprofloxacin (CIP) proficiently and has sturdy bactericidal activity against E. coli. Additionally, TiO_2-Ag-AgCl composites destruct *E. coli* biofilms efficiently. Similarly, biomass and soft/hard templates also present as an efficient of synthesizing biomimetic materials. Notably, at micrometer and nanometer dimensions, natural biomass assemblages commonly display degrees of precision that outperform existing artificial structures and materials. The biomass template mimics many biological processes, including molecular recognition, diffusion, assembly, space confinement, penetration, and template induction. Schnepp *et al.* fabricated pristine, sophisticated, and hierarchical microstructures of magnetic iron carbide (Fe3C) using a leaf skeleton describing a generic approach for the biomimetic preparation through biomass templates to form metal carbide materials [40].

Soft templates (such as microemulsions, liquid crystals, and reverse micelles) and hard templates (such as carbon nanotubes and anodic aluminum oxide) function as effective tools for nanoparticle synthesis using their template-inducing ability and confinement effect, respectively. Hard templates produce nanomaterials with uniform and predictable structures, whereas it is possible to produce various unique and special nanostructures using soft templates. Wu's group encountered egg-shell membranes, celloidin membranes, and commercial polymer films that combine the qualities of both the templates and these combined templates demonstrate a porous structure with functional groups on the pore wall. The structures of these membranes are rigid, and they have space-constricting properties, which are hard template characteristics; meanwhile, these membranes demonstrate modification flexibility and template-inducing properties thanks to their functional assemblies on the pore walls, exhibiting soft template properties [1, 41]. Different methods for the biomimetic synthesis of nanostructure have been compiled in Figure 9.1.

Nanoparticles camouflaged in cell membranes can serve as innovative biomimetic substrates mimicking the functions of cell membranes from which they originated in biological systems [42]. NPs coated with cell membranes have been studied for a wide range of membranes, such as those from immune cells, macrophages, red blood cells, and cancer cells. Jin *et al.*, explained cancer cell membrane coated nanoparticles (CCMCNPs), consisting a nanoparticle core with a cancer cell plasma

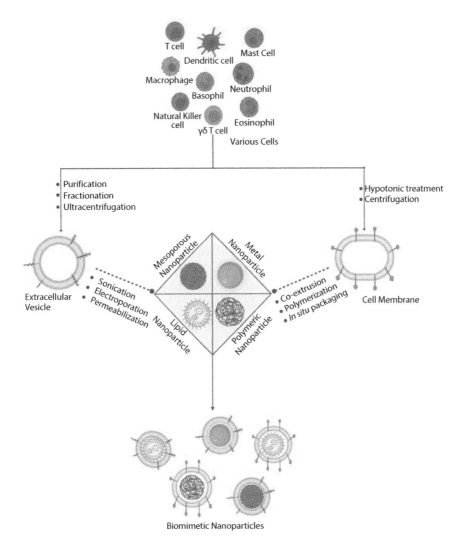

Figure 9.1 Biomimetic synthesis of nanostructures.

membrane coating over it that can deliver tumor-specific receptors and antigens for targeted cancer therapy. CCMCNPs were manufactured by coating the nanoparticles with a lipid bilayer of cancer cell plasma membranes [43]. Organic or inorganic cores can be utilized for the synthesis. Organic nanoparticle cores commonly used are poly lactic-co-glycolic acid (PLGA) [44, 45], poly(caprolactone) (PCL)-pluronic copolymers [46], and semiconducting polymers [47]. Inorganic NP cores can form porous structures to maximize drug-loading capability and exhibit unique magnetic, optical, and electrical properties acclimated to address the specialized needs of biomedical applications. Inorganic nanoparticles can now be formulated with precision over size, shape, and surface chemistry thanks to recent advancements in nanotechnology. These nanoparticle cores function as a cargo to encapsulate therapeutic and imaging moieties. Because of this, CCMC NPs are being explored for cancer theranostics.

The introduction of biomimetic nanodrug delivery systems also presents a clinically viable method for cancer treatment using Noble-metal nanomaterials that exhibit intriguing optical, electromagnetic, and physicochemical properties. Song *et al.* fabricated porous Au@Pt bimetallic nanoparticles by amalgamating chloroauric acid (gold precursor) with the potassium chloroplatinate (platinum precursor) in 1:1 ratio. The Au@Pt NPs revealed good catalytic abilities with superior drug loading efficiency and high photothermal conversion capacity. The resulting Au@Pt nanoparticles were then used to formulate a cell membrane-coated photothermal nanodrug delivery system (DOX/Au@Pt-M-NPs) by incorporating DOX on Au@Pt nanoparticles surface, resulting in greater inhibition of tumor cells. Additionally, the DOX/Au@Pt-M nanoparticle group caused considerably less damage to major organs than the DOX/Au@Pt nanoparticle group and the pure DOX group. Combining chemotherapy and photothermal therapy, nanomedicine carriers are camouflaged by tumor cells, therefore, have the potential to treat tumors [17].

Biomimetic approaches utilizing immune cell membranes like macrophages, neutrophils, or natural killer cell membrane also constitute a promising strategy to suppress cancer progression and metastasis that have the ability to interact with the tumor tissues [48]. Wang *et al.*, [49] reported a groundbreaking "cocktail therapy" technique employing the usage of exosomes (NKEXOs) derived from natural killer cell coupled with biomimetic core-shell nanoparticles (NNs) for targeted tumor treatment. The NNs self-assembled with a hydrophilic NKEXOs shell and a dendrimer core loaded with therapeutic miRNAs. The resulting NN/NKEXO cocktail was observed to possess high targeting efficiency and therapeutic delivery of miRNAs to neuroblastoma cells in vivo, as confirmed by TPEFI (two-photon excited scanning fluorescence imaging) which exhibited dual inhibitory effects on tumor growth. With unique biocompatibility, various fabrication strategies of different biomimetic NPs mentioned above allows for a new avenue for tumor therapy, with budding projections for therapeutic treatments.

9.5 Application of Bioinspired and Biomimetic Structure

Biomimetic and bioinspired structures provide better economic and sustainable applications as compared to synthetic processes and materials. These biomimetic and bioinspired membranes can be used for medical applications and drug delivery cohorts and demonstrate better biocompatibility. The objective of biomimicry is not to construct an exact clone of a biological form, or process, but rather to develop new products and ideas by applying design principles derived from biology. As a result of evolutionary inspiration, prehistoric men designed spears from the teeth of animals, mimicking large predators' sneak-and-pounce hunting techniques. Nevertheless, it has only been recently that a comprehensive methodology is built for investigating biological mechanisms and translating them into medical device designs. Several companies and firms are using biomimicry to arrive at new, efficient and sustainable designs. In general, applied biomimicry can be used in three ways or a combination of these three ways: (i) Form- For instance, creating lightweight structures by mimicking dragonfly wings; (ii) Processes- like capturing solar energy by mimicking photosynthesis; (iii) Systems-For example, developing wall systems that mimic organisms' homeostasis that enables them to maintain their internal parameters, like temperature [50].

Natural processes such as self-assembly and genetic engineering and natural building blocks like proteins, peptides, and lipids have been used to fabricate novel biomimetic structures. Biological entities and processes have recently been used to develop diverse nanostructured materials, such as protein scaffolds and nanofiber peptides [51–53]. Bioengineered protein structures are expected to have a major impact on the facilitation of design of novel materials with tailored morphologies and enhanced properties, including specificity, stimulus sensitivity, and catalytic abilities. There is a huge potential for these biomimetic and functional architectures in biomaterials, including personalized and targeted drug delivery applications. Tabata et al. employed cyclic peptides to prepare peptide nanotubes containing α- and β-amino acids, comprising 1D arrays of anthryl and naphthyl groups. The peptide was synthesized from D-α-anthrylalanine, L-α-naphthylalanine, and four β-alanines (CP6) and molecularly assembled into peptide nanotubes (PNTs). The electronic characteristics emerging from the groups of aromatic groups within PNTs were explored. CP6 coupled with L- and D-α-amino acids were devised to self-assemble into PNTs in a stacking arrangement to increase the intermolecular hydrogen bonds between the cyclic peptides. In the process of PNT development, the side chains (L- and D-α-amino acids) align in a line on the edges of PNTs. Topologically, the close proximity of anthryl groups in the CP6 PNT is explained by induced Cotton effects, enhanced photo-excited energy transfer, and the photo-dimerization occurrence upon PNT development. According to AFM measurements, the PNT bundles of diameters 5 to 15 nm were dielectric microcrystals possessing a piezoelectric coefficient of 2 to 6 pC N−1. Due to 1D array of anthryl groups along PNTs, Kelvin force microscopy observations demonstrate surface potentials of over 100 mV. It is, therefore, an efficient method for designing molecular structures which combines α-amino acids of opposite chirality with cyclic β-peptides to produce nanostructured PNTs exhibiting one-dimensional arrays of chromophores [54]. The applications of biomimetic nanosystems are vast and will become a powerful tool for humans to overcome various diseases in the near future.

Body mucosal surfaces adhere to synthetic and natural macromolecules through bioadhesion which remains a key focus of drug delivery research over the last two decades and enables better control over drug delivery (e.g., site-specific adhesion and prolonged residence times) [55, 56]. Recently, integrins have received a lot of attention, since they are cellular adhesion molecules [57, 58]. A peptide was designed by Yu et al. [59] to study interactions between amino acids, to evaluate its use as a biomaterial coating, and its effectiveness as a delivery device for drugs. Melanoma cells propagated unrestrainedly on carboxyl-coupled Arg-Gly-Asp amphiphiles and did not propagate on amino-coupled Arg-Gly-Asp amphiphiles, demonstrating their effectiveness as cellular recognition agents. To provide immunotherapeutic treatment for cancer, Wu et al. [60] fabricated genetically engineered antibodies (i.e., chimeric structures) based on a single-chain, single-gene approach. Molecular constructs of biomimetic fragmented antibodies with human IgG1 hinges and Fc regions (scFv–Fc dimers) were been developed to design radionucleotide delivery molecules with increased circulation life, effector functions, and reduced immunogenicity. Additionally, with nanoscale drug delivery system, the biological diversity of source cells may be preserved, allowing its use in cancer diagnosis and treatment [61]. Mammalian cells, viruses, extracellular vesicles, pathogens, and more recently, lipoproteins, have been considered as natural nanoparticle candidates for drug delivery.

Wang et al. [29] designed bioinspired actuators with deformable, stimuli-responsive attributes suitable for applications such as in medical devices, diagnostics, intelligent biosensors, and artificial tissues. It is necessary for actuator systems to be biocompatible, biodegradable, mechanically stable, deformable, and reversible to be suitable for these applications. The authors reported a bionic actuator device built from wood-derived cellulose nanofibers (CNFs) and stimuli-responsive genetically engineered silk-elastin-like protein (SELP) hydrogels that were able to detect temperature change and ionic strength under the water through eco-friendly approaches. Programmed site-selective actuation can be calculated and pleated into 3D origami shapes. The reversible deformation capability of the SELP/CNF actuators was evaluated, and the intricate spatial transformations of multi-layer actuators were investigated, including a biomimetic flower structure with particular petal motions. These types of actuators fabricated entirely of biodegradable and biocompatible materials offer the plausibility to construct stimuli-responsive structures for in vivo bionic engineering and soft robotics. In the future, biomimetic structures could be widely used for better biomedical technologies thanks to these groundbreaking developments

Ma et al. [62] manufactured biomimetic nanoerythrosome-coated aptamer–DNA tetrahedron/maytansine Conjugates that are responsive to pH change and have specific cytotoxicity for HER2-Positive Breast Cancer. The work utilizes HER2-targeted DNA-aptamer-modified DNA tetrahedron as a drug delivery system and incorporates maytansine (DM1) to develop HApt-DNA tetrahedron/DM1 conjugate (HTD conjugate) for HER2-positive breast cancer targeted therapy. A biomimetic disguise is utilized to embed HTD to improve its pharmacokinetics and tumor-aggregation properties. The biomimetic camouflage was achieved by fusing the functionalized synthetic liposomes, which are responsive to pH with erythrocyte membranes, providing optimal performance of drug distribution and tumor-stimulated drug release. HER2-positive cancer is better inhibited by hybrid erythrosome-based nanoparticles than by other drug formulations and is more biosafe than other formulations. Hence, HTD is a promising nanomedicine to treat HER2-positive tumors with its benefits of targeted delivery, enhanced drug loading, extended circulation time, and sensitive tumor probing. Significantly, this study introduces dual-targeting nanoparticles by pairing pH-sensitive disguise and HTD conjugates, paving the way for developing and applying biomimetic cell membranes in treatment of cancer and other diseases. Some of the applications of both bioinspired and biomimetic nanostructures are provided in Table 9.2.

Table 9.2 Applications of bioinspired and biomimetic nanostructure.

S. no.	Nanostructure	Application	Reference
Bioinspired			
1	Polydopamine coated molecularly imprinted silicon oxide nanoparticles	Bioanalysis in dietary and nutritional supplement	[63]
2	Nanoporous Prussian blue nanocube heads/titanium oxide wires	Detection of biomolecules	[64]
3	Fluorescent biomimetic quantum dots	Bioanalysis of nutritional and dietary supplements	[65]
4	Castor oil/silica hybrid microparticles	Delivery of poorly water-soluble drugs	[66]
5	Collagen PLGA sandwich	Drug delivery system	[67]
6	Calmodulin incorporated acrylamide hydrogel	Stimuli sensitive drug delivery	[68]
Biomimetic			
7	Albumin templated nanoparticles	Drug delivery	[69]
8	Bovine serum albumin templated nanoparticle	Tumor targeted imaging, ultrasensitive detection	[70]
9	RBC membrane fabricated AuNP	Immunosuppression	[71]
10	Erythrocyte fabricated magnetic nanoparticles	Magnetic resonance imaging, photothermal therapy	[72]
11	Macrophage membrane decorated liposome	Treatment efficacy	[73]

(*Continued*)

Table 9.2 Applications of bioinspired and biomimetic nanostructure. (*Continued*)

S. no.	Nanostructure	Application	Reference
12	Camouflage PLGA on T lymphocyte membrane	Drug delivery	[74]
13	Protease bound DNA origami	Regulate enzyme activity	[75]
14	Engineered exosomes	Repair of cardiac tissue	[76]

9.6 Conclusion

Synthesis of nanoparticles through biomimetic and bioinspired techniques is an effective and versatile method for producing various nanomaterials/nanostructures, which has the benefits of mild reaction conditions (such as neutral or near-neutral pH and ambient temperature and pressure), controllability, high selectivity, atom economy, and sustainability. Biological processes are not mechanically copied in biomimetic synthesis, instead, the researchers add some non-biological processes to combine both biological and non-biological approaches for synthesis that are more convenient and provide superior performance. As far as biomimetic systems go, living organisms are the most intelligent since they are composed of numerous higher life processes. In this chapter, different aspects of mimicking approaches, fabrication strategies, and application of biomimetic structures are discoursed based on the differences in mimicked biological processes. Biomimetic systems based on templates can be used to synthesize sophisticated structures easily and effectively; however, their adjustability is not up to the mark and the finished goods are mostly inefficient. By mimicking one or more natural processes, soft/hard coupled membrane biomimetic systems are considered to be intelligent; therefore, the resultant nanomaterials/nanostructures can be controlled in terms of morphology, size, and shape. Besides being mild and controllable, this method can also solve certain issues arising in conventional methods. For example, the preparation of ultrathin metal films with graphene-like properties is possible under mild conditions, and nanosuperstructures are easily obtained. In addition, it is possible to convert stable crystal forms into metastable crystal forms by abnormal structure conversion. Therefore, bio-inspired and biomimetic nanomaterials form the basis to develop new materials and structures that are more effective in medicine, technology, and biocompatible transplants.

9.7 Future Outlook

Along with the benefits of bioinspired nanostructures, comes the cost of these nanostructures due to the scarcity of available biological agents. Further work on bioinspired structures needs to be focused on the scalability of these structures to the industrial levels with decreased costs. The methods also utilize additives, which are left as impurities in the final steps and the purification steps further add in the cost therefore greener approaches and cost-effective purification methods needs to be worked on for broader application of these nanostructures. The biomimetic nanostructure holds a great potential in market due to the biodegradability and more compatibility with nature.

In the natural world, hybrid materials emerge when multi-scales, multi functionalities, and hierarchies integrate to produce remarkable properties. While natural materials themselves can be used for applications under ambient conditions, the challenge lies in translating the knowledge acquired from investigating complex designs into structural materials that can function in adverse environments, high temperatures, and pressures. The natural structure is usually designed specifically for a particular application, which is another notable observation. As an example, enamel microstructure varies from animal to animal or even tooth to tooth. The applications of synthetic materials, however, tend to be broader. Orthopedic implants, for example, are made from aerospace alloys. If the bioinspired synthetic materials were to become highly specialized, considering the economic value is as important as superior performance. The prospects of marrying a variety of synthetic compounds with the structural control observed in nature may result in the manufacture of improved materials that have an extended range of applications, strength, toughness, and environmental resistance. Achieving those goals would be a major technological breakthrough. Overall, the reviewed aspects of biomimetics show a gradual increase in intelligence and complexity, and studies of biomimetic processes have taken a shift from elementary to smart mimicry. Irrespective of whether bioinspired nanomaterials are widely adopted, skeptics must acknowledge that exploring natural structures and devising ways to replicate them will lead to advances in materials science for years to come.

Acknowledgments

The authors would like to thank Scrivener Publishing and Institute of Nano Science and Technology for their guidance and general support in completing this book chapter.

References

1. Zan, G. and Wu, Q., Biomimetic and bioinspired synthesis of nanomaterials/nanostructures. *Adv. Mater.*, 28, 11, 2099–2147, 2016.
2. Rahamim, V. and Azagury, A., Bioengineered biomimetic and bioinspired noninvasive drug delivery systems. *Adv. Funct. Mater.*, 31, 44, 2102033, 2021.
3. Pradhan, S., Brooks, A.K., Yadavalli, V.K., Nature-derived materials for the fabrication of functional biodevices. *Mater. Today Bio*, 7, 100065, 2020.
4. Troy, E., Tilbury, M.A., Power, A.M., Wall, J.G., Nature-based biomaterials and their application in biomedicine. *Polym. (Basel)*, 13, 19, 3321, 2021.
5. Holzapfel, B.M., Reichert, J.C., Schantz, J.T., Gbureck, U., Rackwitz, L., Nöth, U. *et al.*, How smart do biomaterials need to be? A translational science and clinical point of view. *Adv. Drug Delivery Rev.*, 65, 4, 581–603, 2013.
6. Ortiz, C. and Boyce, M.C., Materials science: Bioinspired structural materials. *Sci.* 319, 5866: 1053–1054, 2008.
7. Begum, G., Reddy, T.N., Kumar, K.P., Dhevendar, K., Singh, S., Amarnath, M. *et al.*, In situ strategy to encapsulate antibiotics in a bioinspired CaCO3 structure enabling pH-sensitive drug release apt for therapeutic and imaging applications. *ACS Appl. Mater. Interfaces*, 8, 34, 22056–22063, 2016.
8. Hench, L.L., Bioceramics: From concept to clinic. *J. Am. Ceram. Soc.*, 74, 7, 1487–1510, 1991.
9. Dorozhkin, S.V., Current state of bioceramics. *J. Ceram. Sci. Technol.*, 9, 4, 353–370, 2018.
10. Seidi, K., Ayoubi-Joshaghani, M.H., Azizi, M., Javaheri, T., Jaymand, M., Alizadeh, E. *et al.*, Bioinspired hydrogels build a bridge from bench to bedside. *Nano Today*, 39, 101157, 2021.
11. Wang, M., Liu, Y., Ren, G., Wang, W., Wu, S., Shen, J., Bioinspired carbon quantum dots for sensitive fluorescent detection of vitamin B12 in cell system. *Anal. Chim. Acta*, 1032, 154–162, 2018.
12. Mavridi-Printezi, A., Guernelli, M., Menichetti, A., Montalti, M., Bio-applications of multifunctional melanin nanoparticles: From nanomedicine to nanocosmetics. *Nanomaterials*, 10, 11, 2276, 2020.
13. Knight, A.S., Zhou, E.Y., Francis, M.B., Zuckermann, R.N., Sequence programmable peptoid polymers for diverse materials applications. *Adv. Mater.*, 27, 38, 5665–91, 2015.

14. Li, X., Xiong, Y., Qing, G., Jiang, G., Li, X., Sun, T. et al., Bioinspired saccharide-saccharide interaction and smart polymer for specific enrichment of sialylated glycopeptides. *ACS Appl. Mater. Interfaces*, 8, 21, 13294–302, 2016.
15. Narayanan, D., Jayakumar, R., Chennazhi, K.P., Versatile carboxymethyl chitin and chitosan nanomaterials: A review. *Wiley Interdiscip. Rev. Nanomed. Nanobiotechnol.*, 6, 6, 574–598, 2014.
16. Wegst, U.G.K., Bai, H., Saiz, E., Tomsia, A.P., Ritchie, R.O., Bioinspired structural materials. *Nat. Mater.*, 14, 1, 23–36, 2015.
17. Song, Q., Zhao, H., Zheng, C., Wang, K., Gao, H., Feng, Q., A. et al., Bioinspired versatile spore coat nanomaterial for oral probiotics delivery. *Adv. Funct. Mater.*, 31, 41, 2104994, 2021.
18. Ouyang, J., Chen, M., Bao, W.J., Zhang, Q.W., Wang, K., Xia, X.H., Morphology controlled poly(aminophenylboronic acid) nanostructures as smart substrates for enhanced capture and release of circulating tumor cells. *Adv. Funct. Mater.*, 25, 6122–6130, 2015.
19. Li, L., Fu, J., Wang, X., Chen, Q., Zhang, W., Cao, Y. et al., Biomimetic "nanoplatelets" as a targeted drug delivery platform for breast cancer theranostics. *ACS Appl. Mater. Interfaces*, 13, 3, 3605–3621, 2021.
20. He, L., Nie, T., Xia, X., Liu, T., Huang, Y., Wang, X. et al., Designing bioinspired 2D MoSe2 nanosheet for efficient photothermal-triggered cancer immunotherapy with reprogramming tumor-associated macrophages. *Adv. Funct. Mater.*, 29, 30, 1901240, 2019.
21. Li, X., Xiong, Y., Qing, G., Jiang, G., Li, X., Sun, T. et al., Bioinspired saccharide-saccharide interaction and smart polymer for specific enrichment of sialylated glycopeptides. *ACS Appl. Mater. Interfaces*, 8, 21, 13294–13302, 2016.
22. Luo, C., Wu, S., Li, J., Li, X., Yang, P., Li, G., Chitosan/calcium phosphate flower-like microparticles as carriers for drug delivery platform. *Int. J. Biol. Macromol.*, 155, 174–183, 2020.
23. Naeem, M., Choi, M., Cao, J., Lee, Y., Ikram, M., Yoon, S. et al., Colon-targeted delivery of budesonide using dual pH- and time-dependent polymeric nanoparticles for colitis therapy. *Drug Des. Devel. Ther.*, 9, 3789–3799, 2015.
24. Chen, A., Bao, Y., Ge, X., Shin, Y., Du, D., Lin, Y., Magnetic particle-based immunoassay of phosphorylated p53 using protein cage templated lead phosphate and carbon nanospheres for signal amplification. *RSC Adv.*, 2, 11029–11034, 2012.
25. Cao, H., Dan, Z., He, X., Zhang, Z., Yu, H., Yin, Q. et al., Liposomes coated with isolated macrophage membrane can target lung metastasis of breast cancer. *ACS Nano*, 10, 7738–7748, 2016.
26. Wang, Z., Dong, L., Han, L., Wang, K., Lu, X., Fang, L. et al., Self-assembled biodegradable nanoparticles and polysaccharides as biomimetic ECM nanostructures for the synergistic effect of RGD and BMP-2 on bone formation. *Sci. Rep.*, 6, 25090, 2016.

27. Sun, H., Su, J., Meng, Q., Yin, Q., Chen, L., Gu, W. et al., Cancer-cell-biomimetic nanoparticles for targeted therapy of homotypic tumors. *Adv. Mater.*, 28, 33, 9581–3588, 2016.
28. Gong, M., Zhang, Q., Zhao, Q., Zheng, J., Li, Y., Wang, S. et al., Development of synthetic high-density lipoprotein-based ApoA-I mimetic peptide-loaded docetaxel as a drug delivery nanocarrier for breast cancer chemotherapy. *Drug Delivery*, 26, 708–716, 2019.
29. Wang, Y., Huang, W., Huang, W., Wang, Y., Mu, X., Ling, S. et al., Stimuli-responsive composite biopolymer actuators with selective spatial deformation behavior. *Proc. Natl. Acad. Sci. U.S.A.*, 117, 25, 14602–14608, 2020.
30. Tong, Q., Qiu, N., Ji, J., Ye, L., Zhai, G., Research progress in bioinspired drug delivery systems. *Expert Opin. Drug Delivery*, 17, 9, 1269–1288, 2020.
31. Thanuja, M.Y., Anupama, C., Ranganath, S.H., Bioengineered cellular and cell membrane-derived vehicles for actively targeted drug delivery: So near and yet so far. *Adv. Drug Delivery Rev.*, 132, 57–80, 2018.
32. Xu, X., Rosi, N.L., Wang, Y., Huo, F., Mirkin, C.A., Asymmetric functionalization of gold nanoparticles with oligonucleotides. *J. Am. Chem. Soc.*, 2006.
33. Mastroianni, A.J., Claridge, S.A., Paul Alivisatos, A., Pyramidal and chiral groupings of gold nanocrystals assembled using DNA scaffolds. *J. Am. Chem. Soc.*, 2009.
34. Park, S.Y., Lytton-Jean, A.K.R., Lee, B., Weigand, S., Schatz, G.C., Mirkin, C.A., DNA-programmable nanoparticle crystallization. *Nature*, 2008.
35. Zheng, J., Constantinou, P.E., Micheel, C., Alivisatos, A.P., Kiehl, R.A., Seeman, N.C., Two-dimensional nanoparticle arrays show the organizational power of robust DNA motifs. *Nano Lett.*, 2006.
36. Liu, K. and Jiang, L., Multifunctional integration: From biological to bio-inspired materials. *ACS Nano*, 5, 6786–6790, 2011.
37. Hong, J., Lee, M., Lee, B., Seo, D.H., Park, C.B., Kang, K., Biologically inspired pteridine redox centres for rechargeable batteries. *Nat. Commun.*, 5, 5335, 2014.
38. Favi, P.M., Yi, S., Lenaghan, S.C., Xia, L., Zhang, M., Inspiration from the natural world: From bio-adhesives to bio-inspired adhesives. *J. Adhes. Sci. Technol.*, 28, 290–319, 2014.
39. Xu, X., Wu, C., Guo, A., Qin, B., Sun, Y., Zhao, C. et al., Visible-light photocatalysis of organic contaminants and disinfection using biomimetic-synthesized TiO2-Ag-AgCl composite. *Appl. Surf. Sci.*, 2022.
40. Schnepp, Z., Yang, W., Antonietti, M., Giordano, C., Biotemplating of metal carbide microstructures: The magnetic leaf. *Angew. Chem.-Int. Ed.*, 2010.
41. Zhang, H., Xu, H., Wu, M., Zhong, Y., Wang, D., Jiao, Z., A soft-hard template approach towards hollow mesoporous silica nanoparticles with rough surfaces for controlled drug delivery and protein adsorption. *J. Mater. Chem. B*, 2015.
42. Kroll, A.V., Fang, R.H., Zhang, L., Biointerfacing and applications of cell membrane-coated nanoparticles. *Bioconjug. Chem.*, 2017.

43. Jin, J. and Bhujwalla, Z.M., Biomimetic nanoparticles camouflaged in cancer cell membranes and their applications in cancer theranostics. *Front. Oncol.*, 9, 1560, 2020.
44. Yang, R., Xu, J., Xu, L., Sun, X., Chen, Q., Zhao, Y. *et al.*, Cancer cell membrane-coated adjuvant nanoparticles with mannose modification for effective anticancer vaccination. *ACS Nano*, 2018.
45. Chen, M., Chen, M., He, J., Cancer cell membrane cloaking nanoparticles for targeted co-delivery of doxorubicin and PD-L1 siRNA. *Artif. Cells Nanomed. Biotechnol.*, 2019.
46. Fuentes, I., Blanco-Fernandez, B., Alvarado, N., Leiva, Á., Radić, D., Alvarez-Lorenzo, C. *et al.*, Encapsulation of antioxidant gallate derivatives in biocompatible poly(ϵ-caprolactone)-b-pluronic-b-poly(ϵ-caprolactone) micelles. *Langmuir*, 32, 3331–3339, 2016.
47. Li, J., Zhen, X., Lyu, Y., Jiang, Y., Huang, J., Pu, K., Cell membrane coated semiconducting polymer nanoparticles for enhanced multimodal cancer phototheranostics. *ACS Nano*, 2018.
48. Oroojalian, F., Beygi, M., Baradaran, B., Mokhtarzadeh, A., Shahbazi, M.A., Immune cell membrane-coated biomimetic nanoparticles for targeted cancer therapy. *Small*, 2021.
49. Wang, G., Hu, W., Chen, H., Shou, X., Ye, T., Xu, Y., Cocktail strategy based on NK cell-derived exosomes and their biomimetic nanoparticles for dual tumor therapy. *Cancers (Basel)*, 2019.
50. Jamei, E. and Vrcelj, Z., Biomimicry and the built environment, learning from nature's solutions. *Appl. Sci.*, 2021.
51. Naresh, V. and Lee, N., A review on biosensors and recent development of nanostructured materials-enabled biosensors. *Sensors (Switzerland)*, 2021.
52. Singh, T.V. and Shagolsem, L.S., Biopolymer based nano-structured materials and their applications. In *Nanostructured Materials and Their Applications, Materials Horizons: From Nature to Nanomaterials*, Bibhu Prasad Swain (ed.) Singapore: Springer, 337–366.
53. Harish, V., Tewari, D., Gaur, M., Yadav, A.B., Swaroop, S., Bechelany, M. *et al.*, Review on nanoparticles and nanostructured materials: Bioimaging, biosensing, drug delivery, tissue engineering, antimicrobial, and agro-food applications. *Nanomaterials*, 2022.
54. Tabata, Y., Uji, H., Imai, T., Kimura, S., Two one-dimensional arrays of naphthyl and anthryl groups along peptide nanotubes prepared from cyclic peptides comprising α- and β-amino acids. *Soft Matter*, 2018.
55. Asati, S., Jain, S., Choubey, A., Bioadhesive or mucoadhesive drug delivery system: A potential alternative to conventional therapy. *J. Drug Deliv. Ther.*, 2019.
56. Yu, L., Luo, Z., Chen, T., Ouyang, Y., Xiao, L., Liang, S. *et al.*, Bioadhesive nanoparticles for local drug delivery. *Int. J. Mol. Sci.*, 2022.
57. da Silva, E.C., Dontenwill, M., Choulier, L., Lehmann, M., Role of integrins in resistance to therapies targeting growth factor receptors in cancer. *Cancers (Basel)*, 2019.

58. Harjunpää, H., Asens, M.L., Guenther, C., Fagerholm, S.C., Cell adhesion molecules and their roles and regulation in the immune and tumor microenvironment. *Front. Immunol.*, 10, 1078, 2019.
59. Yu, Y.C., Berndt, P., Tirrell, M., Fields, G.B., Self-assembling amphiphiles for construction of protein molecular architecture. *J. Am. Chem. Soc.*, 1996.
60. Wu, A.M., Tan, G.J., Sherman, M.A., Clarke, P., Olafsen, T., Forman, S.J. *et al.*, Multimerization of a chimeric anti-CD20 single-chain Fv-Fc fusion protein is mediated through variable domain exchange. *Protein Eng.*, 2001.
61. Nuñez-Prado, N., Compte, M., Harwood, S., Álvarez-Méndez, A., Lykkemark, S., Sanz, L. *et al.*, The coming of age of engineered multivalent antibodies. *Drug Discovery Today*, 2015.
62. Ma, W., Yang, Y., Zhu, J., Jia, W., Zhang, T., Liu, Z. *et al.*, Biomimetic nanoerythrosome-coated aptamer–DNA tetrahedron/maytansine conjugates: pH-responsive and targeted cytotoxicity for HER2-positive breast cancer. *Adv. Mater.*, 2022.
63. Zhang, J., Wang, Y., Lu, X., Molecular imprinting technology for sensing foodborne pathogenic bacteria. *Anal. Bioanal. Chem.*, 413, 4581–4598, 2021.
64. Kong, B., Tang, J., Wu, Z., Selomulya, C., Wang, H., Wei, J. *et al.*, Bio-inspired porous antenna-like nanocube/nanowire heterostructure as ultra-sensitive cellular interfaces. *NPG Asia Mater.*, 6, e117, 2014.
65. Pan, M., Xie, X., Liu, K., Yang, J., Hong, L., Wang, S., Fluorescent carbon quantum dots-synthesis, functionalization and sensing application in food analysis. *Nanomaterials*, 10, 930, 2020.
66. Doufène, K., Lapinte, V., Gaveau, P., Félix, G., Cacciaguerra, T., Chopineau, J. *et al.*, Tunable vegetable oil/silica hybrid microparticles for poorly water-soluble drug delivery. *Int. J. Pharm.*, 567, 118478, 2019.
67. Chen, D.W., Hsu, Y.H., Liao, J.Y., Liu, S.J., Chen, J.K., Ueng, S.W.N., Sustainable release of vancomycin, gentamicin and lidocaine from novel electrospun sandwich-structured PLGA/collagen nanofibrous membranes. *Int. J. Pharm.*, 430, 335–341, 2012.
68. Fox, C.S., Berry, H.A., Pedigo, S., Development and characterization of calmodulin-based copolymeric hydrogels. *Biomacromolecules*, 21, 2073–2086, 2020.
69. Tian, L., Chen, Q., Yi, X., Chen, J., Liang, C., Chao, Y. *et al.*, Albumin-templated manganese dioxide nanoparticles for enhanced radioisotope therapy. *Small*, 13, 2017.
70. Wang, J. and Zhang, B., Bovine serum albumin as a versatile platform for cancer imaging and therapy. *Curr. Med. Chem.*, 25, 2938–2953, 2017.
71. Shi, Y., Qian, H., Rao, P., Mu, D., Liu, Y., Liu, G. *et al.*, Bioinspired membrane-based nanomodulators for immunotherapy of autoimmune and infectious diseases. *Acta Pharm. Sin. B*, 12, 1126–1147, 2022.
72. Rao, L., Cai, B., Bu, L.L., Liao, Q.Q., Guo, S.S., Zhao, X.Z. *et al.*, Microfluidic electroporation-facilitated synthesis of erythrocyte membrane-coated

magnetic nanoparticles for enhanced imaging-guided cancer therapy. *ACS Nano*, 11, 3496–3505, 2017.
73. Cao, H., Dan, Z., He, X., Zhang, Z., Yu, H., Yin, Q. *et al.*, Liposomes coated with isolated macrophage membrane can target lung metastasis of breast cancer. *ACS Nano*, 10, 7738–7748, 2016.
74. Yaman, S., Ramachandramoorthy, H., Oter, G., Zhukova, D., Nguyen, T., Sabnani, M.K. *et al.*, Melanoma peptide MHC specific TCR expressing T-cell membrane camouflaged PLGA nanoparticles for treatment of melanoma skin cancer. *Front. Bioeng. Biotechnol.*, 8, 943, 2020.
75. Zhao, Z., Fu, J., Dhakal, S., Johnson-Buck, A., Liu, M., Zhang, T. *et al.*, Nanocaged enzymes with enhanced catalytic activity and increased stability against protease digestion. *Nat. Commun.*, 7, 10619, 2016.
76. Bheri, S., Hoffman, J.R., Park, H.J., Davis, M.E., Biomimetic nanovesicle design for cardiac tissue repair. *Nanomedicine*, 15, 1873–1896, 2020.

10
A Feasibility Study of the Bioinspired Green Manufacturing of Nanocomposite Materials

Arpita Bhattacharya

Amity Institute of Nanotechnology, Amity University, Uttar Pradesh, Noida, India

Abstract

Nanocomposites are one the most important and widely used engineered materials in the field of nanotechnology. Naturally occurring biological nanocomposites are highly inspiring for their unique mechanical properties with light weight. Wood, nacre, spider silk, teeth, bones, etc. are few examples where unique arrangement of nanomaterials in matrices provide synergistic properties of composite materials. A deep insight in bioinspired material helps to produce high performance engineered nanocomposites. Nowadays, the green nanocomposites are attractive due to their low cost, lightweight, biocompatible and environment-friendly nature. Plants are the main source of green composite material. Selection of proper synthesis technique to mix nanofibrils in biodegradable polymer matrices is one of the important criteria to achieve improved properties in nanocomposites. This chapter provides the recent advancements in bioinspired nanocomposites that are used in various sectors like biomedical, packaging, electronics, durable goods, etc. These nanocomposites possess much improved properties in terms of mechanical, thermal, barrier, and many more as compared to previously used natural or synthetic polymers. Cellulose, the most common natural polymer from plant origin has been used for production of sustainable bioinspired nanocomposites. Simple ball milling of cellulose produce some mechanoradical, which reduce metal ions like gold, silver, palladium, etc., to neutral metal nanoparticles, which can deposit on cellulose matrix. Thus, prepared metal-cellulose composites show catalytic activity, as well as anti-microbial activity. This method avoids many toxic chemicals and directly produce metal-cellulose compounds and blends. In a different way, microwave-assisted and UV-irradiated bioinspired nanocomposites of silver-reduced graphene oxide can be synthesized using lemon juice, which has been reported to have great antimicrobial activity. In agriculture field also, many bacterial diseases of crops can be controlled by bioinspired green chitosan-ZnO–based nanocomposites with

Email: abhattacharya@amity.edu

Mousumi Sen and Monalisa Mukherjee (eds.) Bioinspired and Green Synthesis of Nanostructures: A Sustainable Approach, (231–262) © 2023 Scrivener Publishing LLC

strong antibacterial activity. The recent trends in nanofunctional materials and renewable materials for the preparation of bioinspired nanocomposites, which are specially used in agricultural, biomedical, and healthcare sector are discussed in this chapter.

Keywords: Biological nanocomposites, green synthesis, biopolymers, bioinspired materials, orthopedic applications, tissue engineering

10.1 Introduction

Nanotechnology is the engineering of materials at the atomic or molecular level (from 1 to 100 nm). Nanomaterials are those where the size exists in between 1 and 100 nm at least in one dimension. Nanomaterials show unusual and exotic properties that are not present in traditional bulk materials. The special feature of nanomaterial is large specific surface area. The unusual features of nanomaterials can significantly influence the physical, chemical, biological, mechanical, and electrical properties [1, 2]. The important applications of nanotechnology are in energy sector, information technology, medical purposes, agriculture, food aspect, etc.

Nanocomposites are one of the unique class of materials where nano-sized fillers are added in order to improve the properties of the resulting materials. Nanocomposite is defined as two-phase material where one phase has nanometer size range [3]. Nanocomposites are composed of at least two constituents having different physical and chemical properties. The major component which is present in large quantity is known as matrix. The other constituent, which is in less quantity and dispersed into the matrix material is called nanofillers. The property improvement in nanocomposites is obtained by adding little amount of nanofillers compared to conventional composites. So nanocomposites are much lighter in weight as compared to conventional composites.

Nanocomposites are classified according to the types of matrix used in their composition. As per the type of matrix material, nanocomposites are classified as follows:

1. Polymer Nanocomposites
2. Ceramic Nanocomposites
3. Metal Nanocomposites

As a matrix material, polymers have several advantages like lightweight, high durability, easy processing, corrosion resistance, ductility and low cost. But there are some disadvantages also for polymer materials in comparison to ceramics and metals matrices. Polymers have relatively poor mechanical, thermal and electrical properties. Barrier properties of polymers are not good and they are highly flammable. The drawback of polymers can be overcome by incorporating proper nanofiller in it. The nanofillers can be 1-dimensional (nanotubes and fibers), 2-dimensional (layered materials like clay) or 3-dimensional/0-dimensional (spherical particles). The other excellent properties which can be achieved in polymer nanocomposites by adding various nanofillers are - barrier resistance, flame retardancy, wear resistance, magnetic, electrical and optical properties.

Now-a-days, utilization of nanocomposites in various biological applications have become very promising and popular. Bionanocomposites are the combination of biopolymers such as polysaccharides, proteins, and nucleic acids with inorganic solids at the nanometric scale [4]. Biomimetics is the process to build new materials that mimic biological system and function as natural materials [5, 6]. These bioinspired nanocomposites are useful in tissue engineering, orthopaedics, drug delivery, ocular therapy etc.

10.2 Biopolymers

These are naturally occurring materials and obtained from plants, animals and microorganisms. They mainly contain cellulose, hemicellulose and lignin. The biopolymers are biocompatible and biodegradable, and useful in different biomedical applications like drug delivery material, medical implants, wound healing, tissue scaffolds, and dressing materials in pharmaceutical industries.

Sources of bio-based polymers are given in Figure 10.1.

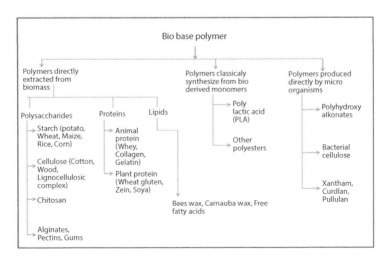

Figure 10.1 Sources of biomaterials.

10.2.1 Cellulose

This is a polysaccharide mostly found in plants. It is most abundant natural polymer on earth. Cellulose nanofiber (CNF) [7] can be prepared from cellulose which is a natural biodegradable nanofillers.

It is a polymer chains of glucose units joined by a beta-acetal linkage which is shown in Figure 10.2. Cellulose is highly biodegradable, abundant, renewable and require less processing steps during manufacturing.

10.2.2 Chitosan

This is a polysaccharide having large number of amino and hydroxyl groups. Chemical structure of chitosan is given in Figure 10.3. It can be synthesized by deacetylation of chitin where at least 50% free amino group(-NH2) is present [8]. It has a heterogeneous chemical structure made up of 1-4 linked 2-acetamido-2-deoxy-β-D-glucopyranose as well as 2-amino-2-deoxy-β-D-glucopyranose.

In medical applications, it is used as bandage and absorbable sutures.

10.2.3 Starch

Rice, potato, maize and wheat are the major source of starch [9]. It has two components amylose and amylopectin. Starch is a natural polymeric

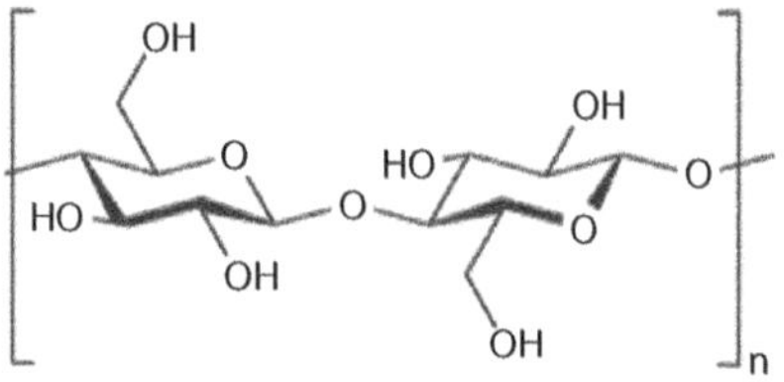

Figure 10.2 Repeating unit of cellulose.

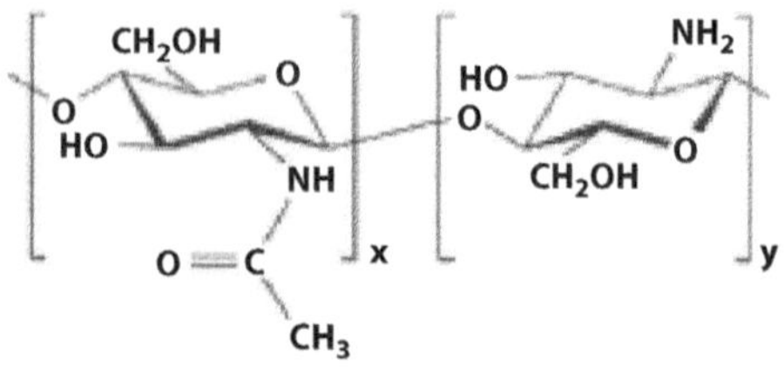

Figure 10.3 Chemical structure of chitosan.

carbohydrate which contains large number of glucose units joined by glycosidic bonds. This is extensively used in pharmaceutical and adhesive industry.

10.2.4 Chitin

After cellulose, chitin is the second most abundant biopolymer. It is a glucose derivative synthesized from long-chain polymer of N-acetylglucosamine. It is mostly used as an effective binder in dyes and fabrics [10].

10.2.5 Polyhydroxyalkanoates (PHA)

Polyhydroxyalkanoates (PHA) belong to naturally occurring biocompatible, hydrophobic and biodegradable polyesters [11]. They can be either thermoplastic or elastomeric materials with melting point ranging from 40-180°C [12].

10.2.6 Polylactic Acid (PLA)

Polylactic acid (PLA) is a biodegradable and aliphatic polyester derived from various renewable resources. It is also called as Poly lactide. It is bioactive thermoplastic polymer. It is used in 3D printers, plaster like moulding materials, and implants. Due to its bio-compatibility and biodegradability properties, it is used as a polymeric scaffold for drug delivery. Figure 10.4 shows monomer, oligomer and polymer of PLA.

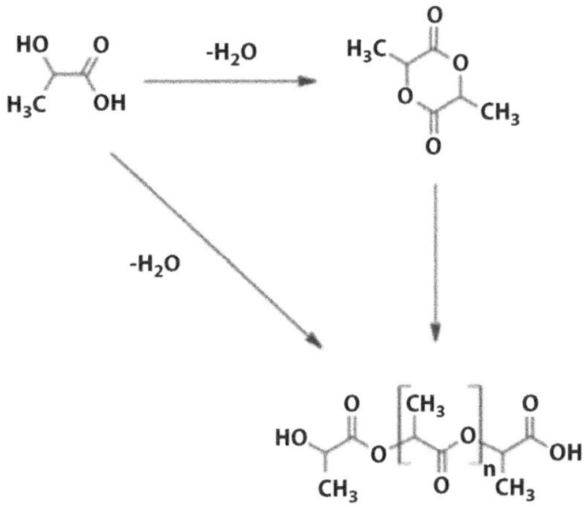

Figure 10.4 Monomer, oligomer, and polymer of polylactic acid.

10.3 Different Types of Bioinspired Nanocomposites

10.3.1 Polymer-HAp Nanoparticle Composites

Bioinspired nanohydroxyapatite (nHAp) is a natural mineral in hard tissue [13]. HAP is difficult to shape due to its brittleness and lack of flexibility. It is a major component of bone and teeth enamel. Polysaccharides and polypeptides can be mixed with HAP for nanocomposite formation [14].

Structure of HAp and HAp-chitosan composite structure are given in Figure 10.5 here.

Nano-HAp and calcium phosphate are major constituents of hard tissues like bone, cartilage and teeth. If a small amount of nano-HAp is added in polymer matrix, the mechanical properties like modulus and the tensile strength of the polymer matrix can be enhanced. Bio-nanocomposite scaffolds was fabricated by various research group from collagen and

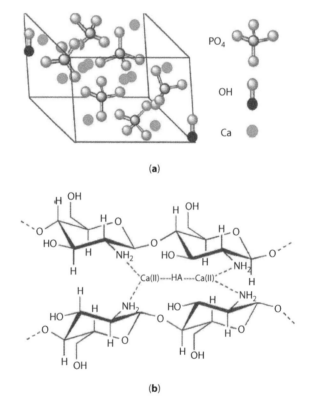

Figure 10.5 (a) Structure of HAp and (b) Chitisan-HAp nanocomposite.

nano-HAp using co-electrospinning. With increase in nano-HAp concentration, fiber diameter of fibrous scaffold increases and interconnected pore structure is obtained in scaffold. An addition of 10% HAp improves tensile strength almost three times and also increase modulus in four fold. The strong interaction between collagen and nano-HAp results the enhancement of mechanical properties. Further increase in the mechanical properties can be achieved by using crosslinker glutaraldehyde which chemically cross-links the collagen network. But as glutaraldehyde is not biocompatible, it may affect immune response.

PLA-co-PEG–HAp nanocomposites with good mechanical strength is obtained when glutamic acid (Glu), a negatively charged peptide is added as a cross-linker. The addition of the cross-linker resulted in a 100% rise in the shear modulus of the nanocomposite material formed. This change in mechanical properties does not happen when micro-HAp is used.

10.3.2 Nanowhisker-Based Bionanocomposites

Cellulose nanowhiskers are rodlike highly crystalline nanofiber with diameter less than 100 nm with a rectangular cross section. They have high aspect ratio The rod-like whisker structure is shown in Figure 10.6. They are formed by the acid hydrolysis of cellulose fibers using sulfuric or hydrochloric acid. The protocol is given in Figure 10.7.

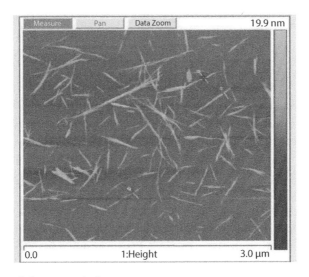

Figure 10.6 Cellulose nanowhisker.

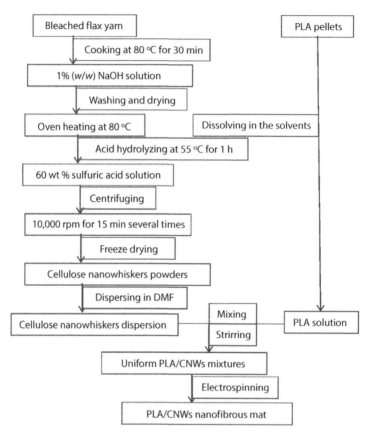

Figure 10.7 Protocol for nanowhisker preparation from cellulose.

10.3.3 Clay-Polymer Nanocomposites

Clay is a naturally occurring material, which is inert, non-toxic, abundant and low cost mineral. Chemically clay is hydrous aluminium silicates and sometimes they contain variable amount of calcium, magnesium. Potassium, sodium etc. Clay mineral is made of two distinct structural units- tetrahedral (silica) and octahedral (alumina) sheets, which are shown in Figure 10.8.

Depending on the number of layer units, clay can be classified as 1:1 and 2:1 type. In 1:1 type one tetrahedral silica and one octahedral alumina layer are present, where in 2:1 type two tetrahedral and one octahedral layer are present. Common clay materials are montmorillonite, bentonite, kaolinite, hectorite, and halloysite.

GREEN SYNTHESIS OF BIOINSPIRED NANOCOMPOSITES 239

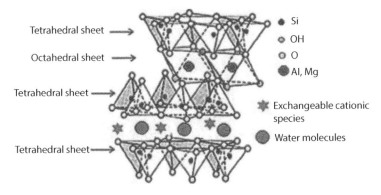

Figure 10.8 Structure of nanoclay (2:1 type).

When polymer is mixed with clay material, the interaction of polymer and layered silicate clays is described in Figure 10.9 mostly by the following nanocomposite structures:

(i) conventional composite,
(ii) intercalated nanocomposite
(iii) exfoliated nanocomposite (lamellae homogeneously dispersed in the polymer matrix)

By incorporating nanoclay in polymer matrix, several property improvements can be achieved like mechanical properties, thermal properties,

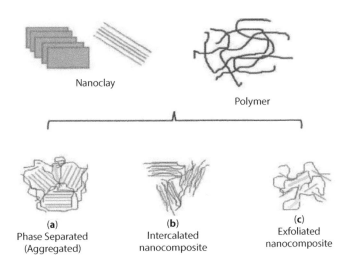

Figure 10.9 Various forms of nanoclay-polymer nanocomposites.

flammability reduction and dimensional stability. These type of nanocomposites can be used in biomedical applications, food packaging, coating system etc.

10.4 Fabrication of Bionanocomposites

The unique properties of biopolymer like biodegradability, low cost, abundant occurrence make them a promising candidate for fabrication of bio-nanocomposites [15]. Among them cellulose based nanocomposite is most important. However, fabrication process is challenging due to its hydrophilic nature. It can be dissolved in a mixture of dimethylsulfoxide (DMSO) and N,N-Dimethylformamide (DMF) with small amount of water by sonication. Homogenous dispersion of polymer matrix can be achieved by functionalization of the surface hydroxyl group [16].

The main techniques for bioinspired nanocomposites are:

    electrospinning,
    cast drying,
    freeze drying,
    vacuum assisted filtration,
    wet spinning, layer by layer assembly and several micropatterning techniques are the predominant methods [17].

10.4.1 Electrospinning

Electrospinning is a common technique used for the manufacture of nanofibers. Cellulose nanowhiskers can be prepared from plant cellulose or bacterial cellulose by treating with strong sulphuric acid. Nanowhiskers undergo blending with suitable polymers and electrospun to form nanocomposites. Electrospun composite nanofibers possess high mechanical strength as cellulose nanowhiskers are embed and aligned inside the nanofibers [17]. Since nanocellulose is a material having no cytotoxic effect, these can have applications as a tissue engineering scaffold [18].

10.4.2 Solvent Casting

In solvent casting method, aqueous suspension of cellulose nanoparticles and the aqueous solution of hydro polymer are mixed thoroughly. Then the resulting mixture can be casted and evaporated to obtain nanocomposites as solid films [19]. Various studies have reported the preparation of cellulose nanofiber reinforced starch [20, 21], silk fibroin [22], poly(oxyethylene) [23, 24], polyvinyl alcohol [25], hydroxypropyl cellulose and soy protein isolate [26].

The improved mechanical properties result from the homogenous dispersion of nanowhiskers. A barrier membrane was developed by Paralikar *et al.* using poly(vinyl alcohol) (PVOH), nanocellulose and poly(acrylic acid) (PAA) via casting/evaporation technique [27]. PAA and PVA powders were dissolved in water and the solution obtained was mixed with the dispersion of nanocellulose. Whole mixture was sonicated before casting to disperse the agglomerates. Then film was casted by pouring the solution in flat bottom dish, then air dried and finally heated in oven. In cast drying technique, the colloidal mixture gets deposited on the clean substrate surface to produce solid films after evaporation of solvent. Different coating procedures are: bar coating or spray coating which require complete evaporation of solvent.

Polylactide (PLA)-cellulose nanocomposite can be fabricated by blending PLA and modified CNW using solution casting method. PLA was allowed to dry for 20 h and mixed with modified CNW in dichloromethane at 25 °C for about 12 h. The mixture was casted into films, evaporated, and dried completely at 50 °C in vacuum oven for 24 h [28]. The fabricated nanocomposites exhibit enhanced mechanical properties by acting as a nucleating agent [29, 30]. Similarly, solvent casted CNW nanocomposite wet films were prepared with the carboxyl methyl cellulose having enhanced strength [31].

Homogeneous distribution of HAp fibres throughout the PLGA scaffold could only be obtained by using a solvent-casting technique to prepare a composite material of HAp fibres, PLGA and gelatin.

10.4.3 Melt Moulding

This process involves filling a Teflon mould with PLGA powder and gelatine microspheres, of specific diameter, and then heating the mould for the regeneration of hard and soft tissues above the glass-transition temperature of PLGA while applying pressure to the mixture [32]. On application of heat, PLGA particles bind together. When the mould is removed, the gelatin is leached out by dipping it into water. Then scaffold is dried and thus produced scaffold takes the shape of the mould. Short fibers of HAp can be incorporated in polymer matrix by melt moulding process.

10.4.4 Freeze Drying

Freeze drying/lyophilization involves removal of water from hydrogel structure to produce aerogel. Here, initially freezing of samples remove solvents via sublimation. This method maintains pore size and aerogel particle distribution. To get aerogels, the ice-template method is mostly used where crystal growth formation results in nanostructure assembly. Chazeau et al. (2000) investigated the plastic behaviour of poly(vinyl chloride) (PVC) reinforced by the CNW and the nanocomposites were processed using freeze drying [33]. The cellulose whisker suspension was mixed with the aqueous suspension of the PVC and then freeze dried. The produced freeze-dried powder, di-ethylhexyl phthalate (DOP) and stearic acid based lubricant were mixed together and the mixture was hot-pressed by compression moulding to make the nanocomposite films or sheets.

10.4.5 3D Printing

3D printing method is one of the efficient method to prepare nanocomposites by incorporating nanomaterials into the polymer matrices. The nanomaterial can be added to the host matrix initially and then the 3D printing of the mixture can be done for nanocomposite fabrication [34]. To produce composite materials for biomedical applications like regenerative medicine and tissue engineering, 3D printing method can be employed. It involves layer by layer printing of different biopolymers like polysaccharides and proteins as bio-inks. Cellulose nanomaterials can be either used as a substrate or ink constituent for 3D printing [35, 36]. 3D printed cellulose nanocomposites can be used as adhesive bandage to aid wound healing process. Electrode printed with silver ink has been inserted onto the bandage to measure the temperature of the wound. The same technique was employed for printing of cartilage using alginate/hyaluronic acid. The 3D printing can also have applications in the making of nanocellulose derived protective clothing and textiles [37]. Sultan and Mathew (2019) researched about the synthesis of porous cubic scaffold with uniform pore structure using 3D printing of the CNC based hydrogel ink with sodium alginate and gelatin [38]. Thus, prepared 3D scaffolds possess similar porosity and compression modulus values, which are required for cartilage regeneration purposes. To control the pore size and compression modulus of scaffolds, the composition of hydrogel ink and 3D processing conditions can be changed. Scaffold synthesis as per patient demand can be done using this technique. Kajsa *et al.* showed the printing of various pieces using CNF/alginate dispersion [39]. The moulds obtained by printing were soaked in a $CaCl_2$ solution which act as cross-linker for sodium alginate. 3D structures with improved storage modulus, compressive strength and shear modulus were obtained by using the mixture of 80% CNF/20% alginate and thereafter cross-linking.

The CNW nanocomposites-polyvinyl alcohol filaments containing various amounts of the CNW for bone engineering application was fabricated where 10% CNW is used as functional additive to improve the 3D printing performance [40]. The 3D printing of the CNW nanocomposite hydrogel via stereo-lithography is used in tissue engineering applications [41].

10.4.6 Ball Milling Method

High Energy Ball Milling is another alternative method for obtaining bionanocomposites. This method is an efficient mixing method of solid state organic and inorganic compounds by mechanical grinding. A composite was prepared with apple peel pectin and montmorillonite clay by using ball milling method [42]. The ball milled powder was mixed with water and casted as a film. The characterization of the film by X-ray diffraction showed absence of peak which corresponds to basal spacing of the clay. The result shows proper dispersion of clay in pectin matrix takes place with exfoliation of clay sheets. The thermal and mechanical property improvements were achieved with this nanocomposites.

Metal-cellulose nanocomposites was prepared by ball mill method using cellulose and metal salts [43]. During grinding in ball mill, cellulose produce radicals which reduce metal ions to metal nanoparticles. These metal nanoparticles are deposited and stabilized in cellulose matrix to form bionanocomposites. This is a clean process as here toxic chemical reducing and stabilizing agents are not used and number of steps for preparing nanocomposites is also less. The property of thus prepared metal-cellulose nanocomposites depends on nature of metal nanoparticles present in composite materials. For example, Au-cellulose nanocomposites show good catalytic actinity whereas Ag-cellulose nanocomposites show unique antimicrobial activity. This ball milling method is also useful to make blend of synthetic and natural polymers and their composites. These type of metal-polymer composites have applications in diagnostics and other catalytic applications. Figure 10.10 shows schematic for the synthesis of cellulose-metal nanocomposites by ball milling method.

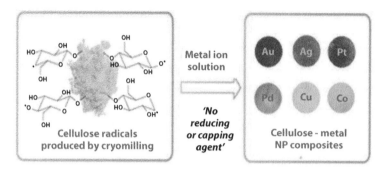

Figure 10.10 Cellulose-metal nanocomposite by ball milling method.

10.4.7 Microwave-Assisted Method for Bionanocomposite Preparation

The microwave-assisted synthesis is a simple and rapid technique with higher reproducibility. It is a green and effective method for the synthesis of various materials. Microwave irradiations have been used to synthesize biological nanocomposites. The microwaves refer to electromagnetic radiation with wavelength ranges from 0.01 to 1 m which is in between the infrared region and radio waves. This corresponds to a frequency of 30 to 0.3 GHz and are able to start nucleation or growth of materials. Normally, microwaves interact with polar molecules. This interaction leads to the rotational and vibrational motion of polar molecules which generates heat within the material. The main advantage of this technique is uniform heating within the material.

In a study, Karaya gum was modified by polyacrylic acid by microwave radiation and mixed with copper and nickel nanoparticles [44]. Nanoparticles were embedded in modified Karaya gum by microwave radiation for very small time, 20 S at 100 W, when bionanocomposites were formed. This nanocomposite showed efficient removal of organic dyes.

In another report, Simvastatin, a poorly soluble oral drug was incorporated in modified natural polymer gum ghatti to improve solubility and dissolution rate of the drug [45, 46]. Water-soluble natural polymers like guar gum, gum ghatti, etc. show high viscosity. Heating these natural polymers for a particular time and temperature in microwave reduces the viscosity and changes in surface property. It has been found that modified natural polymer gum ghatti has great potential for enhancement of solubility, dissolution rate. So it improves bioavailability of poorly soluble simvastatin. Microwave irradiation reduces the particle size and improves the wettability of the drug. The optimum ratio of drug to modified natural polymer was found to be 1:3 w/w.

Table 10.1 Difference between conventional heating and microwave-assisted heating.

Sr. no.	Conventional heating	Microwave-assisted heating
1	Thermal reaction begins from surface of the material	Thermal reaction initiates simultaneously from surface, as well as at bulk of the material.
2	Physical contact between materials' surface and vessel is required for heat transfer	Physical contact between materials' surface and vessel is not required for heat transfer.
3	Heat can be supplied by electric or thermal source	Heating takes place by microwaves.
4	Thermal conduction mechanism is involved in conventional heating	Dielectric polarization of material occurs in microwave-assisted heating.
5	Heating rate is low	Higher heating rate.

In another study, linseed oil based bionanocomposite was prepared with organo-modified montmorillonite (OMMT) clay by microwave assisted heating technique for protective coating [47]. The composites can serve as environment-friendly waterborne protective coatings at high temperature up to 290°C. Table 10.1 highlights the difference between conventional and microwave-assisted heating.

10.4.8 Ultraviolet Irradiation Method

Silver-montmorillonite-chitosan nanocomposites were synthesized by UV-irradiation method [48]. Precursor used was silver nitrate, montmorillonite was used as nanofiller and chitosan was natural binder in bionanocomposite formation. No further reducing agent or heat treatment was employed here. The properties of nanocomposites depend on the UV-irradiation time. The nanocomposite prepared showed effective anti-microbial activity against gram-positive and gram-negative bacteria.

Similarly, alginate-silver nanoparticle-chitosan based bio-nanocomposites was prepared using UV-irradiation method [49]. Thus, prepared green bio-nanocomposites exhibited very good antibacterial activity.

Other research group developed the methodology for the synthesis of silver/reduced graphene oxide (Ag/RGO) using lemon juice under microwave and UV irradiation [50]. The properties of prepared nanocomposites depend on several parameters like precursor concentration, irradiation time. Optimum time is 20 minutes for microwave irradiation and 60 seconds for UV irradiation. Thus, prepared green nanocomposites showed very good anti-bacterial property.

10.5 Application of Bionanocomposites

10.5.1 Orthopedics

Natural bone is a nanostructured composite composed of collagen along with other non-collagenous proteins, proteoglycans, lipids, peptides, and water. This matrix is reinforced with nanosized ceramic particles hydroxyapatite (HAp). In bone matrix, there are bone cells, namely osteoclasts and osteoblasts. The bone tissue faces many challenges during lifetime, such as osteoporosis, osteosarcoma, osteoarthritis, and traumas, i.e., fractures, microfractures [51]. Millions of people all over the world suffer from bone illness annually [52]. Due to ageing, loss of various bone structures occurs, which can be replaced by the bone substitutes. This leads to develop a new class of bio-nanocomposites for orthopaedic applications.

The development of organic–inorganic bio-composites improve the efficacy of medical science implants. Mainly, bioinspired nanocomposites optimize these implants and their interaction with the host tissue. The mechanical and physicochemical properties of nanocomposite materials depend on the materials, as well as the fabrication technique, which ultimately affect the performance and safety of these materials. These also improve their biodegradability as well as biocompatibility. In bone regeneration, better osteo-conductivity can be achieved if synthetic materials are fabricated to resemble bone in terms of its nanoscale features [53, 54]. Du *et al.* synthesized nano-HAp/collagen composites, which has a porous microstructure similar to bone. These materials enhanced the deposition of a new bone matrix with osteoblasts within this biologically inspired composite. Webster *et al.* showed in their research significant increase in initial protein adsorption and subsequent osteoblast adhesion on the nanosized ceramic materials like titania. Besides that, nanosized ceramic materials possess enhanced surface wettability due to greater surface roughness and larger numbers of grain boundaries on their surfaces. Their improved mechanical properties and biocompatibility provides wider orthopedic applications.

Surface roughness of nanomaterials influence adsorption of selected proteins and subsequent cell adhesion [55–57]. Liu et al. reported that nanotitania/PLGA composites, which had closest surface roughness to natural bone, demonstrated greatest osteoblast adhesion and subsequent calcium containing mineral deposition [58]. Wei et al. demonstrated greater initial protein adsorption important for osteoblast adhesion on nano-HAp/PLA porous scaffolds than on respective micro-HA/PLA scaffolds [59]. Some studies in this field showed mechanisms of improved osteoblast activity on nanomaterials. In vitro studies showed that grain size in the nanometer range as the major parameter for improving ceramic cyto-compatibility. As compared to larger grain size enhanced adhesion of osteoblasts were observed on nanophase alumina, titania, and HAp. In fact, decreasing alumina grain size from 167 to 24 nm increased osteoblast adhesion 51% within 4 hours [60]. Nanobiocomposites require mechanical properties like hardness, bending, compressive and tensile strengths that are significantly different from conventional materials but similar to those of natural bone. McManus et al. reported that the bending moduli of composites of PLA with 40 and 50 wt.% nanoalumina, titania and HAp were significantly greater than corresponding composite formulations with coarser grained ceramic materials. Specifically, compared to a bending modulus of 60 ± 3 MPa for plain PLA and 870 ± 30 MPa for conventional titania/PLA composites with the weight ratio of 50/50, the bending modulus of nanophase titania/PLA composites with the weight ratio of 50/50 was 1960 ± 250 MPa, which were on the same order of magnitude of healthy trabecular bone [61]. Nanophase ceramics possess increased surface roughness and decreased surface pore diameter as compared to conventional ceramics.

Collagen which is a main cell constituent can be obtained using standard methods with more than 90 % purity. Chondroitin sulphate is a promising material for mimicking natural bone composition. The other major constituents are calcium phosphate minerals like calcium hydroxyapatite and tricalcium phosphate, which are commonly used for biomedical applications. These calcium-based materials have the ability to enhance new bone formation. Protein adsorption increases osteoblast adhesion when their crystal size is in nanometric size range [62]. Bone apatite contains several trace elements, such as magnesium, copper, zinc, silicate, fluoride which are necessary for having good bone quality [63]. For this purpose, several collagen-calcium phosphate composite materials have been tailored which are used as temporary scaffolds in studies on animals and humans for tissue regeneration [64–66]. The main advantage of collagen-calcium phosphate nanocomposite is the excellent biocompatibility. The main advantage to use these composites is that they are degraded and reabsorbed easily in the body. They also do not have enough strength after hydration in comparison to bone [67]. That can be solved by chemical cross-linking. Poly(ε-caprolactone) (PCL) is used in bone tissue engineering [68]. Recent studies showed the potential of silicate nanocomposites due to their strong interaction with stem cells, which could further induce osteogenesis without the use of any additional growth factors [69]. Silicate materials are disc shaped and negatively charged material and are easily internalized by endocytosis. Due to internalization, these nanoparticles activate various osteo related genes and proteins, like osteocalcin and osteopontin and produce mineralized ECM. Hydroxyapatite-based nanocomposites were fabricated with other polymer scaffolds also for bone regeneration, and the main objective in all cases is to mimic the natural structures. These nanocomposites have been used for osteoconductive applications [70], as well as the addition of nHAp has increased the compressive strength of poly(ethylene glycol) (PEG) hydrogels. According to these, nanocomposites show high viscoelastic quality due to covalent crosslinking between PEG, poly(ε-caprolactone) (PCL) used in bone tissue engineering [68]. The scaffolds give the obligatory support for the cells to multiply and discriminate for the final shape of new bones [71].

10.5.2 Dental Applications

A range of nanomaterials such as electrospun nanofiber, nanotextured surfaces, self-assembled nanoparticles and nanocomposites are used to mimic mechanical, chemical, and biological properties of natural tissues [72, 73]. Silica/poly(ethylene glycol) (PEG) nanocomposite resulted in a significant increase in mechanical strength and bioactivity compared to PEG hydrogels [74, 75]. These bioactive nanocomposite hydrogels with high mechanical strength is useful for injectable matrix in orthopedic and dental applications. A number of nanoparticles were used to include bioactive properties and to enhance biological activities of nanocomposites [76, 77]. The various layers of dental tissue are as follows: enamel, dentin, and dental pulp. Enamel is one of the hardest tissues found in the body. It contains inorganic hydroxyapatite and a small amount of unique non-collagenous proteins, resulting in a composite structure [78]. Human teeth do not have the capacity to regenerate after eruption. If the tooth is damaged, the only treatment option is the use of biocompatible synthetic materials [79]. For dental tissue regeneration, stem cells from teeth and periodontium have been used as potential source. But these approaches do not provide protection against future dental diseases. Nanomaterials with their unique properties provide a wider range of dental restorations with greater abrasion resistance, high mechanical properties, improved esthetics, and better controlled cellular environment [80, 81]. For dental restoration, two major types of nanomaterials bioinert and bioactive nanomaterials are used. Common examples of bioactive nanomaterials include hydroxyapatite, tricalcium phosphate and bioglass nanomaterials, whereas bioinert nanomaterials include alumina, zirconia, titanium, vitreous carbon, etc.

Hydroxyapatite-based nanocomposites are used mostly for dental applications. As HAp has bone-bonding ability, it is mainly used as a coating material for different types of dental implants and grafts. Hydroxyapatite is significantly biocompatible and can osteo-integrate with bone tissue. But HAp is highly brittle and cannot be used for load-bearing applications [82]. By adding polymer to HAp produces bionanocomposites where mechanical toughness of this HAp can be developed [83]. This type of hybrid nanocomposites is used to design bioactive coatings on dental implants. A porous nanocomposites were designed by Brostow et al. with hydroxyapatite (150 μm) and polyurethane to obtain implants with high mechanical properties. Polyurethane has tunable mechanical hardness and thus hybrid nanocomposites with optimum mechanical strength with porous structure can be obtained.

The other important nanofiller for dental application is bioactive glass. It contains various oxides, like silicon dioxide, sodium oxide, calcium oxide, and phosphorous pentoxide in specific proportions. It has very good bone-bonding ability [84–87]. When the bioglass is subjected to an aqueous environment, it results in the formation of hydroxycarbonate apatite/hydroxyapatite layers on the surface [88, 89]. Like HAp, bioglass is also brittle and has a low wear resistance. So it is not suitable for load carrying applications. Several methods were developed to improve the mechanical properties of bioglasses. Ananth et al. incorporated yttria-stabilized zirconia in bioglass. This yttria stabilized zirconia bioglass is deposited on the titanium implant (Ti_6Al_4V) using electrophoretic deposition [90]. This composite materials show good mechanical strength and biocompatibility for dental restoration.

Zirconium dioxide (zirconia) has several advantages like it is biocompatible, possess low reactivity, high wear resistance, and good optical properties. Due to these properties, it can be used extensively in dental implants and restorations [91, 92]. The mechanical properties of zirconia can be improved by phase transformation toughening using stabilizers, such as yttria, magnesia, calcium, and ceria, to stabilize the tetragonal phase of zirconia [93–96]. Nanosized zirconia can be used as a reinforcing agent in various dental fillers. Zirconia nanoclusters can be incorporated in different polymer matrix to obtain dental fillers. When zirconia nanoparticles are added in polymers, mechanical strength is increased and tissue adhesion also improves. Lohbauer *et al.* showed that a zirconia-based nanoparticle system can be used as dental adhesive [97].

Silver nanoparticle is well known about its antimicrobial property and has been used for various biomedical applications. Recently, silver nanoparticles are being used in dental medicine, mainly for the treatment of oral cavities due to the antimicrobial property of silver nanoparticles [98]. The antimicrobial property of silver occurs by the release of silver ions which are highly reactive. This reactive silver interacts with the bacterial cell wall, penetrates inside the cell and ultimately causing cell death [99]. Several research groups evaluated antimicrobial activity of silver against *Streptococcus mutans* that are responsible for lesions and tooth decay [100]. These silver nanoparticles can be utilized in hard and soft tissue lining. Near the soft linings, *Candida albicans* is the commonly found fungal colony. Chladek at al. modified soft silicone linings of dentures with silver nanoparticles [101] which showed very good antifungal properties and can be used for dental infections. In other study, Torres *et al.* also evaluated antifungal efficiency of silver nanoparticles by incorporating it within denture resins [102]. They prepared denture resins by incorporating silver nanoparticles within polymethyl methacrylate (PMMA). Mitsunori *et al.* showed that the addition of silver nanoparticles to porcelain significantly enhances mechanical toughness [103]. The mechanical properties of ceramic porcelain were increased with uniform distribution of silver nanoparticles. The addition of silver nanoparticles resists crack propagation on the implants and improves fracture toughness. Smaller crystallites in porcelain ceramic are formed with silver nanoparticles which resist crack propagation. The mechanical properties as well as antimicrobial property of dental implants can be improved with silver nanoparticles.

Synthetic silicates are an important category of nanofillers which interact physically with both synthetic and natural polymers and can be used as injectable matrices for cellular therapies [76, 104]. These silicate-based polymer nanocomposites can be used for repairing and regeneration of mineralized tissue, including bone and dental tissue. The addition of silicates in long chain poly(ethylene oxide) (PEO) enhanced the mechanical strength and stability of the nanocomposite network formed. When these nanocomposites are dried in a sequential manner, a hierarchical structure is formed [76]. These hierarchical structures contain highly organized layered structure composed of silicates and polymers which show controlled cell adhesion and spreading characteristics.

10.5.3 Tissue Engineering

Tissue engineering is the use of a combination of cells on properly designed scaffold that focus on the biochemical and physicochemical properties, which can replace many natural biological tissues [105]. Tissue engineering is based on three-dimensional (3D) biocompatible scaffolds and biologically active, appropriate bioreactor condition which is resemble to extracellular matrix (ECM) of the tissue of various organs [106]. The ECM nanocomposite provides the mechanical base for implanted cells and interacts to promote and regulate cellular adhesion, migration, proliferation, differentiation, and morphogenesis [107]. Both natural and synthetic polymer materials have been utilized as scaffolds for tissue engineering. Natural polymeric materials that are usually used are as follows: collagen [108], silk protein [109], agarose [110], alginate [111], and chitosan [112]. Although these materials show promise in tissue repair, some issues regarding mechanical properties are there. On the other hand, synthetic materials can be prepared with improved biocompatibility, controlled degradation and tunable mechanical properties [113, 114]. Important synthetic polymers are : polylactic acid (PLA), polyamide, PLA–glycolic acid copolymer (PLGA), polycaprolactone polyester, polyanhydride, polyglycolic acid (PGA), polysaccharides, polyurethane, polyacrylate, and proteins [115] which all have great biocompatibility and have been employed as degradable scaffolds for a variety of tissues and organs [100]. Besides that, nanomaterials, bioactive moieties, and functional groups can be readily incorporated into the polymeric system, giving rise to smart and responsive materials [116–119].

Incorporation of various nanomaterials in bionanocomposites additionally improves several properties, which can be utilized in various types of tissue engineering. Carbon-based nanomaterials play critical roles in the field of neural tissue engineering. Fullerenes, carbon nanotubes, and graphene (G) are main examples of carbon-based nanomaterials. Electrical stimulation is beneficial to the regeneration of neurons and graphene based material which has good electrical conductivity, flexibility and mechanical strength, is important for this purpose which also help in neuron cell differentiation and proliferation [120]. The combination of functional GO nanosheets and nanofibers modulated the physicochemical and biological properties of scaffold used in repairing neural tissue [121]. Other than graphene, carbon nanotube is another popular material in neural tissue engineering. Many biomaterials like Chitosan, alginate, PLGA, PLA, PEG, etc. have been used for nerve diseases related research. These materials showed promising results in nerve tissue engineering. Silver nanoparticle-based bio-nanocomposites also show excellent application in skin tissue regeneration.

10.6 Conclusion

Various bionanocomposites are available in nature like bone, teeth, wood, etc. with specific structural and functional properties. To mimic these structures, researchers tried to develop bio-nanocomposites with biocompatible natural or synthetic polymers and nanofillers. These developed nanocomposites must possess certain properties, like non-toxicity, biodegradability, biocompatibility, etc. The selection of nanofillers also dictates the structural, mechanical and biological properties of bio-inspired nanocomposites formed. Environmental friendly approach of synthesis methodologies of bio-inspired nanocomposites have been discussed in this chapter, which will produce green bio-nanocomposites. Mostly, in these methods, harmful chemicals are avoided. The applications of these bio-inspired composites are discussed here mainly in three major areas namely orthopedics, dental applications, and tissue engineering. The other applications of general bio-nanocomposites like food packaging, drug delivery, agricultural applications, etc. are out of the scope of this chapter. Although scientists achieved so far a certain level of success in the field of bio-inspired nanocomposites to replace or regenerate natural materials still in many cases the complexity of natural materials are undiscovered. Hence, there is a scope of further research to understand the structural arrangement and interactions between the components of natural materials to mimic them artificially.

References

1. Sajid, M., Nanomaterials: Types, properties, recent advances, and toxicity concerns. *Curr. Opin. Environ. Sci. Health*, 25, 100319, 2022.
2. Khan, I., Saeed, K., Khan, I., Nanoparticles: Properties, applications and toxicities. *Arabian J. Chem.*, 12, 7, 908–931, 2019.
3. Zhu, R., Yadama, V., Liu, H., Lin, R.J.T., Harper, D.P., Fabrication and characterization of Nylon 6/cellulose nanofibrils melt-spun nanocomposite filaments. *Compos. Part A*, 97, 111–119, 2017.
4. Okpala, C.C., Nanocomposites-An overview. *Int. J. Eng. Res. Dev.*, 8, 17–23, 2013.

5. Zhou, H., Fan, T., Zhang, D., Biotemplated materials for sustainable energy and environment: Current status and challenges. *ChemSusChem*, 4, 10, 1344–87, 2011.
6. Ji, B. and Gao, H., Mechanical properties of nanostructure of biological materials. *J. Mech. Phys. Solids*, 52, 1963–90, 2004.
7. Herrick, F.W., Casebier, R.L., Hamilton, J.K., Sandberg, K.R., Microfibrillated cellulose: Morphology and accessibility. *J. Appl. Polym. Sci.: Appl. Polym. Symp., (United States)*, 37, 1983.
8. Huang, Z., Chen, H., Yip, A., Ng, G., Guo, F., Chen, Z.K., Roco, M.C., Longitudinal patent analysis for nanoscale science and engineering: Country, institution and technology field. *J. Nanopart. Res.*, 5, 3-4, 333–63, 2003.
9. Chen, M., Chen, B., Evans, J.R.G., Novel thermoplastic starch-clay nanocomposite foams. *Nanotechnology*, 16, 10, 2334, 2005.
10. Oksman, K., Aitomäki, Y., Mathew, A.P., Siqueira, G., Zhou, Q., Butylina, S., Tanpichai, S., Zhou, X., Hooshmand, S., Review of the recent developments in cellulose nanocomposite processing. *Compos. Part A: Appl. Sci. Manuf.*, 83, 2–18, 2016.
11. Mauter, M.S. and Elimelech, M., Environmental applications of carbon-based nanomaterials. *Environ. Sci. Technol.*, 42, 16, 5843–59, 2008.
12. Lohse, S.E. and Murphy, C.J., Applications of colloidal inorganic nanoparticles: From medicine to energy. *J. Am. Chem. Soc.*, 134, 38, 15607–20, 2012.
13. Gaharwar, A.K., Dammu, S.A., Canter, J.M., Wu, C.-J., Schmidt, G., Highly extensible, tough, and elastomeric nanocomposite hydrogels from poly(ethylene glycol) and hydroxyapatite nanoparticles. *Biomacromolecules*, 12, 5, 1641–50, 2011.
14. Tang, S.H., Lee, H.H., Wang, P., Kim, H.J., Collagen/hydroxyapatite composite nanofibers by electrospinning. *Mater. Lett.*, 62, 17-18, 3055–3058, 2008.
15. Pandey, J.K., Nakagaito, A.N., Takagi, H., Fabrication and applications of cellulose nanoparticle-based polymer composites. *Polym. Eng. Sci.*, 53, 1–8, 2013.
16. Wang, N., Ding, E.Y., Cheng, R.S., Surface modification of cellulose nanocrystals. *Front. Chem. Eng. China*, 1, 228–232, 2007.
17. Sharma, A., Thakur, M., Bhattacharya, M., Mandal, T., Goswami, S., Commercial application of cellulose nano-composites—A review. *Biotechnol. Rep.*, 21, e00316, 2019.
18. Qiu, K.Y. and Netravali, A.N., A review of fabrication and applications of bacterial cellulose based nanocomposites. *Polym. Rev.*, 54, 598–626, 2014.
19. Kalia, S., Dufresne, A., Cherian, B.M., Kaith, B.S., Avérous, L., Njuguna, J., Nassiopoulos, E., Cellulose-based bio- and nanocomposites: A review. *Int. J. Polym. Sci.*, 2011, 1–35, 2011.
20. Anglès, M.N. and Dufresne, A., Plasticized starch/tunicin whiskers nanocomposite materials. 2. mechanical behaviour. *Macromolecules*, 34, 2921–2931, 2001.

21. Liu, D.G., Zhong, T.H., Chang, P.R., Li, K.F., Wu, Q.L., Starch composites reinforced by bamboo cellulosic crystals. *Bioresour. Technol.*, 10, 2529–2536, 2010.
22. Noishiki, Y., Nishiyama, Y., Wada, M., Kuga, S., Magoshi, J., Mechanical properties of silk fibroin-microcrystalline cellulose composite films. *J. Appl. Polym. Sci.*, 86, 3425–3429, 2002.
23. Azizi Samir, M.A.S., Alloin, F., Sanchez, J.Y., Dufresne, A., Cellulose nanocrystals reinforced poly(oxyethylene). *Polymer*, 45, 4149–4157, 2004.
24. Azizi Samir, M.A.S., Alloin, F., Dufresne, A., High performance nanocomposite polymer electrolytes. *Compos. Interfaces*, 13, 545–559, 2006.
25. Zimmermann, T., Pöhler, E., Schwaller, P., Mechanical and morphological properties of cellulose fibril reinforced nanocomposites. *Adv. Eng. Mater.*, 7, 1156–1161, 2005.
26. Wang, Y.X., Cao, X.D., Zhang, L.N., Effects of cellulose whiskers on properties of soy protein thermoplastics. *Macromol. Biosci.*, 6, 524–531, 2006.
27. Paralikar, S.A., Simonsen, J., Lombardi, J., Poly(vinyl alcohol)/cellulose nanocrystal barrier membranes. *J. Membr. Sci.*, 320, 248–258, 2008.
28. Chai, H.B., Chang, Y., Zhang, Y.C., Chen, Z.Z., Zhong, Y., Zhang, L.P., Sui, X.F., Xu, H., Mao, Z.P., The fabrication of polylactide/cellulose nanocomposites with enhanced crystallization and mechanical properties. *Int. J. Biol. Macromol.*, 155, 1578–1588, 2020.
29. Qian, S.P., Zhang, H.H., Yao, W.C., Sheng, K.C., Effects of bamboo cellulose nanowhisker content on the morphology, crystallization, mechanical, and thermal properties of PLA matrix biocomposites. *Compos. Part B: Eng.*, 133, 203–209, 2018.
30. Kian, L.K., Jawaid, M., Ariffin, H., Karim, Z., Isolation and characterization of nanocrystalline cellulose from roselle-derived microcrystalline cellulose. *Int. J. Biol. Macromol.*, 114, 54–63, 2018.
31. Oksman, K., Aitomäki, Y., Mathew, A.P., Siqueira, G., Zhou, Q., Butylina, S., Tanpichai, S., Zhou, X.J., Hooshmand, S., Review of the recent developments in cellulose nanocomposite processing. *Compos. Part A*, 83, 2–18, 2016.
32. Allaf, R.M., *Functional 3D tissue engineering scaffolds; Materials, technologies and applications*, pp. 75–100, Woodhead Publishing, USA, 2018.
33. Chen, X.Y., Low, H.R., Loi, X.Y., Merel, L., Mohd Cairul Iqbal, M.A., Fabrication and evaluation of bacterial nanocellulose/poly(acrylic acid)/graphene oxide composite hydrogel: Characterizations and biocompatibility studies for wound dressing. *J. Biomed. Mater. Res. Part B*, 107, 2140–2151, 2019.
34. Campbell, T.A. and Ivanova, O.S., 3D printing of multifunctional nanocomposites. *Nano Today*, 8, 119–120, 2013.
35. Li, Y.Y., Zhu, H.L., Wang, Y.B., Ray, U., Zhu, S.Z., Dai, J.Q., Chen, C.J., Fu, K., Jang, S.H., Henderson, D., Li, T., Hu, L.B., Cellulose-nanofiber-enabled 3D printing of a carbon-nanotube microfiber network. *Small Methods*, 1, 1700222, 2017.

36. Joseph, B., James, J., Grohens, Y., Kalarikkal, N., Thomas, S., Material aspects during additive manufacturing of nano-cellulose composites, in: *Structure and Properties of Additive Manufactured Polymer Components*, pp. 409–428, Elsevier, Amsterdam, 2020.
37. Tenhunen, T.M., Moslemian, O., Kammiovirta, K., Harlin, A., Kääriäinen, P., Österberg, M., Tammelin, T., Orelma, H., Surface tailoring and design-driven prototyping of fabrics with 3D-printing: An all cellulose approach. *Mater. Des.*, 140, 409–419, 2018.
38. Sultan, S. and Mathew, A.P., 3D printed porous cellulose nanocomposite hydrogel scaffolds. *J. Vis. Exp.*, 2019.
39. Kajsa, M., Athanasios, M., Ivan, T., Héctor, M.Á., Daniel, H., Paul, G., 3D Bioprinting human chondrocytes with nanocellulose-alginate bioink for cartilage tissue engineering applications. *Biomacromolecules*, 16, 1489–1496, 2015.
40. Wang, Q.Q., Sun, J.Z., Yao, Q., Ji, C.C., Liu, J., Zhu, Q.Q., 3D printing with cellulose materials. *Cellulose*, 25, 4275–4301, 2018.
41. Palaganas, N.B., Mangadlao, J.D., de Leon, A.C.C., Palaganas, J.O., Pangilinan, K.D., Lee, Y.J., Advincula, R.C., 3D printing of photocurable cellulose nanocrystal composite for fabrication of complex architectures via stereolithography. *ACS Appl. Mater. Interfaces*, 9, 34314–34324, 2017.
42. Mangiacapra, P., Gorrasi, G., Sorrentino, A., Vittoria, V., Biodegradable nanocomposites obtained by ball milling of pectin and montmorillonites. *Carbohydr. Polym.*, 64, 4, 516–523, 2006.
43. Kwiczak-Yiğitbaşı, J., Laçin, Ö., Demir, M., Ahan, R.E., Şeker, U Ö Ş, Baytekin, B., A sustainable preparation of catalytically active and antibacterial cellulose metal nanocomposites *via* ball milling of cellulose. *Green Chem.*, 22, 2, 455–464, 2020.
44. Gupta, D., Jamwal1, D., Rana, D., Katoch, A., Microwave synthesized nanocomposites for enhancing oral bioavailability of drugs, in: *Applications of Nanocomposite Materials in Drug Delivery*, Inamuddin, A.M. Asiri, A. Mohammad (Eds.), Woodhead Publishing series in Biomaterials, USA, 2018.
45. Saruchi, Kumar, V., Ghfar, A.A., Pandey, S., Microwave synthesize karaya gum-cu, Ni nanoparticles based bionanocomposite as an adsorbent for malachite green dye: Kinetics and thermodynamics. *Front. Mater., Sec. Polymeric and Composite Materials*, 9, 2022.
46. Diliprao, D., Husain, M., Mohini, B., Nikita, D., Vaibhav, I., Microwave generated bionanocomposites for solubility and dissolution rate enhancement of poorly water soluble drug simvastatin. *Indo Am. J. Pharm. Res.*, 7(3), 8020–8031, 2017.
47. Zafar, F., Sharmin, E., Zafar, H., Yaseen Shah, M., Nishat, N., Ahmad, S., Facile microwave-assisted preparation of waterborne polyesteramide/OMMT clay bio-nanocomposites for protective coatings. *Ind. Crops Prod.*, 67, 484–491, 2015.

48. Shameli, K., Ahmad, M.B., Md Zin Wan Yunus, W., Rustaiyan, A., Ibrahim, N.A., Zargarand, M., Green synthesis of silver/montmorillonite/chitosan bionanocomposites using the UV irradiation method and evaluation of antibacterial activity. *Int. J. Nanomed.*, 5, 875–887, 2010.
49. Bousalem, N., Benmansour, K., Ziani Cherif, H., Synthesis and characterization of antibacterial silver-alginate - chitosan bionanocomposite films using UV irradiation method. *Adv. Perform. Mater.*, 32, 367–377, 2017.
50. Alsharaeh, E., Alazzam, S., Ahmed, F., Arshi, N., Al-Hindawi, M., Sing, G.K., Green synthesis of silver nanoparticles and their reduced graphene oxide nanocomposites as antibacterial agents: A bio-inspired approach. *Acta Metall. Sin. (Engl. Lett.)*, 30, 45–52, 2017.
51. Mishra, S., Sharma, S., Javed, Md. N., Pottoo, F.H., Barkat, Md. A., Harshita, Alam, Md. S., Amir, Md., Sarafroz, Md., Bioinspired nanocomposites: Applications in disease diagnosis and treatment. *Pharm. Nanotechnol.*, 7, 1–15, 2019.
52. Rodrigues, C.V.M., Serricella, P., Linhares, A.B.R., Guerdes, R.M., Borojevic, R., Rossi, M.A., Duarte, M.E.L., Farina, M., Characterization of a bovine collagen-hydroxyapatite composite scaffold for bone tissue engineering. *Biomaterials*, 24, 27, 4987–97, 2003.
53. Du, C., Cui, F.Z., Zhu, X.D., de Groot, K., Three-dimensional nanoHAP/collagen matrix loading with osteogenic cells in organ culture. *J. Biomed. Mater. Res.*, 44, 4, 407–415, 1999.
54. Webster, T.J., Siegel, R.W., Bizios, R., Nanoceramic surface roughness enhances osteoblast and osteoclast functions for improved orthopaedic/dental implant efficacy. *Scr. Mater.*, 44, 8-9, 1639–1642, 2001.
55. Webster, T.J., Ergun, C., Doremus, R.H., Siegel, R.W., Bizios, R., Specific proteins mediate enhanced osteoblast adhesion on nanophase ceramics. *J. Biomed. Mater. Res.*, 51, 3, 475–483, 2000.
56. Yamasaki, H. and Sakai, H., Osteogenic response to porous hydroxyapatite ceramics under the skin of dogs. *Biomaterials*, 13, 5, 308–312, 1999.
57. Yuan, H., Kurashina, K., de Bruijin, J.D., Li, Y., de Grout, K., Zhang, X., A preliminary study on osteoinduction of two kinds of calcium phosphate ceramics. *Biomaterials*, 20, 19, 1799–1806, 1999.
58. Liu, H., Slamovich, E.B., Webster, T.J., Increased osteoblast functions on nanophase titania dispersed in poly-lactic-co-glycolic acid composites. *Nanotechnology*, 16, 7, S601–S608, 2005.
59. Wei, G. and Ma, P.X., Structure and properties of nano-hydroxyapatite/polymer composite scaffolds for bone tissue engineering. *Biomaterials*, 25, 19, 4749–4757, 2004.
60. Webster, T.J., Siegel, R.W., Bizios, R., Osteoblast adhesion on nanophase ceramics. *Biomaterials*, 20, 13, 1221–1227, 1999.
61. McManus, A.J., Doremus, R.H., Siegel, R.W., Bizios, R., Evaluation of cytocompatibility and bending modulus of nanoceramic/polymer composites. *J. Biomed. Mater. Res.*, 72A, 1, 98–106, 2005.

62. Nelson, M., Balasundaram, G., Webster, T.J., Increased osteoblast adhesion on nanoparticulate crystalline hydroxyapatite functionalized with KRSR. *Int. J. Nanomed.*, 1, 3, 339–49, 2006.
63. Camargo, P.H.C., Satyanarayana, K.G., Wypych, F., Nanocomposites: Synthesis, structure, properties and new application opportunities. *Mater. Res.*, 12, 1, 1–39, 2009.
64. Yi, H., Rehman, F.U., Zhao, C., Liu, B., He, N., Recent advances in nano scaffolds for bone repair. *Bone Res.*, 4, 16050, 2016.
65. Henkel, J., Woodruff, M.A., Epari, D.R., Steck, R., Glatt, V., Dickinson, I.C., Choong, P.F.M., Schuetz, M.A., Hutmacher, D.W., Bone regeneration based on tissue engineering conceptions — A 21st century perspective. *Bone Res.*, 1, 3, 216–48, 2013 Sep 25.
66. Gao, C., Deng, Y., Feng, P., Mao, Z., Li, P., Yang, B., Deng, J., Cao, Y., Shuai, C., Peng, S., Current progress in bioactive ceramic scaffolds for bone repair and regeneration. *Int. J. Mol. Sci.*, 15, 3, 4714–32, 2014 Mar 18.
67. Matsuno, T., Uchimura, E., Ohno, T., Satoh, T., Sogo, Y., Ito, A., Yamazaki, A., Ishikawa, Y., Kondo, N., Ichinose, N., Hydroxyapatite containing immobilized collagen and fibronectin promotes bone regeneration. *Int. Congr. Ser.*, 1284, 330–1, 2005 Sep 1.
68. Kumar, P., Sandeep, K.P., Alavi, S., Truong, V.D., Gorga, R.E., Preparation and characterization of bionanocomposite films based on soy protein isolate and montmorillonite using melt extrusion. *J. Food Eng.*, 100, 3, 480–9, 2010 Oct 1.
69. Mihaila, S.M., Gaharwar, A.K., Reis, R.L., Khademhosseini, A., Marques, A.P., Gomes, M.E., The osteogenic differentiation of SSEA-4 sub-population of human adipose derived stem cells using silicate nanoplatelets. *Biomaterials*, 35, 33, 9087–99, 2014 Nov.
70. Gaharwar, A.K., Dammu, S.A., Canter, J.M., Wu, C.-J., Schmidt, G., Highly extensible, tough, and elastomeric nanocomposite hydrogels from poly(ethylene glycol) and hydroxyapatite nanoparticles. *Biomacromolecules*, 12, 5, 1641–50, 2011 May 9.
71. Fisher, J.P., Vehof, J.W.M., Dean, D., van der Waerden, J.P.C.M., Holland, T.A., Mikos, A.G., Jansen, J.A., Soft and hard tissue response to photocrosslinked poly(propylene fumarate) scaffolds in a rabbit model. *J. Biomed. Mater. Res.*, 59, 3, 547–56, 2002 Mar 5.
72. Xavier, J.R., Desai, P., Varanasi, V.G., Al-Hashimi, I., Gaharwar, A.K., Advanced nanomaterials: Promises for improved dental tissue regeneration, in: *Nanotechnology in Endodontics: Current and Potential Clinical Applications*, vol. 5, Springer International Publishing, Switzerland, 2015,
73. Thomas, J., Peppas, N., Sato, M., Webster, T., *Nanotechnology and biomaterials*, CRC Taylor and Francis, Boca Raton, 2006.
74. Balazs, A.C., Emrick, T., Russell, T.P., Nanoparticle polymer composites: Where two small worlds meet. *Science*, 314, 5802, 1107–10, 2006.

75. Gaharwar, A.K., Rivera, C., Wu, C.-J., Chan, B.K., Schmidt, G., Photocrosslinked nanocomposite hydrogels from PEG and silica nanospheres: Structural, mechanical and cell adhesion characteristics. *Mater. Sci. Eng. C*, 33, 3, 1800–7, 2013.
76. Gaharwar, A.K., Kishore, V., Rivera, C., Bullock, W., Wu, C.J., Akkus, O. et al., Physically crosslinked nanocomposites from silicate-crosslinked PEO: Mechanical properties and osteogenic differentiation of human mesenchymal stem cells. *Macromol. Biosci.*, 12, 6, 779, 2012.
77. Gaharwar, A.K., Rivera, C.P., Wu, C.-J., Schmidt, G., Transparent, elastomeric and tough hydrogels from poly (ethylene glycol) and silicate nanoparticles. *Acta Biomater.*, 7, 12, 4139–48, 2011.
78. Yen, A.H. and Yelick, P.C., Dental tissue regeneration – A mini-review. *Gerontology*, 57, 1, 85–94, 2011.
79. Ratner, B.D., Replacing and renewing: Synthetic materials, biomimetics, and tissue engineering in implant dentistry. *J. Dent. Educ.*, 65, 12, 1340–7, 2001.
80. Piva, E., Silva, A.F., Nor, J.E., Functionalized scaffolds to control dental pulp stem cell fate. *J. Endod.*, 40, 4 Suppl, S33–40, 2014, PubMed PMID: 24698691.
81. Mendonça, G., Mendonca, D., Aragao, F.J.L., Cooper, L.F., Advancing dental implant surface technology– from micron- to nanotopography. *Biomaterials*, 29, 28, 3822–35, 2008.
82. Cao, W. and Hench, L.L., Bioactive materials. *Ceram. Int.*, 22, 6, 493–507, 1996.
83. Brostow, W., Estevez, M., Lobland, H.E.H., Hoang, L., Rodriguez, J.R., Vargar, S., Porous hydroxyapatitebased obturation materials for dentistry. *J. Mater. Res.*, 23, 06, 1587–96, 2008.
84. Kudo, K., Miyasawa, M., Fujioka, Y., Kamegai, T., Nakano, H., Seino, Y. et al., Clinical application of dental implant with root of coated bioglass: Short-term results. *Oral. Surg. Oral. Med. Oral. Pathol.*, 70, 1, 18–23, 1990, J.R. Xavier et al. 21.
85. Hench, L.L. and Paschall, H.A., Direct chemical bond of bioactive glass-ceramic materials to bone and muscle. *J. Biomed. Mater. Res.*, 7, 3, 25–42, 1973.
86. Stanley, H.R., Hench, L., Going, R., Bennett, C., Chellemi, S.J., King, C. et al., The implantation of natural tooth form bioglasses in baboons: A preliminary report. *Oral. Surg. Oral. Med. Oral. Pathol.*, 42, 3, 339–56, 1976.
87. Piotrowski, G., Hench, L.L., Allen, W.C., Miller, G.J., Mechanical studies of the bone bioglass interfacial bond. *J. Biomed. Mater. Res.*, 9, 4, 47–61, 1975.
88. Oonishi, H., Hench, L.L., Wilson, J., Sugihara, F., Tsuji, E., Kushitani, S. et al., Comparative bone growth behavior in granules of bioceramic materials of various sizes. *J. Biomed. Mater. Res.*, 44, 1, 31–43, 1999.
89. Nganga, S., Zhang, D., Moritz, N., Vallittu, P.K., Hupa, L., Multi-layer porous fiber-reinforced composites for implants: *In vitro* calcium phosphate formation in the presence of bioactive glass. *Dent. Mater.*, 28, 11, 1134–45, 2012.
90. Ananth, K.P., Suganya, S., Mangalaraj, D., Ferreira, J., Balamurugan, A., Electrophoretic bilayer deposition of zirconia and reinforced bioglass system

on Ti6Al4V for implant applications: An in vitro investigation. *Mater. Sci. Eng. C*, 33, 7, 4160–6, 2013.
91. Deville, S., Gremillard, L., Chevalier, J., Fantozzi, G., A critical comparison of methods for the determination of the aging sensitivity in biomedical grade yttria stabilized zirconia. *J. Biomed. Mater. Res. B Appl. Biomater.*, 72, 2, 239–45, 2005.
92. Chevalier, J., What future for zirconia as a biomaterial? *Biomaterials*, 27, 4, 535–43, 2006, 2 Advanced Nanomaterials: Promises for Improved Dental Tissue Regeneration 22.
93. Piconi, C. and Maccauro, G., Zirconia as a ceramic biomaterial. *Biomaterials*, 20, 1, 1–25, 1999.
94. Chevalier, J., Deville, S., Münch, E., Jullian, R., Lair, F., Critical effect of cubic phase on aging in 3 mol% yttria-stabilized zirconia ceramics for hip replacement prosthesis. *Biomaterials*, 25, 24, 5539–45, 2004.
95. Piconi, C., Burger, W., Richter, H.G., Cittadini, A., Maccauro, G., Covacci, V. et al., Y-TZP ceramics for artificial joint replacements. *Biomaterials*, 19, 16, 1489–94, 1998.
96. Bao, L., Liu, J., Shi, F., Jiang, Y., Liu, G., Preparation and characterization of TiO2 and Si-doped octacalcium phosphate composite coatings on zirconia ceramics (Y-TZP) for dental implant applications. *Appl. Surf. Sci.*, 290, 48–52, 2014.
97. Lohbauer, U., Wagner, A., Belli, R., Stoetzel, C., Hilpert, A., Kurland, H.-D. et al., Zirconia nanoparticles prepared by laser vaporization as fillers for dental adhesives. *Acta Biomater.*, 6, 12, 4539–46, 2010.
98. García-Contreras, R., Argueta-Figueroa, L., MejíaRubalcava, C., Jiménez-Martínez, R., Cuevas-Guajardo, S., Sánchez-Reyna, P. et al., Perspectives for the use of silver nanoparticles in dental practice. *Int. Dent. J.*, 61, 6, 297–301, 2011.
99. Chladek, G., Barszczewska-Rybarek, I., Lukaszczyk, J., Developing the procedure of modifying the denture soft liner by silver nanoparticles. *Acta Bioeng. Biomech.*, 14, 1, 23–9, 2012.
100. Magalhães, A.P.R., Santos, L.B., Lopes, L.G., Estrela, C.R.A., Estrela, C., Torres, É.M. et al., Nanosilver application in dental cements. *ISRN Nanotechnol.*, 2012, 365438, 6, 2012.
101. Chladek, G., Mertas, A., Barszczewska-Rybarek, I., Nalewajek, T., Żmudzki, J., Król, W. et al., Antifungal activity of denture soft lining material modified by silver nanoparticles—A pilot study. *Int. J. Mol. Sci.*, 12, 7, 4735–44, 2011.
102. Acosta-Torres, L.S., Mendieta, I., Nuñez-Anita, R.E., Cajero-Juárez, M., Castano, V.M., Cytocompatible antifungal acrylic resin containing silver nanoparticles for dentures. *Int. J. Nanomed.*, 7, 4777, 2012.
103. Uno, M., Kurachi, M., Wakamatsu, N., Doi, Y., Effects of adding silver nanoparticles on the toughening of dental porcelain. *J. Prosthet. Dent.*, 109, 4, 241–7, 2013.

104. Gaharwar, A.K., Mukundan, S., Karaca, E., DolatshahiPirouz, A., Patel, A., Rangarajan, K. et al., Nanoclayenriched poly (ε-caprolactone) electrospun Scaffolds for osteogenic differentiation of human mesenchymal stem cells. *Tissue Eng. Part A*, 20, 15–16, 2088–2101, 2014.
105. Khademhosseini, A., Vacanti, J.P., Langer, R., Progress in tissue engineering. *Sci. Am.*, 300, 5, 64–71, 2009 May.
106. Nakanishi, J., Takarada, T., Yamaguchi, K., Maeda, M., Recent advances in cell micropatterning techniques for bioanalytical and biomedical sciences. *Anal. Sci.*, 24, 1, 67–72, 2008 Jan.
107. Goldberg, M., Langer, R., Jia, X., Nanostructured materials for applications in drug delivery and tissue engineering. *J. Biomater. Sci. Polym. Ed.*, 18, 3, 241–68, 2007.
108. Freyman, T.M., Yannas, I.V., Yokoo, R., Gibson, L.J., Fibroblast contraction of a collagen-GAG matrix. *Biomaterials*, 22, 21, 2883–91, 2001 Nov.
109. Wang, Y., Kim, U.-J., Blasioli, D.J., Kim, H.-J., Kaplan, D.L., In vitro cartilage tissue engineering with 3D porous aqueous-derived silk scaffolds and mesenchymal stem cells. *Biomaterials*, 26, 34, 7082–94, 2005 Dec.
110. Mauck, R.L., Yuan, X., Tuan, R.S., Chondrogenic differentiation and functional maturation of bovine mesenchymal stem cells in long-term agarose culture. *Osteoarthr. Cartil.*, 14, 2, 179–89, 2006 Feb.
111. Marijnissen, W.J.C.M., van Osch, G.J.V.M., Aigner, J., van der Veen, S.W., Hollander, A.P., Verwoerd-Verhoef, H.L., Verhaar, J.A.N., Alginate as a chondrocyte-delivery substance in combination with a non-woven scaffold for cartilage tissue engineering. *Biomaterials*, 23, 6, 1511–7, 2002 Mar.
112. Ciardelli, G. and Chiono, V., Materials for peripheral nerve regeneration. *Macromol. Biosci.*, 6, 1, 13–26, 2006 Jan 5.
113. Wang, Y., Ameer, G.A., Sheppard, B.J., Langer, R., A tough biodegradable elastomer. *Nat. Biotechnol.*, 20, 6, 602–6, 2002 Jun.
114. Lendlein, A. and Langer, R., Biodegradable, elastic shapememory polymers for potential biomedical applications. *Science*, 296, 5573, 1673–6, 2002 May 31.
115. Nicolas, J., Mura, S., Brambilla, D., Mackiewicz, N., Couvreur, P., Design, functionalization strategies and biomedical applications of targeted biodegradable/biocompatible polymer-based nanocarriers for drug delivery. *Chem. Soc. Rev.*, 42, 1147–1235, 2013.
116. Lutolf, M.P., Raeber, G.P., Zisch, A.H., Tirelli, N., Hubbell, J.A., Cell-responsive synthetic hydrogels. *Adv. Mater.*, 15, 11, 888–92, 2003 Jun 5.
117. Barrera, D.A., Zylstra, E., Lansbury, P.T., Langer, R., Synthesis and RGD peptide modification of a new biodegradable copolymer: Poly(lactic acid-co-lysine). *J. Am. Chem. Soc.*, 115, 23, 11010–1, 1993 Nov 1.
118. Cook, A.D., Hrkach, J.S., Gao, N.N., Johnson, I.M., Pajvani, U.B., Cannizzaro, S.M., Langer, R., Characterization and development of RGD-peptide-modified poly(lactic acid-co-lysine) as an interactive, resorbable biomaterial. *J. Biomed. Mater. Res.*, 35, 4, 513–23, 1997.

119. Parrish, B. and Emrick, T., Aliphatic polyesters with pendant cyclopentene groups: Controlled synthesis and conversion to polyester-graft-PEG copolymers. *Macromolecules*, 37, 16, 5863–5, 2004 Aug 1.
120. Fabbro, A., Bosi, S., Ballerini, L., Prato, M., Carbon nanotubes: Artificial nanomaterials to engineer single neurons and neuronal networks, *ACS Chem. Neurosci.*, 3, 611–618, 2012.
121. Girão, A.F., Sousa, J., Domínguez-Bajo, A., González-Mayorga, A.,Bdikin, I., Pujades-Otero, E., Casañ-Pastor, N., Hortigüela, M.J., Otero-Irurueta, G., Completo, A., Serrano, M.C., Marques, P.A.A.P., 3D reduced graphene oxide scaffolds with a combinatorial fibrous-porous architecture for neural tissue engineering, *ACS Appl. Mater. Interfaces*, 12, 38962–38975, 2020.

11

Bioinspiration as Tools for the Design of Innovative Materials and Systems Bioinspired Piezoelectric Materials: Design, Synthesis, and Biomedical Applications

Santu Bera

University Bordeaux, CNRS, CBMN, UMR 5248, Institut Européen de Chimie et Biologie, Pessac, France

---

Abstract

Bioinspiration represents the phenomenon of utilizing principles from nature to stimulate research in various disciplines, including chemistry, biology, physics or engineering, and develop modern technology. For many decades, research in science and technology has been greatly inspired by nature in terms of millions of well-coordinated and engineered processes, materials, and designs. Biomaterials are optimized tools that are skillfully designed by nature to flourish in the challenging environment. The inherent biocompatibility, high biodegradability and affinity to easily integrate with natural systems make biomaterials attractive systems for investigating for a large range of applications starting from drug delivery platforms, tissue engineering, medical implants, biosensors. Besides many other characteristics, piezoelectricity, the phenomenon to generate electrical signal in regards to mechanical deformation, has been identified recently in many biomaterials that build the human body or natural systems. Yet, the origin and significance of piezoelectricity in diverse biomaterials have not fully understood; however, these are assumed to play crucial function in the health conditions. In this chapter, we systematically discuss the recent development of biopiezoelectric materials based on natural or nature-inspired biomolecules—design strategy, synthesis, integration into biopiezoelectric platforms, and finally, their deployment in the latest biomedical applications has been emphasized.

Email: santubera49@gmail.com

Mousumi Sen and Monalisa Mukherjee (eds.) *Bioinspired and Green Synthesis of Nanostructures: A Sustainable Approach*, (263–290) © 2023 Scrivener Publishing LLC

Keywords: Bioinspiration, biomaterials, piezoelectricity, energy harvesting, biomedical application, biosensing, disease treatment

11.1 Bioinspiration and Sophisticated Materials Design

The large amount of diversity of structural biological systems exits in nature and their structure-function relationship often provide the idea to design and engineer systems for developing modern technology [1–5]. A most remarkable feature of these biomaterials is their simplicity in composition but high degree of efficiency in performance [6, 7]. The level of sophistication and miniaturization of biomaterials always amazed researchers. For that reason, nature and natural materials has been remained as endless source of inspiration to mankind for developing new ideas and design advanced materials. As day-by-day global challenges are raising the bar toward more complexity, scientists are more and more looking toward biological systems for bringing new solutions. Until now, biomaterials have been comprehensively explored for a variety of applications in biomedicine because of their ease of integration with the body. Among the biomaterials, cellulose, collagen, gelatin, and chitosan are placed on the top of the priority list for study owing to their coherence with the living systems. In-depth understanding the working principles of these biomaterials and establishing their structure-function relation pave the way for modular design of bio-inspired artificial materials with acute control over desired functionalities.

Besides other functions, piezoelectricity [8, 9], the potential of materials to transform mechanical forces into electricity, has been identified recently in many natural materials [10–13] that construct the human body or natural systems, like bone, hair, skin, virus, cellulose, protein, amino acid, DNA, etc. (Figure 11.1). Piezoelectricity is the property of crystals without inversion symmetry where piezoelectricity appears due to arrangement of ions under mechanical agitation [14, 15]. Upon application of stress, the internal dipole moments of the material alter linearly and results into the development of electric filed across the material boundary. The piezoelectric properties of biomaterials are found to play crucial role in the health conditions. As for example, the piezoelectric property of bone and collagen has found to exert significant role on the mechanism of constituent tissue functions like bone growth, healing, regeneration, and remodeling according to the so-called Wolff's law [16–19]. In line with this assumption, several *in vitro* and *in vivo* researches have revealed the importance

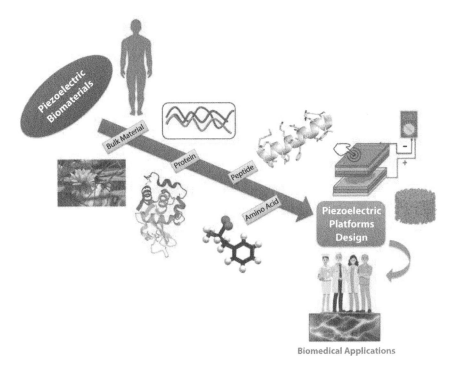

Figure 11.1 Schematic of nature and nature-inspired materials based biopiezoelectric platforms, and their biomedical applications.

of electrical stimulation therapy to enhancing and stimulating osteogenic activity [20, 21]. The discovery of piezoelectricity in natural materials and their role in tissue function represents a paradigm shift in bio-inspired material research, attracts the widespread attention into designing biocompatible, nontoxic, self-powered biopiezoelectric platforms [22–27]. A significant effort by the biomedical research community to create biopiezoelectric materials has unveiled that these materials can produce physiologically relevant electrical clue from the regular physical functions of body like body movement, muscle construct, heartbeat, blood circulation, etc. Besides providing the advantage of replacement of short-lived battery in the current state-of-the-art implantable medical devices, these materials are capable of regulating array of physiological behaviors of cells, for example proliferation, differentiation, and apoptosis, thus demonstrating remedial effects on diseases. At the early stage of construction such materials, the focus was to mimicking the fundamental constructs of our body like amino acids, peptides, and proteins by copying the exact sequences. Later, the attention has been diverted toward creating biopiezoelectric materials from *de novo* design sequences that might achieve efficiency beyond the natural counterparts. In this chapter, I will summarize the recent progress in the upliftment of biopiezoelectric materials based on natural or synthetic biomolecules—their systematic design strategy, synthesis, incorporation into piezoelectric devices, and finally their utilization in biomedicine will be discussed. At the end, current challenges of fabricating biopiezoelectric platforms suitable for biomedical applications and the promising pathway to minimize the risks will be highlighted.

11.1.1 Piezoelectricity in Natural Bulk Materials

The piezoelectric effect is identified in many natural components like bones, ligaments, tendons, skins, hairs, bacteria, wood, etc. [10–13, 28–32]. These natural materials generate physiologically relevant electric clue upon supplying mechanical constrain, and thus facilitated tissue functions like tissue growth, wound healing, and regeneration. Therefore, these are promising aspirants for application in biomedicine. The basis of piezoelectricity in natural components is their constituent proteins. Fukada and Yasuda first discovered the electromechanical coupling in bone and revealed bone to act as bulk piezoelectric entity having a piezoelectric tensor about one-tenth to that of quartz [10]. It has been identified that the source of piezoelectricity in bone originates from the constituent collagen which possesses piezoelectric property. The generation of mechanoelectrical signal through piezoelectric effect of collagen has been established to be crucial for bone regeneration [16, 17]. As piezoelectric response of a materials is directly related to degree of polarization, alignment of building units is vital to negate the charge cancellation and achieve high piezoelectricity. As a result of nice orientation of collagen in tendon, piezoelectricity in cow and horse's tendons is found to have significantly improved than that of bone [33]. A detailed experimental study on rat tail tendon by accounting the angle of measurement using PFM evidently showed the significance of oriented collagen for acquiring the high piezoelectricity [33]. The skin is another tissue in the living system that is known to show piezoelectricity. Shamos *et al.* in the year of 1967 examined the piezoelectric characteristics of skin collected from different living bodies, like human, cat, pig [34]. Later, the epidermis of the human skin was made and examined for the piezoelectricity. In all these cases, the origin of piezoelectricity was believed to be due to the presence of oriented collagen [35].

Many plant tissues also show the piezoelectric property [36]. As for example, the piezoelectric response of wood acquires from the major component of many plants—lignocellulosic molecules, which are piezoactive

[37, 38]. Similarly, *Sonchus asper* (SA), environment-friendly and stiffest natural polymer fiber composed of ascorbic acid, carotenoids, and fatty acids, exhibited piezoelectricity [39, 40]. Spider silk has been also reported to show structure-dependent piezoelectric response [41].

11.1.2 Piezoelectricity in Proteins

Proteins are the long chains of amino acids and main structural component of living systems. Collagen is the most abundant proteins in mammals by weight, building up for 25% to 35% of the total protein components [42]. The extracellular matrix (ECM) of various tissues, like bone, cartilage, skin is mainly composed of collagen. At the atomic level composition of collagen, three peptide helices entwined around a common axis to fabricate the triple helical conformation [43, 44]. The polypeptide chain of collagen consist of a characteristic tripeptide repeat: $(G-X-Y)_n$, where G stands for glycine (Gly), X generally symbolizing proline (Pro), and Y is hydroxyproline (Hyp). The requirement of Gly residue at every third position is essential. This triplet repeats organize for favorable non-covalent interactions, stabilizing the three polyproline peptide chains into a triple- helix structure. This helical conformation of collagen backbone was found to be crucial for its piezoelectricity [45, 46]. Zhou *et al.* used full atomistic simulation to understand the mechanism of piezoelectricity in experimental "super-twisted" model collagen [47]. Their investigation exhibited a organized dipole moment of collagen fiber toward the elongated axis. Moreover, their results clearly demonstrated that the degree of piezoelectricity in solvent encapsulated building blocks is ruled by the mechanical stress-induced reorganization of constituent building unit's dipoles. From the molecular dynamics simulation, a longitudinal, d_{33} coefficient was obtained in the range of 1 to 2 pm/V (pC/N). Piezoresponse force microscopy (PFM) was employed to quantitative analysis of piezoelectricity. Yu *et al.* measured the piezoelectricity of type I bovine collagen single fiber having diameter around 100 nm [48]. Single collagen fibrils were identified to comprise with oriented dipole organization toward the length of the fibrils and acted majorly as shear piezoelectric materials with a piezoelectric coefficient, d_{15} = pm V^{-1}. Resonance measurement of the piezoelectric coefficient, d_{14} of collagen films solubilized from bovine serosa gave a value around 0.096 pC N^{-1} [49]. Piezoelectricity of collagen has found to be affected by many factors like extraction procedure from tissue, their preparation, p^H of the solution, as well as by humidity [50]. As the degree of piezoelectricity of collagen strongly depends on the favorable hydrogen bonding, crystallinity, proper orientation of collagen monomers toward the long axis of fiber, each of above mentioned condition can strongly alter these and so affects the resultant piezoelectricity of collagen.

Recently, development of collagen-based piezoelectric green energy harvester has become an attractive area of research for biomedical scientist. Ghosh *et al.* fabricated a proficient bio-piezoelectric nanogenerator (BPNG) based on the fish swim bladder (FSB) which is constructed of oriented collagen fibers [51]. To fabricate a robust and flexible nanogenerator, first gold was deposited on both sides of FSB and attached with fine copper wires. Finally, the system was sealed with polydimethylsiloxane (PDMS). Human finger was used to provide a gentle force that could able to stimulate the BPNG resulted with 10 V open-circuit voltages and 51 nA short-circuit current. In a similar way, they also fabricated BPNG with "bio-waste" transparent fish scale (FSC), which is also made up with unidirectional collagen fibrils [52]. The BPNG was capable of scratching around mechanical energies from different kind of sources which are abundant in living system like body motion, machine and also through sound vibrations, wind flow. The voltage and current produced by the nanogenerator were 4 V and 1.5 l A, respectively. Another BPNG was fabricated by these authors based on fish skin which was highly sensitive (sensitivity ~27 mV N^{-1}), durable (over 75,000 cycles) and also displayed rapid response time (~4.9 ms) [53]. An open-circuit voltage, V_{oc} ~ 2 V and short-circuit current, I_{sc} ~ 20 nA was obtained while pressing the device by an external pressure of 1.8 MPa. Fish scale composed of oriented collagen fibrils was also utilized as bio-waste together with polymer to produce composite materials and fabricate efficient energy harvester [54]. Human finger generated gentle force was found to be sufficient for the operation of device besides different common forms of mechanical compression, such as twisting, bending, walking, foot tapping. The device demonstrated very high voltage output and energy density of 22 V and 28.5 $\mu W/cm^2$, respectively. Vivekananthan *et al.* built similar BPNG by using a market obtainable cotton fabric composed of collagen which was attached with electrode from both the sides [55]. Copper wires built the connection and finally the fabrication was laminated with the polypropylene film. It was found that collagen-based BPNG could generate 45 V/250 nA output under applied pressure of 5 N.

Many other proteins like keratin, prestin, and lysozyme also exhibit piezoelectricity. Keratin is predominantly observed in skin, hair, and nails. The major constituent of keratin backbone is alanine, leucine, arginine, cyteine which organizes into the right-handed α-helical structure and fabricates fibrillar structure. A.J.P Martin first observed the piezoelectric charge generation by wool fibers (or human hairs) while applying mechanical force through rubbing together [56]. It was also established that the origin of piezoelectricity in keratin assimilated from the organized dipole moment due to oriented conformation of keratin [57, 58]. Piezoelectricity was also characterized by lysozyme. The two different crystalline forms of lysozyme, monoclinic and tetragonal exhibited significantly different piezoelectricity with later one showed up to piezoelectric coefficient of 19.3 pm V^{-1} measured by PFM [59–61].

By using electrospinning machinery, Ghosh *et al.* demonstrated fabrication of energy harvester based on fish gelatin nanofibers (GNFs) [62]. Fish gelatin is known to be produced from the partial denaturation of collagen. The device was fabricated by using the copper–nickel plated fine knit polyester fabric as top and bottom electrodes for gelatin nanofibers. PDMS was employed to encapsulate the whole sandwich structure for providing mechanical robustness and moisture protection. PFM characterization revealed a piezoelectric coefficient, d_{33}=20 pm/V with tremendous functional steadiness (over 108,000 cycles) and anti-fatigue (over 6 months) characteristics. The basis of molecular dipole and piezoelectricity in gelatin also expected to acquire from the polar –C=O….H-N H-bonding units among the peptide chains. Biowaste pomegranate peel was utilized in combination of polymer to fabricate green energy harvester [63]. The device was fabricated by sandwiching biowaste-polymer composite film with silver paste from both the sides. Copper electrodes were fond of to both electrodes for output measurements. The sandwich system was finally wrapped up with polypropylene tape. To protect the whole system, the fabrication was covered with PDMS. Scavenging mechanical energy from body motion like walking, twisting, and bending, the device was able to generate electricity with an open circuit voltage of 65 V and power density of 84 µW/cm^2. Organized M13 bacteriophage film was also used for fabrication of piezoelectric devices [64]. First, 10 nm chromium and then 30 nm gold were deposited on Thermanox films to prepared flexible gold-coated platforms. Then, bacteriophage solution was put on the substrate and covered up by another gold-coated flexible substrate. To achieve structural stability, the whole fabrication was embedded between two PDMS matrices. The nanogenerator exhibited maximum output current and voltage of 6 nA and 400 mV, respectively. The structural alignment of the M13 bacteriophages was expected to improve the polarization and output significantly. Shin *et al.* proved this by constructing BPNG using vertically aligned M13 bacteriophage nanopillars that demonstrated a 2.6-fold increased of current and voltage in comparison to the output of device made up with laterally oriented viruses [65]. Lee *et al.* also demonstrated polarized phage-based BPNG that upon pressing by a force of 17 N showed voltage, current and potential of 2.8 V, 120 nA, and 236 nW, respectively [66].

11.1.3 Piezoelectric Ultra-Short Peptides

Peptides are the short chains composed of amino acids connected through peptide bond. Piezoelectricity has been identified in a numbers of α-helical polypeptides [67–70]. The key origin of piezoelectricity in these systems is the aligned hydrogen bonds along the axis of the helices. Fukada et al. reported the piezoelectric property in poly-γ-methyl-L-glutamate (PMLG) and poly-γ-benzyl-L-glutamate (PBLG) synthetic polypeptides [71–73]. Scientists put significant effort to generate aligned structure as this is the main feature that controls the efficiency of piezoelectric materials [74–76]. Magnetic field was used as an important tool to orient the structures which resulted a strong piezoelectric coefficient of d_{14} = 26 pC N^{-1} for the PBLG membrane [74]. The piezoelectric coefficient was found to increase linearly with applied magnetic field. Corona discharging method to produce external electric field was also employed to produce oriented PBLG-PMMA film [75]. The alignment exhibited a strong piezoelectric coefficient of d_{33} = 23 pC N^{-1}. Electrospinning method was also employed for these purposes. The oriented PBLG fibers produced through electrospinning showed a strong piezoelectric coefficient of d_{33} = 25 pC N^{-1}, and high thermal stability capable of functional after keeping at 100°C for more than 24 hours [76].

Recently, we have discovered the formation of helical structure from shortest peptide sequence built on natural amino acids only, Pro-Phe-Phe and Hyp-Phe-Phe [77, 78]. The tripeptides showed excellent mechanical rigidity and piezoelectricity which was maximized in case of Hyp-Phe-Phe due to additional number of hydrogen bonds. Theoretical study based on DFT calculation revealed that due to additional hydroxyl group in Hyp-Phe-Phe, piezoelectric coefficient increased by an order of magnitude (d_{16} and d_{35} = 27 pm/V). To experimentally validate the piezoelectricity of tripeptides, we employed PFM characterization. The results exhibited the vertical coefficient d_{33} of Pro-Phe-Phe was 2.15 ± 0.86 pm V^{-1} and for Hyp-Phe-Phe, it was 4.03 ± 1.96 pm V^{-1}. The shear piezoelectricity of Hyp-Phe-Phe was d_{34}=16.12 ± 2.3 pm V^{-1}, which was higher than LiNbO$_3$ (13 pm V^{-1}), ZnO (12 pm V^{-1}), γ-glycine (10 pm V^{-1}), M13 virus (6–8 pm V^{-1}) and collagen film (1 pm V^{-1}). Finally, we fabricated tripeptide based BPNG by utilizing two Ag electrodes to sandwiching peptide assemblies. Copper wires were used to make the connection. The whole system was wrapped up with Kapton tape to protect from temperature, external dust and humidity. The Pro-Phe-Phe based device exhibited a voltage and current output of 1.4 V and 52 nA, respectively under a force of 55 N. Almost similar output current was obtained from Hyp-Phe-Phe based device while pressing with a mechanical force of 23 N, half to that used previously. Moreover, the output voltage and current showed a linear changes with the applied force (4 to 23 N), representative of true piezoelectric characteristics of tripeptide-based nanogenerators. The sustainability of the devices under cyclic forced was investigated, and it was found that the output electricity did not show any deprivation over working for 1000 press/release cycles indicated significant stability of the nanogenerator. We also designed collagenous polyproline II-structured peptide from a single triplet motif, Fmoc-Gly-Pro-Hyp [79]. This tripeptide exhibited a maximum predicted strain constant d_{14} with a value of 8.6 pC/N and induced collagenous trait into non-collagenous hydrogel scaffold.

Among the dipeptide, diphenylalanine (FF) was the most attractive piezoelectric building block, comprehensively investigated by many researchers [80–87]. Kholkin et al. was the first to observe piezoelectricity in FF nanotubes by using PFM technique [80]. Depending on the diameter of tubes, a strong shear piezoelectric responses of d_{15} (35 and 60 pm V^{-1} for 100 and 200 nm, respectively), were measured. Later, Vasilev et al. also calculated the coefficients d_{33}, d_{31}, d_{15}, and d_{14} for FF nanostructures through the PFM, and identified value of 18 ± 5, 4 ± 1, 80 ± 15, and 10 ± 1 pm V^{-1}, respectively [81]. These strong piezoelectric co-efficient of FF nanostructures attracted the attention of researchers to build FF-based biocompatible green energy harvesters. However, initially, a very limited success was achieved through device design as the method of controlling the polarization through uniform orientation of FF nanotubes was unknown. In 2016, Yang et al. put forward a procedure founded on application of external electric filed to achieve oriented array of FF nanotubes (Figure 11.2) [88]. At first, peptide solution was prepared by dissolving the powder in 1,1,1,3,3,3-hexafluoro-2-propanol (HFP) and then drop casted over Au-coated silicon substrate to create an amorphous and transparent film. Then this was used as the seeding platform to grow the vertically oriented FF rods. An electric filed was employed during the growing process to control the polarization directions of FF nanostructures that was possible to alter by changing the direction of electric filed. PFM characterization demonstrated a high piezoelectric coefficient of $d_{33} \approx 17.9$ pm V^{-1} for the oriented FF tubes. The aligned nanotubes were then utilized to prepare energy harvester. Two gold coated plates were used to tightly sandwich the FF assemblies and acted as two electrodes. The electrodes were connected with lead wires to complete the device fabrication. The nanogenerator produced a voltage of 1.4 V, current of 39.2 nA and power density of 3.3 nW cm^{-2} upon compression with 60 N force. In a similar way, a different technique was employed to produce horizontally oriented nanotubes in a large

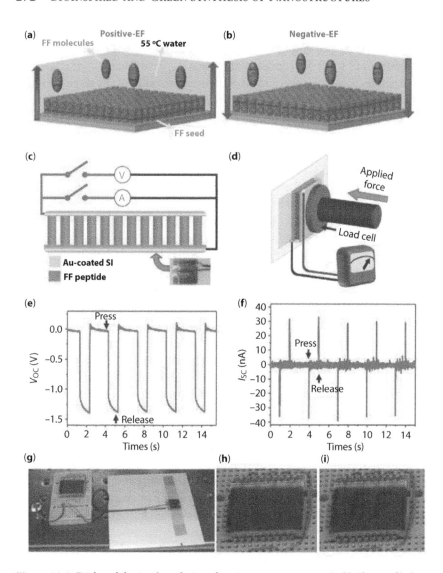

Figure 11.2 Diphenylalanine-based piezoelectric power generator. (a, b) Electric filed induced vertical growth of FF nanotubes with oriented polarization, (c) Schematic representation of the fabricated device along with mode of connection to the measuring unit, (d) Application of the mechanical force on the top electrode of the device (e, f). Upon application of a periodic force, F = 60 N to the FF-based device, the output open-circuit voltage (V_{oc}) (e) and short-circuit current (I_{sc}) (f), (g–i) FF-based generator as a direct power source for an LCD. Photograph of the LCD (g), before (h) and after (i) pressing by a human finger. Reproduced with permission [88]. Copyright 2016, Springer, Nature.

area [89]. A substrate with line scratch contained FF seeds was put into the concentrated peptide solution, and a horizontally array of FF nanostructures was produced on the substrate by slow pulling it vertically from the solution. The morphology of FF tubes could be controlled by several factors like solvent, concentration of peptide as well as the puling rate. A nanogenerator was fabricated using these uniformly polarized FF nanotubes. The prepared nanogenerator exhibited an output voltage of 2.8 V, current of 37.4 nA and power of 8.2 nW.

Recently, we have designed a number of FF derivatives such as Dip-Dip, cyclo-Dip-Dip, and *tert*-butyloxycarbonyl (Boc)-Dip-Dip [90]. Excitingly, Boc-Dip-Dip displayed numerous interesting characteristics like organized dipole moment and strong piezoelectric polarization as compared to the parent FF dipeptide. Theoretical calculation based on density function theory (DFT) of Boc-Dip-Dip single crystal showed a piezoelectric coefficient of $d_{33} \approx 73.1$ pC N^{-1}. Fabrication of mechanical energy harvesting device built on this aromatic-rich dipeptide exhibited that upon applying 40 N mechanical forces, a maximum output voltage and current could reach to 1 V and 60 nA.

Our group also designed a cyclic dipeptide, glycine-tryptophan (cyclo-GW) which self-assembled into supramolecular β-sheet structure stabilized through hydrogen bonds and aromatic-aromatic connections [91]. The cyclic peptide showed high thermal stability ($\approx$ 643 K) and rigidity (elastic modulus $\approx$ 24.0 GPa). The noncentrosymmetric crystals of cyclo-GW were used to fabricate green energy harvester by sandwiching between two electrodes. The fabrication was then wrapped up with Kapton to guard from external dust and moisture. The device demonstrated a voltage of 1.2 V and current of 1.75 nA, while compressed with force of 65 N.

The design was further extended to produce a series of tryptophan-based aromatic cyclic dipeptides [92]. The cyclo phenylalanine–tryptophan (cyclo-FW) and cyclo-tryptophan–tryptophan (cyclo-WW) also showed excellent mechanical rigidity and strong thermal stability. Then, a peptide based nanogenerator was similarly fabricated. Upon compressed with 56 N mechanical force, the device exhibited a voltage of 1.4 V and current of 1.75 nA.

11.1.4 Single Amino Acid Assembly and Coassembly-Based Piezoelectric Materials

Amino acids are the basic structural units of proteins and peptides. Self-assembled amino acids are interesting biomaterials to study their piezoelectric properties as they produce high polarization due to inherent dipole moment. Except of Lysine and Arginine, all other amino acids are capable of producing single crystals which could be piezoactive while associated into 20 non-centrosymmetric crystal classes. Using nuclear quadruple resonance spectrometer, Vasilescu *et al.* examined the piezoelectricity of amino acids powder in various forms, like pure (L or D) and recemic mixtures (DL) [93]. Their study showed good agreement between experimental results and theoretical studies based on the discovered crystallographic classes. Guerin *et al.* extensively studied the piezoelectric property of amino acids both theoretically and experimentally. In 2018, they reported strong longitudinal piezoelectricity by L-amino acid films growth on conductive substrate that exhibited similar range of piezoelectricity with larger biomolecules [94]. The existence of longitudinal piezoelectricity in the L-amino acid films were verified using the DFT calculations, and the aggregates also displayed natural polarization. They also used DFT based simulations to measure the piezoelectric coefficient of the major components of collagen: hydroxyproline, proline, and alanine [95]. The simple presence of an additional –OH group in Hyp was capable of increasing the piezoelectric coefficient significantly (d_{25} = 25 ± 5 pC N^{-1}) as compared to that of Pro (d_{14} = 3.35 pC N^{-1}), indicated that piezoelectricity of amino acids can be fine-tuned by playing with simple side chain chemistry. They also studied the direct piezoelectric effect in L-leucine by growing the bioelectret crystal films on conductive substrates [96]. Their results were varied through DFT calculation, and it revealed that L-leucine is a potential candidate for the development of biocompatible, body-friendly, efficient power generator. The group further extended their investigation to understand the role of chirality on piezoelectricity of amino acids by studying DL-alanine [97]. L-alanine did not exhibit any piezoelectric property due to lack of polarization as the sum of the dipole vectors was zero. However, dipole vectors for DL-alanine pointed toward same direction and thus displayed longitudinal piezoelectric properties. Experimental measurement of piezoelectric properties of DL-alanine showed longitudinal piezoelectric coefficient of 4.8 pC N^{-1} for thin films, and a copper electrode based energy harvester produced a voltage of 0.8 V under manual compression.

Glycine, the only nonchiral amino acid, is the most studied amino acid for piezoelectricity [98–101]. Glycine crystallizes in three different polymorphic phases—α, β and γ. Among these three forms, α-glycine belongs to centrosymmetric point group, thus excludes from piezoelectric study. However, the β and γ-phases are belong to noncentrosymmetric point group and attracted widespread attention for their strong piezoelectric characteristics. Initial investigation of glycine piezoelectricity showed a very lower range of value without knowledge about their origin. Recently, Guerin et al. extensively studied the piezoelectric phenomenon of the different phases of glycine utilizing both theoretical and experimental techniques, and found both forms exhibited considerable piezoelectricity [102]. Theoretical calculation revealed that piezoelectric coefficient d_{16} for β-glycine reached to 200 pm V^{-1}, which is higher than many conventional piezoelectric inorganic, organic and polymer based materials. The result was further confirmed through experimental validation. However, the β-phase of glycine is well known to have very high tendency to gradually transform into the α- or γ-phase at the atmospheric conditions. A piezoelectric nanogenerator was prepared based on γ-glycine crystals by sandwiching between two copper electrodes that produced an open-circuit voltage of 0.45 V at a manual pressure of 0.17 N.

We studied a number of aromatic amino acids like L-phenylalanine (L-Phe), L-tyrosine (L-Tyr), and 3,4-dihydroxyphenylalanine (L-DOPA), and found formation of order supramolecular architectures [103]. The noncentrosymmetric crystal of L-Tyr demonstrated rigid assemblies through formation of hydrogen bonded dimeric aggregates. This structural pattern provided many advantages to L-Tyr assemblies like elevated thermal stability (≈581 K), good mechanical rigidity, and strong power generation as compared to analogous aromatic amino acids. L-Tyr based power generator was fabricated by using the assemblies in between two electrodes which was then wrapped up with PDMS. An output voltage and current of 0.5 V and 35 nA, respectively were obtained from this nanogenerator while pressing periodically with a force of 31 N.

We extended our design strategy to include co-assembly approach between aromatic 4,4′-bipyridine (4,4′-Bpy) and nonaromatic amino acid derivative of N-acetyl-L-alanine (AcA) to understand the role of co-assembly in piezoelectricity [104]. The co-assembly induced a transition of packing mode from H-aggregation (face-to-face stacking) in pure form of 4,4′-Bpy to J (staggered stacking) and X (crossed stacking) aggregation in co-assembly system. While pure 4,4′-Bpy crystals displayed good piezoelectric property as a result of low symmetry in crystal packing, the hybrid system exhibited improved conductivity. The noncentrosymmetric nature of these crystals revealed their potential as piezoelectric material which was studied both through DFT calculation and experimental analysis. For 4,4′-Bpy, AcA, and 4,4′-Bpy/AcA crystals, the calculated piezoelectric coefficient (d_{max}) and piezoelectric voltage constant (g_{max}) were 20.9, 9.5 and 14.9 pC N^{-1}, and 1.15, 0.48, and 0.67 V m N^{-1}, respectively. Inspired from theoretical data, 4,4′-Bpy and 4,4′-Bpy/AcA crystals based nanogenerators were fabricated using two electrodes. An output voltage for 4,4'-Bpy and 4,4'-Bpy/AcA crystals based nanogenerators showed up to values of 0.65 and 0.35 V, respectively under a mechanical compression of 56 N. Next, we further extended our study to understand the impact of chirality in organic co-crystals for the favorable packing and piezoelectric properties [105]. Using acetylated alanine (AcA) and a nonchiral bipyridine derivative (BPE), it was demonstrated that chirality played a significant role in dictating several physical properties like macroscale chirality, thermal stability, mechanical rigidity, and piezoelectricity. Piezoelectric coefficient of pure single crystals and co-crystals were studied by DFT calculation. The two racemic single crystal and co-crystal (DL-AcA and BPE/DL-AcA) were belongs to centrosymmetric space groups (P21/c and P2/n), and thus excluded for piezoelectric characterization. BPE showed the highest piezoelectric coefficient with d_{24}=10.9 pC/N among the three pure single crystals. However, the piezoelectricity increased significantly for both enantiopure cocrystals of BPE/L-AcA and BPE/D-AcA, which exhibited values up to 34.9 pC/N and 22.9 pC/N, respectively.

11.2 Biomedical Applications

The discovery of biomaterials based piezoelectric self-powered nanogenerators has opened door for design platforms for long-term power supply to biomedical or skin-contact electronic devices as these scavenge the mechanical energy from sustainable sources like ultrasonic waves, pressure, body motion and transform into electrical signal. At the same time, inherent polarization based mechanoelectronic coupling of biomaterials has found to control a numbers of physiological behaviors of cells, such as proliferation, migration, and differentiation, thus displaying therapeutic effects on diseases. Due to their low cost, easy synthesis, inherent biocompatibility and environmental friendliness, biomedical research community have put significant efforts to explore the ability of biomolecular nanogenerators in many areas like energy harvesting, health monitoring, *in vivo* implanting biomedical applications, tissue engineering, etc.

11.2.1 Piezoelectric Sensors

The current state-of-the-art capacitor- and inductor-based biosensors have several drawbacks like big size, rigid structure, limited time energy supply by battery, and low biocompatibility, which significantly restrict their practical applications. Due to the potential of converting mechanical energy from regular physical activities into electrical signal, biopiezoelectric materials are promising candidates for monitoring many body signals like the heartbeat or breathing. With the progress of development of biopiezoelectric materials, a numbers of miniature and intelligent self-powered biosensors have fabricated by the scientists for detecting different physiological signals. As these devices are composed of self-power operation, these are eliminating the problem of short battery life and repeating envisions therapies for replacing the battery.

Recently, flexible electronic devices are developed based on piezoelectric materials as wearable and attachable health monitoring platforms. These platforms are negating the painful surgery for implementation as these can easily attached to the skin or tissue. Ghosal *et al.* developed a biocompatible piezoelectric platform by combining environmental bacterial strain with the microstructure of organic polymer, PVDF [106]. Incorporation of bacterial strain was found to result with significantly improve properties like biocompatibility, piezoelectric properties of PVDF film. While the pure PVDF showed a piezoelectric coefficient, d_{33}= -0.85 pC/N, the porous structures of hybrid bioorganic film exhibited the value up to -43 pC/N, indicated a significant improvement. The output power from the fabricated biopiezoelectric device was 640 mW/m^2 and showed very good pressure sensitivity (1.26 and 0.86 V kPa^{-1} in the pressure range of 0.01–1 kPa and 2–15 kPa, respectively). The good biocompatibility and pressure sensitivity encouraged the researchers to investigate potential for real-time healthcare monitoring. The device was therefore deployed toward biomedical applications of healthcare monitoring. The same group also developed a self-powered wearable biopiezoelectric device based on oriented collagen from fish skin (Figure 11.3) [53]. With the efficiency like current and voltage output, the fabricated biocompatible energy harvester was capable of monitoring several physiological signals such as arterial pulses, vocal cord vibration without requirement of supplying energy from outside. They also used fish gelatin nanofibers (GNFs) that was oriented through electrospinning technology and utilized for the fabrication of a wearable pressure sensor (i.e., bio-e-skin) [62]. The excellent sensitivity of designed bio-e-skin as self-powered device was capable of mimicking spatiotemporal human perception and demonstrated for monitoring real-time physiological signals. A low (~0.3 kPa) to medium (~25 kPa) pressure was employed to realize the control sensing performance by the device. Due to its higher sensitivity (S~0.8 V kPa^{-1}) in low pressure region, the bio-e-skin was found to be strongly capable of physiological signal monitoring in low pressure regime. In medium pressure regime, the production of high output voltage suggested its application as low power portable electronics. Overall,

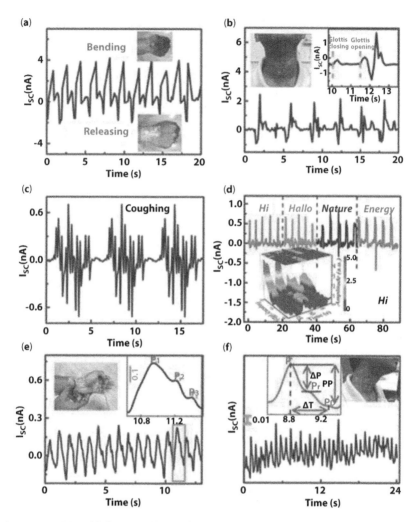

Figure 11.3 Wearable biopiezoelectric devices based on biomaterials. The output current from the fabricated device in the course of (a) periodic movement of the wrist, (b) swallowing motions, (c) periodic coughing actions, (d) while someone is talking", (e) monitoring radial artery and (f) carotid artery pulses with enlarged view of one cycle in the insets. Reproduced with permission [53]. Copyright 2017, American Chemical Society.

the efficiency of this biomaterials based bio-e-skin was found to be similar with frequently used polymer based piezoelectric devices.

Joseph *et al.* developed a unique approach to orient silk films that could potentially be used for the fabrication of different kinds of wearable sensors [107]. With the help of PFM analysis, the piezoelectric coefficient of the silk film was evaluated and found an average d_{33} value of 56.7 pm/V. Using a simple process flow, a pressure sensor was fabricated based on piezoelectricity of silk thin film that did not require any silk patterning step. Microstructured surface intertextures of silk was also used as moulds for preparation of organized PDMS films which was utilized to constrict flexible pressure sensors (E-skins) [108]. The hybrid system based developed E-skin exhibited advanced characteristics like superior sensitivity (1.80 kPa^{-1}), very low detectable pressure limit (0.6 Pa), fast response time (<10 ms), and high stability (>67 500 cycles), that provided its capability to detection feather-light pressures. By using this E-skins, it was possible to monitoring human physiological signals (PS) like wrist pulse and muscle movement while a person was speaking. This indicated the potential of the developed E-skins for different biomedical application like disease diagnosis and voice recognition.

11.2.2 Tissue Regeneration

A piezoelectric-based scaffold produces electrical potential differences while subjected to mechanical stress *via* regular body motion, heartbeats, breathing, or blood circulation. In general the conversion of mechanical energy into physiologically relevant electrical signals regulates voltage gated ion channels and tunes the intracellular ion level, and thus accelerating cell proliferation and differentiation [109]. Noris-Suárez *et al.* examined the influence of piezoelectricity through elastic deformation of bone collagen [110]. The study revealed that piezoelectric effect induced the mineralization mainly on the pressed side of bone. The results from combined experimental techniques showed that, even in the absence of osteoblast cells, the deformed collagen could produce hydroxyapatite by the effect of piezoelectric dipoles, in contrast to the classical biomimetic deposition process which occurred in presence of a catalytic converter. Panwar *et al.* reported to prepare a numbers of conductive hydrogels (Ch–CMC–PDA) by mixing chitosan, cellulose (CMC) and dopamine (DA), without addition of conductive component [111]. The hydrogel scaffold showed very good ion conductivity (0.01–3.4 × 10^{-3} S cm^{-1}) and injectability which facilitated its biomedical application. The hydrogel scaffold was utilized to build nanogenerator which converted mechanical energy from finger tapping to electrical signal. The system was also examined in various bioelectronics applications like for gathering of stable real-time signals. The high mechanical rigidity and superior piezoelectric properties of hydrogel scaffold inspired the author to examine if the platform could have potential to accelerate tissue regeneration. *In vivo* tissue regeneration was studied by using a rat model. The potential of the conductive hydrogel scaffold to promote cell growth and tissue regeneration was established from the result of both *in vitro* and *in vivo* studies. Poly(l-lactic acid) (PLLA) was used together with glass-reinforced hydroxyapatite granules (gHA) particles to develop composite electrospun membrane to accelerate the bone regeneration capability of PLLA [109]. To compare the effect of gHA on cell growth, osteoblast cells were seeded both on PLLA-gHA hybrid system and PLLA control materials. After 1 day, it was found that the growth of cells over hybrid system was much faster as compared to the control system. Although both systems revealed the nucleation of HA crystals, an increases alkaline phosphatase activity for the hybrid system made it promising candidates for promoting bone regeneration applications.

Wound healing is also a branch of research on tissue regeneration. Current state-of-the-art therapies employed for wound repair treatment mostly concentrates on two factor-reducing wound infection and increasing tissue rehydration [112, 113]. In contrast to this passive treatment strategies, biomaterials based self-powered piezoelectric platforms can converts sustainable mechanical energy into electrical signal to accelerate would repair process. The three important stages of wound healing-inflammation, proliferation, and remodeling were reported to influence significantly by the external electric filed [114, 115]. Increasing blood flow and tissue oxygenation supply by electric filed in the inflammation stage protects the bacterial growth and minimizes wound edema. Wound contraction, fibroblast proliferation, angiogenesis, and collagen deposition is increased by external electric filed in the proliferative phase. In the last stage, the remodeling stage, wound contraction is accelerating through electrical stimulation driven speedy maturation and remodeling of collagens. Ghosh and coworkers demonstrated the use of piezoelectric scaffolds that was capable of supply electrical stimulation to accelerate faster wound healing [116]. The conductive hydrogel scaffold was prepared by combining polydopamine/polydopamine/polyacrylamide and electroactive electrospun PVDF membrane which was competent of producing electrical clue through biomechanical activities. The mechanically robust hydrogel platform could tolerate different mechanical forces provided through walking, stretching, etc. The prospective of the piezoelectric dressing scaffold to accelerate the healing process was validated through examination of an *in vivo* excisional skin wound model. The results revealed that electrical simulation *via* scaffold improved the re-epithelialization, vascularization resulted in faster healing.

11.3 Conclusion and Future Perspectives

In recent years, biomaterials based piezoelectric self-powered platforms become a research hotspot for biomedicine due to their easy synthesis, low cost, biocompatibility, ease of fabrication, light-weight nature, mechanical flexibility, and strong piezoelectric properties. The characteristics of piezoelectricity enable to produce electricity from the regular biological activities like body movement, muscle movement, blood circulation, heartbeats and breathing. The biomaterials based piezoelectric platforms are likely to impact three major areas of research. First, they can be used as functional materials for long term and stable power supply to the implantable medical devices due to their mechanical agitation and inherent biocompatibility. Secondly, because of strong flexibility and sensitivity toward small mechanical pressure, these can also be utilized as platforms for monitoring a variety of vital signs such as heart rate, breathing and blood pressure. Lastly, piezoelectric materials can be used to fabricate mechanically stable scaffolds that can act as tissue stimulator and accelerate the differentiation and growth of specific tissues. Many biomedical devices like cardiac pacemakers, cochlear implants, artificial retinas, electronic skins can be design based on biopiezoelectric materials based biosensors. Although initial studies on biopiezoelectric materials mainly concentrated on developing bioelectronics devices, later the focused has been shifted toward exploring new opportunities to use these self-powered systems in biomedicine. The characteristics of electromechanical coupling enable piezoelectric materials to generate reactive oxygen species for cancer treatment. The high efficiency and sensitivity of these systems make them fascinating platforms for diverse applications, such as tissue regeneration, electronic skins, drug delivery, cancer treatment, etc.

In spite of tremendous potential of biomaterials based piezoelectric platforms, there are several critical issues that need to be resolved for accelerating integration of these platform into biomedicine: (1) To develop implantable medical devices based on biopiezoelectric materials, several characteristics features like good biointegrability, high flexibility and excellent efficiency for electromechanical coupling need to be considered seriously to integrate with the fabricated devices. Moreover, in case of disease treatment, biodegradability, and immunogenicity of platforms along with the capability of faster tissue accumulation should be considered. To fulfill these points, sophisticated lead-free new piezoelectric materials need to be designed based on the principle of nature. (2) Although researcher focused to uncover the fundamental principle of biological materials' piezoelectricity, a lots need to be done to fully understand the phenomenon. Recently, tremendous progress of using theoretical studies based on DFT, MD simulations allows their use to understand different phenomenon and apply them to design new materials. Along with analyzing the single crystal structure of piezoelectric materials, theoretical study could be used as a vital tool for new materials development. (3) Many piezoelectric material-based scaffolds have showed to promote the cell growth and tissue regeneration. However, the produced safe range of electricity compatible with the *in vivo* condition has not been studied systematically. It is well known that overdoses of electricity can exert a negative effect on cell growth. So the proper range of electricity useful for particular cells and tissues need to understand clearly *via* systematic exploration. Biodegradability with desired time frame is another important issue for these piezoelectric platforms for biodegradable sensors and scaffold applications. Biodegradability of these materials need to be examined and also could be engineered through different techniques like temperature, electric poling, etc. (4) As biomedical applications is the ultimate goal of developing biopiezoelectric platforms, their biocompatibility, long-term toxicity, biosafety, immunogenicity, controlled biodegradability are crucial features that need to be thoroughly explored to speed up their practical application for healthcare.

Although biopiezeoelectric materials are at their early stage and a huge efforts requires to resolving all the current challenging issues. However, based on the significant research interests of multiple disciplines, it can be envisioned that biopiezoelectric platforms will continue their rapid growth in recent future and bring exciting new solutions for biomedicine.

Acknowledgment

This work was supported by the European Union's Horizon 2020 research and innovation program under Marie Sklodowska Curie fellowship (grant agreement no. 101032317).

References

1. Ortiz, C. and Boyce, M.C., Bioinspired structural materials. *Science*, 319, 1053, 2008.
2. Sanchez, C., Arribart, H., Guille, M.M.G., Biomimetism and bioinspiration as tools for the design of innovative materials and systems. *Nat. Mater.*, 4, 277, 2005.

3. Whitesides, G.M., Bioinspiration: Something for everyone. *Interface Focus*, 5, 20150031, 2015.
4. Fratzl, P., Sauer, C., Razghandi, K., Special issue: Bioinspired architectural and architected materials. *Bioinspir. Biomim.*, 17, 040401, 2022.
5. Das, S., Ahn, B.K., Martinez-Rodriguez, N.R., Biomimicry and bioinspiration as tools for the design of innovative materials and systems. *Appl. Bionics Biomech.*, 2018, 6103537, 1, 2018
6. Katiyar, N.K., Goel, G., Hawi, S., Goel, S., Nature-inspired materials: Emerging trends and prospects. *NPG Asia Mater.*, 13, 56, 2021.
7. Wang, Y., Naleway, S.E., Wang, B., Biological and bioinspired materials: Structure leading to functional and mechanical performance. *Bioact. Mater.*, 5, 745, 2020.
8. Wu, W.Z., Wang, L., Li, Y.L., Zhang, F., Lin, L., Niu, S.M., Chenet, D., Zhang, X., Hao, Y.F., Heinz, T.F., Hone, J., Wang, Z.L., Piezoelectricity of single-atomic-layer MoS_2 for energy conversion and piezotronics. *Nature*, 514, 470, 2014.
9. Li, F., Lin, D.B., Chen, Z.B., Cheng, Z.X., Wang, J.L., Li, C.C., Xu, Z., Huang, Q.W., Liao, X.Z., Chen, L.Q., Shrout, T.R., Zhang, S.J., Ultrahigh piezoelectricity in ferroelectric ceramics by design. *Nat. Mater.*, 17, 349, 2018.
10. Fukada, E. and Yasuda, I., On the piezoelectric effect of bone. *J. Phys. Soc. Jpn.*, 12, 1158, 1957.
11. Lang, S.B., Pyroelectric effect in bone and tendon. *Nature*, 212, 705, 1966.
12. Derossi, D., Pastacaldi, P., Domenici, C., Piezoelectric properties of dry human skin. *IEEE Trans. Electr. Insul.*, 21, 511, 1986.
13. Marino, A.A. and Becker, R.O., Piezoelectricity in hydrated frozen bone and tendon. *Nature*, 253, 627, 1975.
14. Mason, W.P., Piezoelectricity, its history and applications. *J. Acoust. Soc. Am.*, 70, 1561, 1981.
15. Katzir, S., The discovery of the piezoelectric effect. *Arch. Hist. Exact Sci.*, 57, 61, 2003.
16. Korostoff, E., A linear piezoelectric model for characterizing stress generated potentials in bone. *J. Biomech.*, 12, 335, 1979.
17. Guzelsu, N. and Saha, S., In theoretical analysis of piezoelectric wave propagation in bone. *American Society of Mechanical Engineers 101st Winter Annual Meeting*, 1980.
18. Ahn, A.C. and Grodzinsky, A.J., Relevance of collagen piezoelectricity to "Wolff's Law": A critical review. *J. Biomed. Eng.*, 31, 733, 2009.
19. Martin, R.B., Burr, D.B., Sharkey, N.A., *Skeletal tissue mechanics*, vol. 190, Springer, New York, NY, 1998.
20. Kim, M.S., Lee, M.H., Kwon, B.-J., Seo, H.J., Koo, M.-A., You, K.E., Kim, D., Park, J.-C., Development of decellularized scaffolds for stem cell-driven tissue engineering. *J. Tissue Eng. Regener. Med.*, 11, 862, 2017.
21. Bolander, J., Chai, Y.C., Geris, L., Schrooten, J., Lambrechts, D., Roberts, S.J., Luyten, F.P., Early BMP, Wnt and Ca2+/PKC pathway activation predicts the

bone forming capacity of periosteal cells in combination with calcium phosphates. *Biomaterials*, 86, 106, 2016.
22. Kim, D., Han, S.A., Kim, J.H., Lee, J.-H., Kim, S.-W., Lee, S.-W., Biomolecular piezoelectric materials: From amino acids to living tissues. *Adv. Mater.*, 32, 1906989, 2020.
23. Xu, Q., Gao, X., Zhao, S., Liu, Y.-N., Zhang, D., Zhou, K., Khanbareh, H., Chen, W., Zhang, Y., Bowen, C., Construction of bio-piezoelectric platforms: From structures and synthesis to applications. *Adv. Mater.*, 33, 2008452, 2021.
24. Vijayakanth, T., Liptrot, D.J., Gazit, E., Boomishankar, R., Bowen, C.R., Recent advances in organic and organic–inorganic hybrid materials for piezoelectric mechanical energy harvesting. *Adv. Funct. Mater.*, 32, 2109492, 2022.
25. Shin, D.-M., Hong, S.W., Hwang, Y.-H., Recent advances in organic piezoelectric biomaterials for energy and biomedical applications. *Nanomaterials*, 10, 123, 2020.
26. Chorsi, M.T., Curry, E.J., Chorsi, H.T., Das, R., Baroody, J., Purohit, P.K., Ilies, H., Nguyen, T.D., Piezoelectric biomaterials for sensors and actuators. *Adv. Mater.*, 31, 1802084, 2019.
27. Wang, Y.-M., Zeng, Q., He, L., Yin, P., Sun, Y., Hu, W., Yang, R., Fabrication and application of biocompatible nanogenerators. *iScience*, 24, 102274, 2021.
28. Bassett, C.A.L., Biologic significance of piezoelectricity. *Calcif. Tissue Res.*, 1, 252, 1968.
29. Fukada, E., History and recent progress in piezoelectric polymers. *IEEE Trans. Ultrason. Ferroelectr.*, 47, 1277, 2000.
30. Fukada, E., The effect of blood viscoelasticity on pulsatile flow in stationary and axially moving tubes. *Biorheology*, 33, 95, 1996.
31. Marino, A.A., Becker, R.O., Soderholm, S.C., Origin of the piezoelectric effect in bone. *Calcif. Tissue Res.*, 8, 177, 1971.
32. Jayasuriya, A.C., Scheinbeim, J.I., Lubkin, V., Bennett, G., Kramer, P., Piezoelectric and mechanical properties in bovine cornea. *J. Biomed. Mater. Res. Part A*, 66a, 260, 2003.
33. Denning, D., Kilpatrick, J.I., Fukada, E., Zhang, N., Habelitz, S., Fertala, A., Gilchrist, M.D., Zhang, Y., Tofail, S.A.M., Rodriguez, B.J., Piezoelectric tensor of collagen fibrils determined at the nanoscale. *ACS Biomater. Sci. Eng.*, 3, 929, 2017.
34. Shamos, M.H. and Lavine, L.S., Piezoelectricity as a fundamental property of biological tissues. *Nature*, 213, 267, 1967.
35. Athenstaedt, H., Claussen, H., Schaper, D., Pyroelectric and piezoelectric sensor layer. *Science*, 216, 1018, 1982.
36. Yamashiro, T., Piezoelectric effect of plant leaf. *Ferroelectrics*, 171, 211, 1995.
37. Csoka, L., Hoeger, I.C., Rojas, O.J., Peszlen, I., Pawlak, J.J., Peralta, P.N., Piezoelectric effect of cellulose nanocrystals thin films. *ACS Macro Lett.*, 1, 867, 2012.

38. Fukada, E., Piezoelectricity as a fundamental property of wood. *Wood Sci. Technol.*, 2, 299, 1968.
39. Sarkar, D., Das, N., Saikh, Md. M., Biswas, P., Das, S., Das, S., Hoque, N.A., Basu, R., Development of a sustainable and biodegradable *sonchus asper* cotton pappus based piezoelectric nanogenerator for instrument vibration and human body motion sensing with mechanical energy harvesting applications. *ACS Omega*, 6, 28710, 2021.
40. Afolayan, A.J. and Jimoh, F.O., Nutritional quality of some wild leafy vegetables in South Africa. *Int. J. Food Sci. Nutr.*, 60, 424, 2009.
41. Karan, S.K., Maiti, S., Kwon, O., Paria, S., Maitra, A., Si, S.K., Kim, Y., Kim, J.K., Khatua, B.B., Nature driven spider silk as high energy conversion efficient bio-piezoelectric nanogenerator. *Nano Energy*, 49, 655, 2018.
42. Bielajew, B.J., Hu, J.C., Athanasiou, K.A., Collagen: Quantification, biomechanics, and role of minor subtypes in cartilage. *Nat. Rev. Mater.*, 5, 730, 2020.
43. Bella, J., Eaton, M., Brodsky, B., Berman, H.M., Crystal and molecular structure of a collagen-like peptide at 1.9 Å resolution. *Science*, 266, 75, 1994.
44. Berisio, R., Vitagliano, L., Mazzarella, L., Zagari, A., Crystal structure of a collagen-like polypeptide with repeating sequence Pro–Hyp–Gly at 1.4 Å resolution: Implications for collagen hydration. *Biopolymers*, 56, 8, 2001.
45. Silva, C.C., Pinheiro, A.G., Figueiro, S.D., Goes, J.C., Sasaki, J.M., Miranda, M.A.R., Sombra, A.S.B., Piezoelectric properties of collagen-nanocrystalline hydroxyapatite composites. *J. Mater. Sci.*, 37, 2061, 2002.
46. Ravi, H.K., Simona, F., Hulliger, J., Cascella, M., Molecular origin of piezo- and pyroelectric properties in collagen investigated by molecular dynamics simulations. *J. Phys. Chem. B*, 116, 1901, 2012.
47. Zhou, Z., Qian, D., Minary-Jolandan, M., Molecular mechanism of polarization and piezoelectric effect in super-twisted collagen. *ACS Biomater. Sci. Eng.*, 2, 929, 2016.
48. Minary-Jolandan, M. and Yu, M.F., Nanoscale characterization of isolated individual type I collagen fibrils: Polarization and piezoelectricity. *Nanotechnology*, 20, 085706, 2009.
49. Goes, J.C., Figueiro, S.D., De Pavia, J.A.C., Sombra, A.S.B., Piezoelectric and dielectric properties of collagen films. *Phys. Status Solidi*, 176, 1077, 1999.
50. Denning, D., Paukshto, M.V., Habelitz, S., Rodriguez, B.J., Piezoelectric properties of aligned collagen membranes. *J. Biomed. Mater. Res. Part B*, 102, 284, 2014.
51. Ghosh, S.K. and Mandal, D., Efficient natural piezoelectric nanogenerator: Electricity generation from fish swim bladder. *Nano Energy*, 28, 356, 2016.
52. Ghosh, S.K. and Mandal, D., High-performance bio-piezoelectric nanogenerator made with fish scale. *Appl. Phys. Lett.*, 109, 103701, 2016.
53. Ghosh, S.K. and Mandal, D., Sustainable energy generation from piezoelectric biomaterial for noninvasive physiological signal monitoring. *ACS Sustain. Chem. Eng.*, 5, 8836, 2017.

54. Kumar, C., Gaur, A., Tiwari, S., Biswas, A., Rai, S.K., Maiti, P., Bio-waste polymer hybrid as induced piezoelectric material with high energy harvesting efficiency. *Compos. Commun.*, 11, 56, 2019.
55. Vivekananthan, V., Alluri, N.R., Purusothaman, Y., Chandrasekhar, A., Selvarajan, S., Kim, S.-J., Biocompatible collagen nanofibrils: An approach for sustainable energy harvesting and battery-free humidity sensor applications. *ACS Appl. Mater. Interfaces*, 10, 18650, 2018.
56. Martin, A.J.P., Tribo-electricity in wool and hair. *Proc. Phys. Soc.*, 53, 186, 1941.
57. Fukada, E., Zimmerman, R.L., Mascarenhas, S., Denaturation of horn keratin observed by piezoelectric measurements. *Biochem. Biophys. Res. Commun.*, 62, 415, 1975.
58. Feughelman, M., Lyman, D., Menefee, E., Willis, B., The orientation of the α-helices in α-keratin fibres. *Int. J. Biol. Macromol.*, 33, 149, 2003.
59. Kalinin, S.V., Rodriguez, B.J., Jesse, S., Seal, K., Proksch, R., Hohlbauch, S., Revenko, I., Thompson, G.L., Vertegel, A.A., Towards local electromechanical probing of cellular and biomolecular systems in a liquid environment. *Nanotechnology*, 18, 424020, 2007.
60. Stapleton, A., Noor, M.R., Sweeney, J., Casey, V., Kholkin, A.L., Silien, C., Gandhi, A.A., Soulimane, T., Tofail, S.A.M., The direct piezoelectric effect in the globular protein lysozyme. *Appl. Phys. Lett.*, 111, 142902, 2017.
61. Stapleton, A., Ivanov, M.S., Noor, M.R., Silien, C., Gandhi, A.A., Soulimane, T., Kholkin, A.L., Tofail, S.A.M., Converse piezoelectricity and ferroelectricity in crystals of lysozyme protein revealed by piezoresponse force microscopy. *Ferroelectrics*, 525, 135, 2018.
62. Ghosh, S.K., Adhikary, P., Jana, S., Biswas, A., Sencadas, V., Gupta, S.D., Tudu, B., Mandal, D., Electrospun gelatin nanofiber based self-powered bio-e-skin for healthcare monitoring. *Nano Energy*, 36, 166, 2017.
63. Gaur, A., Tiwari, S., Kumar, C., Maiti, P., Polymer biowaste hybrid for enhanced piezoelectric energy harvesting. *ACS Appl. Electron. Mater.*, 2, 1426, 2020.
64. Lee, B.Y., Zhang, J., Zueger, C., Chung, W.-J., Yoo, S.Y., Wang, E., Meyer, J., Ramesh, R., Lee, S.-W., Virus-based piezoelectric energy generation. *Nat. Nanotechnol.*, 7, 351, 2012.
65. Shin, D.-M., Han, H.J., Kim, W.-G., Kim, E., Kim, C., Hong, S.W., Kim, H.K., Oh, J.-W., Hwang, Y.-H., Bioinspired piezoelectric nanogenerators based on vertically aligned phage nanopillars. *Energy Environ. Sci.*, 8, 3198, 2015.
66. 28. Lee, J.-H., Lee, J.H., Xiao, J., Desai, M.S., Zhang, X., Lee, S.-W., Vertical self-assembly of polarized phage nanostructure for energy harvesting. *Nano Lett.*, 19, 2661, 2019.
67. Imoto, K., Date, M., Fukada, E., Tahara, K., Kamaiyama, Y., Yamakita, T., Tajitsu, Y., Piezoelectric motion of poly (L-lactic acid) film improved by supercritical CO_2 treatment. *Jpn. J. Appl. Phys.*, 48, 09KE06, 2009.

68. Yoshida, T., Imoto, K., Tahara, K., Naka, K., Uehara, Y., Kataoka, S., Date, M., Fukada, E., Tajitsu, Y., Piezoelectricity of poly (L-lactic acid) composite film with stereocomplex of poly (L-lactide) and poly (D-lactide). *Jpn. J. Appl. Phys.*, 49, 09MC11, 2010.
69. Yen, C.C., Tokita, M., Park, B., Takezoe, H., Watanabe, J., Spontaneous organization of helical polypeptide molecules into polar packing structure. *Macromolecules*, 39, 1313, 2006.
70. Namiki, K., Hayakawa, R., Wada, Y., Molecular theory of piezoelectricity of α-helical polypeptide. *J. Polym. Sci.: Polym. Phys. Ed.*, 18, 993, 1980.
71. Furukawa, T. and Fukada, E., Piezoelectric effect and its temperature variation in optically active polypropylene oxide. *Nature*, 221, 1235, 1969.
72. Furukawa, T. and Fukada, E., Piezoelectric relaxation in poly (γ-benzyl-glutamate). *J. Polym. Sci.: Polym. Phys. Ed.*, 14, 1979, 1976.
73. Date, M., Takashita, S., Fukada, E., Temperature variation of piezoelectric moduli in oriented poly(γ-methyl L-glutamate). *J. Polym. Sci. A-2: Polym. Phys.*, 8, 61, 1970.
74. Nakiri, T., Imoto, K., Ishizuka, M., Okamoto, S., Date, M., Uematsu, Y., Fukada, E., Tajitsu, Y., Piezoelectric characteristics of polymer film oriented under a strong magnetic field. *Jpn. J. Appl. Phys.*, 43, 6769, 2004.
75. Farrar, D., West, J.E., Busch-Vishniac, I.J., Yu, S.M., Fabrication of polypeptide-based piezoelectric composite polymer film. *Scr. Mater.*, 59, 1051, 2008.
76. Farrar, D., Ren, K.L., Cheng, D., Kim, S., Moon, W., Wilson, W.L., West, J.E., Yu, S.M., Permanent polarity and piezoelectricity of electrospun α-helical poly(α-amino acid) fibers. *Adv. Mater.*, 23, 3954, 2011.
77. Bera, S., Mondal, S., Xue, B., Shimon, L.J.W., Cao, Y., Gazit, E., Rigid helical-like assemblies from a self-aggregating tripeptide. *Nat. Mater.*, 18, 503, 2019.
78. Bera, S., Guerin, S., Yuan, H., O'Donnell, J., Reynolds, N.P., Maraba, O., Ji, W., Shimon, L.J.W., Cazade, P.-A., Tofail, S.A.M., Thompson, D., Yang, R., Gazit, E., Molecular engineering of piezoelectricity in collagen-mimicking peptide assemblies. *Nat. Commun.*, 12, 2634, 2021.
79. Bera, S., Cazade, P.-A., Bhattacharya, S., Guerin, S., Ghosh, M., Netti, F., Thompson, D., Adler-Abramovich, L., Molecular engineering of rigid hydrogels co-assembled from collagenous helical peptides based on single triplet motif. *ACS Appl. Mater. Interfaces*, 14, 46827, 2022.
80. Kholkin, A., Amdursky, N., Bdikin, I., Gazit, E., Rosenman, G., Strong piezoelectricity in bioinspired peptide nanotubes. *ACS Nano*, 4, 610, 2010.
81. Vasilev, S., Zelenovskiy, P., Vasileva, D., Nuraeva, A., Shur, V.Y., Kholkin, A.L., Piezoelectric properties of diphenylalanine microtubes prepared from the solution. *J. Phys. Chem. Solids*, 93, 68, 2016.
82. Jenkins, K., Kelly, S., Nguyen, V., Wu, Y., Yang, R.S., Piezoelectric diphenylalanine peptide for greatly improved flexible nanogenerators. *Nano Energy*, 51, 317, 2018.

83. Nguyen, V., Jenkins, K., Yang, R.S., Epitaxial growth of vertically aligned piezoelectric diphenylalanine peptide microrods with uniform polarization. *Nano Energy*, 17, 323, 2015.
84. Yuan, H., Lei, T.M., Qin, Y., He, J.H., Yang, R.S., Design and application of piezoelectric biomaterials. *J. Phys. D: Appl. Phys.*, 52, 194002, 2019.
85. Dayarian, S., Kopyl, S., Bystrov, V., Correia, M.R., Ivanov, M.S., Pelegova, E., Kholkin, A., Effect of the chloride anions on the formation of self-assembled diphenylalanine peptide nanotubes. *IEEE Trans. Ultrason. Ferroelectr.*, 65, 1563, 2018.
86. Bosne, E.D., Heredia, A., Kopyl, S., Karpinsky, D.V., Pinto, A.G., Kholkin, A.L., Piezoelectric resonators based on self-assembled diphenylalanine microtubes. *Appl. Phys. Lett.*, 102, 073504, 2013.
87. Nguyen, V., Kelly, S., Yang, R., Piezoelectric peptide-based nanogenerator enhanced by single-electrode triboelectric nanogenerator. *APL Mater.*, 5, 074108, 2017.
88. Nguyen, V., Zhu, R., Jenkins, K., Yang, R., Self-assembly of diphenylalanine peptide with controlled polarization for power generation. *Nat. Commun.*, 7, 13566, 2016.
89. Lee, J.-H., Heo, K., Schulz-Schonbagen, K., Lee, J.H., Desai, M.S., Jin, H.E., Lee, S.W., Diphenylalanine peptide nanotube energy harvesters. *ACS Nano*, 12, 8138, 2018.
90. Basavalingappa, V., Bera, S., Xue, B., O'Donnell, J., Guerin, S., Cazade, P.A., Yuan, H., Haq, E.U., Silien, C., Tao, K., Shimon, L.J.W., Tofail, S.A.M., Thompson, D., Kolusheva, S., Yang, R., Cao, Y., Gazit, E., Diphenylalanine-derivative peptide assemblies with increased aromaticity exhibit metal-like rigidity and high piezoelectricity. *ACS Nano*, 14, 7025, 2020.
91. Tao, K., Hu, W., Xue, B., Chovan, D., Brown, N., Shimon, L.J.W., Maraba, O., Cao, Y., Tofail, S.A.M., Thompson, D., Li, J., Yang, R., Gazit, E., Bioinspired stable and photoluminescent assemblies for power generation. *Adv. Mater.*, 31, 1807481, 2019.
92. Tao, K., Xue, B., Li, Q., Hu, W., Shimon, L.J.W., Makam, P., Si, M., Yan, X., Zhang, M., Cao, Y., Yang, R.S., Li, J.B., Gazit, E., Stable and optoelectronic dipeptide assemblies for power harvesting. *Mater. Today*, 30, 10, 2019.
93. Vasilescu, D., Cornillon, R., Mallet, G., Piezoelectric resonances in amino-acids. *Nature*, 225, 635, 1970.
94. Guerin, S., Tofail, S.A.M., Thompson, D., Longitudinal piezoelectricity in orthorhombic amino acid crystal films. *Cryst. Growth Des.*, 18, 4844, 2018.
95. Guerin, S., Tofail, S.A.M., Thompson, D., Deconstructing collagen piezoelectricity using alanine-hydroxyproline-glycine building blocks. *Nanoscale*, 10, 9653, 2018.
96. O'Donnell, J., Sarkar, S.M., Guerin, S., Borda, G.G., Silien, C., Soulimane, T., Thompson, D., O'Reilly, E., Tofail, S.A.M., Piezoelectricity in the proteinogenic amino acid L-leucine: A novel piezoactive bioelectret. *IEEE Trans. Dielectr. Electr. Insul.*, 27, 5, 1465–1468, October 2020.

97. Guerin, S., O'Donnell, J., Haq, E.U., McKeown, C., Silien, C., Rhen, F.M.F., Soulimane, T., Tofail, S.A.M., Thompson, D., Racemic amino acid piezoelectric transducer. *Phys. Rev. Lett.*, 122, 047701, 2019.
98. Heredia, A., Meunier, V., Bdikin, I.K., Gracio, J., Balke, N., Jesse, S., Tselev, A., Agarwal, P.K., Sumpter, B.G., Kalinin, S.V., Nanoscale ferroelectricity in crystalline γ-glycine, nanoscale ferroelectricity in crystalline γ-glycine. *Adv. Funct. Mater.*, 22, 2996, 2012.
99. Isakov, D., d. M. Gomes, E., Bdikin, I., Almeida, B., Belsley, M., Costa, M., Rodrigues, V., Heredia, A., Production of polar β-glycine nanofibers with enhanced nonlinear optical and piezoelectric properties. *Cryst. Growth Des.*, 11, 4288, 2011.
100. Kumar, R.A., Vizhi, R.E., Vijayan, N., Babu, D.R., Structural, dielectric and piezoelectric properties of nonlinear optical γ-glycine single crystals. *Phys. B*, 406, 2594, 2011.
101. Seyedhosseini, E., Bdikin, I., Ivanov, M., Vasileva, D., Kudryavtsev, A., Rodriguez, B., Kholkin, A., Tipinduced domain structures and polarization switching in ferroelectric amino acid glycine. *J. Appl. Phys.*, 118, 072008, 2015.
102. Guerin, S., Stapleton, A., Chovan, D., Mouras, R., Gleeson, M., McKeown, C., Noor, M.R., Silien, C., Rhen, F.M.F., Kholkin, A.L., Liu, N., Soulimane, T., Tofail, S.A.M., Thompson, D., Control of piezoelectricity in amino acids by supramolecular packing. *Nat. Mater.*, 17, 180, 2018.
103. Ji, W., Xue, B., Arnon, Z.A., Yuan, H., Bera, S., Li, Q., Zaguri, D., Reynolds, N.P., Li, H., Chen, Y., Gilead, S., Rencus-Lazar, S., Li, J., Yang, R., Cao, Y., Gazit, E., Rigid tightly packed amino acid crystals as functional supramolecular materials. *ACS Nano*, 13, 14477, 2019.
104. Ji, W., Xue, B., Bera, S., Guerin, S., Liu, Y., Yuan, H., Li, Q., Yuan, C., Shimon, L.J.W., Ma, Q., Kiely, E., Tofail, S.A.M., Si, M., Yan, X., Cao, Y., Wang, W., Yang, R., Thompson, D., Li, J., Gazit, E., Tunable mechanical and optoelectronic properties of organic cocrystals by unexpected stacking transformation from H-to J-and X-aggregation. *ACS Nano*, 14, 10704, 2020.
105. Ji, W., Xue, B., Bera, S., Guerin, Shimon, L.J.W., Ma, Q., Tofail, S.A.M., Thompson, D., Cao, Y., Wang, W., Gazit, E., Modulation of physical properties of organic cocrystals by amino acid chirality. *Mater. Today*, 42, 29, 2021.
106. Ghosal, C., Ghosh, S.K., Roy, K., Chattopadhyay, B., Mandal, D., Environmental bacteria engineered piezoelectric bio-organic energy harvester towards clinical applications. *Nano Energy*, 93, 106843, 2022.
107. Joseph, J., Singh, S.G., Vanjari, S.R.K., Leveraging innate piezoelectricity of ultra-smooth silk thin films for flexible and wearable sensor applications. *IEEE Sens. J.*, 17, 8306, 2017.
108. Wang, X., Gu, Y., Xiong, Z., Cui, Z., Zhang, T., Silk-molded flexible, ultrasensitive, and highly stable electronic skin for monitoring human physiological signals. *Adv. Mater.*, 26, 1336, 2014.

109. Santos, D., Silva, D.M., Gomes, P.S., Fernandes, M.H., Santos, J.D., Sencadas, V., Multifunctional PLLA-ceramic fiber membranes for bone regeneration applications. *J. Colloid Interface Sci.*, 504, 101, 2017.
110. Noris-Suárez, K., Lira-Olivares, J., Marina Ferreira, A., Feijoo, J.L., Suárez, N., Hernández, M.C., Barrios, E., In vitro deposition of hydroxyapatite on cortical bone collagen stimulated by deformation-induced piezoelectricity. *Biomacromolecules*, 8, 941, 2007.
111. Panwar, V., Babu, A., Sharma, A., Thomas, J., Chopra, V., Malik, P., Rajput, S., Mittal, M., Guha, R., Chattopadhyay, N., Mandal, D., Ghosh, D., Tunable, conductive, self-healing, adhesive and injectable hydrogels for bioelectronics and tissue regeneration applications. *J. Mater. Chem. B*, 9, 6260, 2021.
112. Bhang, S.H., Jang, W.S., Han, J., Yoon, J.-K., La, W.-G., Lee, E., Kim, Y.S., Shin, J.-Y., Lee, T.-J., Baik, H.K., Kim, B.-S., Zinc oxide nanorod-based piezoelectric dermal patch for wound healing. *Adv. Funct. Mater.*, 27, 1603497, 2017.
113. Kapat, K., Shubhra, Q.T.H., Zhou, M., Leeuwenburgh, S., Piezoelectric nano-biomaterials for biomedicine and tissue regeneration. *Adv. Funct. Mater.*, 30, 1909045, 2020.
114. Ashrafi, M., Alonso-Rasgado, T., Baguneid, M., Bayat, A., The efficacy of electrical stimulation in experimentally induced cutaneous wounds in animals. *Vet. Dermatol.*, 27, 235, 2016.
115. Reid, B., Song, B., McCaig, C.D., Zhao, M., Wound healing in rat cornea: The role of electric currents. *FASEB J.*, 19, 379, 2005.
116. Sharma, A., Panwar, V., Mondal, B., Prasher, D., Bera, M.K., Thomas, J., Kumar, A., Kamboj, N., Mandal, D., Ghosh, D., Electrical stimulation induced by a piezo-driven triboelectric nanogenerator and electroactive hydrogel composite, accelerate wound repair. *Nano Energy*, 99, 107419, 2022.

12
Protein Cages and their Potential Application in Therapeutics

Chiging Tupe[1,2] and Soumyananda Chakraborti[1,2]*

[1]*National Institute of Malaria Research, Dwarka Sector 8, New Delhi, India*
[2]*Academy of Scientific and Innovative Research (AcSIR), UP, India*

Abstract

Nature uses self-assembly properties of protein to produce a large variety of complex and highly symmetric protein architectures with a broad range of biological activities. These complexes include filaments, protein lattices and symmetric cages. Within different supra-molecular protein architectures (complexes), protein cages are probably the most sophisticated protein-based architectures. Their self-assembly from a small number of subunits into symmetrical, mono-disperse architectures have inspired thousands of scientists from multidisciplinary fields. Cage architectures are abundant in nature starting from virus capsid to bacterial micro-compartment, such as carboxysome. In the last two decades, protein cages have been developed as extremely useful material for biotechnology and therapeutic applications, mostly because of their remarkable diversity in sizes, shapes and structures. Furthermore, their ease of production in large quantities using biological systems, and biocompatibility makes them a very attractive system for manipulation. Protein cages were also found useful for a variety of other applications, including different nanoparticle synthesis, catalysis, diagnostics and targeted drug delivery. For example, protein cage can make well-defined mono-disperse spheres that can be filled with cargo and the protein cage protects the cargo from harsh external environment. This is a very useful strategy and protein cage-based cargo delivery, is now undergoing clinical trials. It has also been shown that both the interface and outer surface of the protein cage are highly amenable for different types of modifications, so it holds significant promise for targeting applications including tissue and cell specific delivery applications. In this book chapter, we are going to showcase recent findings related to protein cages, including protein cage generation and their potential application in therapeutics.

---

*Corresponding author: soumyabiochem@gmail.com; ORCID: 0000-0002-7384-690X

Keywords: Self-assembly of protein, protein cage, protein architecture, targeted delivery, protein-based therapeutics, drug delivery, vaccine development

12.1 Introduction

Material scientists for a long time are trying to build complex self-assembled protein architecture capable of therapeutic applications utilizing bio-inspired approaches, as there are many complex self-assembled protein structures exist in nature. In nature, most prominent examples of self-assembled three-dimensional material is bone, its excellent mechanical property comes from tightly regulated hierarchical structure of the hydroxyapatite and collagen assembly across macro scales [1]. Other noteworthy example of complex self-assembled structure is muscle, which is constructed of multiple bio-molecular building blocks. The functional properties of muscle emerge as a consequence of multiple interaction partners including actin, myosin and titin [2]. In recent years, nanoparticles, polymers and liquid crystalline molecules have been routinely used to develop complex architectures with delivery capabilities mostly via intermolecular noncovalent interactions such as ionic bonding, hydrophobic interactions and hydrogen bonding [1]. In contrast to these systems, complex delivery vehicles (for e.g. protein cages) made from biological building blocks has several unique advantages specially for biotechnology and therapeutic application mainly because of their high biocompatibility and less toxicity [3].

In nature, one can easily find varieties of self-assembled architectures with diverse biological functions [4]. Amongst these, protein cages are unique and carry one of the most delicate structures [3, 4]. In biology, these protein-based compartments encapsulate natural cargos such as the nucleic acids of virus genomes as well as the nanoparticles of iron oxide that are encapsulated within ferritin [3–5]. The majority of protein cages are hollow and spherical though there are some notable exceptions, and often possess internal symmetry either icosahedral, octahedral or tetrahedral which plays an important role in controlling their inter-subunit interactions. In the last two decades protein cages have been developed as modular platforms for cargo encapsulation and delivery because of their unusual stability and targeting ability [6]. Furthermore, their facile production in large quantities using biological systems, biocompatibility and variety of options for modifications make them attractive choices for drug delivery, protein encapsulation, and delivery application and vaccine development (Figure 12.1) [6, 7].

Currently there are three main sources of protein cages. i) Virus-based protein cages or Virus Like Particles (VLPs) ii) bacteria

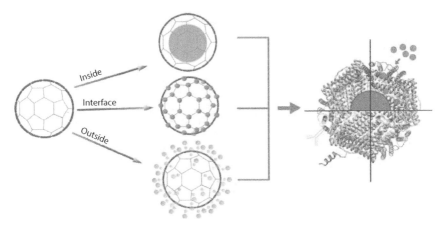

Figure 12.1 Schematic showing possible ways of protein cage modifications and generation of multifunctional protein cage by utilizing all different types of modifications (figure was conceptualized based on earlier work DOI: 10.1126/science.1123223).

micro-compartment–based protein cages and iii) artificial or synthetic protein cages [8]. Among these three classes of protein cages virus-based protein cages are the largest and also most diversified [9]. By definition VLPs are viruses devoid of their genetic materials, structurally they are very robust and biocompatible, and normally show extreme stability under different stress conditions compared to normal protein. Furthermore their ability to self-assemble from single subunits and encapsulate variety of cargoes makes them attractive as modular therapeutic containers [9–11]. Within different virus-based protein cages (or VLPs), Cowpea chlorotic mottle virus (CCMV)-based VLP is widely studied one, mostly because of its broad range of applications in therapeutics, protein delivery and catalysis [12, 13]. CCMV-based VLPs were first developed from cowpea plant (basically CCMV is a plant-based RNA virus) and structurally it has an icosahedral shell consisting of 180 identical 20 kDa subunits (coat protein) [12]. Its outer and inner diameters are 28 and 18 nm, respectively. Recently, CCMV-VLP is found as a useful tool in various biotechnology and therapeutic applications due to its controllable cage assembly/disassembly ability and cargo capture/release features. The CCMV assembly/disassembly is pH dependent and low pH triggers its assembly while high pH disassembles the viral particles [11, 12]. CCMV-VLP encapsulates cargoes with high loading capacities mainly based on the electrostatic interaction, especially towards negatively charged cargoes (e.g. nucleic acid) by interacting with its positively charged N-terminus region of each subunit [14]. Recently it has been shown that CCMV-VLP is capable of encapsulating proteins, such as horseradish peroxidase and fluorescence protein in high efficiency [15, 16]. It has been also found that both internal and external surfaces of CCMV are highly amenable for variety of modifications therefore CCMV holds significant promise for biotechnology and therapeutic application [17].

Beside VLPs, ferritin is undoubtedly the most versatile protein cage, and it's already used for many different applications mainly because of its unique bio-mineralization ability and extraordinary stability [18, 19]. Ferritin is a protein whose *in vivo* role is iron storage and in that way it protect cells from iron related toxicity. Ferritin proteins are found in all domains of life except yeast and some parasites. It is consisted of 24 protein subunits and each subunit is approximately 20 kDa weight. Self-assemblies of individual subunits lead to formation of dimmers and eventually a dodecameric cage with an outer diameter of approximately 12 nm and an inner cavity of 8 nm. Each subunit is roughly cylindrical, with size 50 Å in length and 25 Å in diameter and further it has been found that within each subunit there are extensive side chain interactions that provide high stability to ferritin [20]. The cage is stable up to 120 °C and tolerates relatively high levels of urea, guanidinium chloride, and many other denaturants [21, 22]. Ferritin without the inorganic core is known as apo-ferritin.

Connecting protein cages together to form higher order assembly is a new research area and it is getting attention due to its wide spread biological applications. To date one-dimensional (1D) assembly of protein cages into tube-like structures has been successfully demonstrated through a head-to-head assembly of individual cages. For example, Aida *et al.* have shown a micrometer long self-assembled nanotube from a chemically modified GroEL chaperonin cage [23]. Ordered two-dimensional (2D) arrays of proteins have also been studied in great detail as they are of particular interest in electron crystallography and nanotechnology [24, 25]. Some of the earliest examples of self-assembled 2D arrays comes from ferritin and was prepared on lipid monolayers at the air–water interface [24, 25] where interaction of protein cages with polar head groups of lipids was key to manipulate array formation. Construction of three-dimensional (3D) superlattices from functional nanoparticle building blocks in a controlled manner is the most challenging, as most of the time it leads to amorphous aggregates. Nevertheless, assembling a 3D superlattice is of significant interest among materials scientists because they have the potential for making a new class of materials. A range of methods have been explored to construct 3D assemblies of nanoparticles, peptides and in most cases, linker molecules have been utilized to direct and control assembly. So far, it has been found that electrostatic interactions plays the most crucial role to induce assembly [26, 27]. For example, Kostiainen *et al.* have demonstrated the formation of a protein cage (ferritin and CCMV) based superlattice materials through electrostatic interactions between the protein cages and oppositely charged mediator particles [28]. Although higher order protein cage–based architecture shows lots of promises, but their development as a modular delivery vehicle is still in the stage of infancy.

Among different protein cages, synthetic protein cages have evolved most recently. The concept of synthetic protein cage was first surfaced by *Yates et al.* with a pioneering study published in Science [29], afterward, there were several publications showing practicality of *in-silico* designed highly programmable protein cages [8, 30]. These synthetic protein cages are highly modular and can withstand multiple modifications in their structure that makes them very unique for targeted delivery application. Biological cargo loading is also very easy within synthetic protein cage cavity compared to any naturally occurring protein cages and that is one of the main reasons for their popularity within the community [30–32]. Recently it has been also shown that symmetry specific sites of synthetic protein cage could be utilized for cargo loading [33].

12.2 Different Methods of Cage Modifications and Cargo Loading

In the above section, we already discussed different types of protein cage, their genesis, and their pros and cons. One of the main applications of protein cage in biotechnology and therapeutics is controlled cargo encapsulation and targeted delivery. Scientists are quite active in this field especially in the last two decades and developed several interesting methods for cargo loading inside protein cages. Encapsulations of cargo inside protein cages were mainly achieved through passive diffusion, electrostatic interaction, or by using direct genetic fusion to cage subunits (in our recent review we already discussed them in details) [34]. In the proceeding section we briefly discussed about few specialized techniques available for protein cage modification and cargo (mainly protein) encapsulation.

i) By Supercharging: One of the sophisticated methods available for protein cage modification and cargo loading. In the supercharging method either protein cage or cargo molecules are modified to make them highly positive or negatively charged molecules, and several methods available for the modification, first and most convenient one is to generate supercharged protein cages [35]. Historically, protein supercharging was done to suppress protein amorphous aggregation [36]. As electrostatic force plays a very important role in any kind of cargo loading inside protein cages, [37] we believe supercharging will be a very fruitful strategy in future. So far Beck et al. generated supercharged version of ferritin protein cages from human H-chain of ferritin, and using both positive and negative version of supercharged ferritin cage authors were also successful in generating a binary crystal [38, 39]. In a very similarly way, supercharged version of lumazine synthase protein cage was also produced and supercharged variant was found more efficient in cargo encapsulation and sorting [37, 38]. Currently supercharging strategy was also used in modifying cargo molecules as well. For example, positively supercharged GFP was designed for encapsulation within *Thermotoga maritima* ferritin [11].

ii) Circular permutation: This strategy is so far used in limited number protein cages, and one of the most successful examples of circular permutation is Lumazine synthase protein cage (AaLS). By using this strategy one could easily generate shape variations and alternative cargo loading opportunities inside protein cages [40]. Circular permutation is a strategy to alter the connectivity of secondary structural elements in proteins. Theoretically, the native N- and C-termini are connected with a polypeptide linker, followed by reopening of the circular protein elsewhere in the sequence. In cage-forming proteins, this topological alteration allows relocation of the N- and C-termini on the individual subunits with respect to the overall structure, e.g. from the exterior to the interior or vice versa. Circular permutation also provides a means of fine-tuning the overall tertiary shape of the protein cage. Beside lumazine synthatase, this approach was also used in propanediol utilization microcompartment shell (Pdu) of *Salmonella typhimurium* [41].

iii) Direct evolution: Directed evolution is an extremely elegant technique for generating engineered protein cages for a wide range of applications including cargo loading and release. This technique is based on natural evolution principles, and iterative cycles of mutagenesis [42]. The last step is screening or selection to develop completely new protein and this strategy is extremely useful for both protein cage and cargo molecule modifications and to improve cargo loading efficacy. So far this strategy has been extensively applied in lumazine synthase, MS2 bacteriophage and few other synthetic protein cages including O3-33. Result clearly shows that directed evolution method improves lumazine synthase cargo packing capacity significantly [8, 43].

iv) Genetic incorporation of biologically orthogonal functional groups: Incorporation of unnatural amino acids (uAAs) is a very powerful strategy to incorporate cargo inside protein cages [44]. Incorporation of unnatural amino acids (uAAs) not only provides site-specificity, it also allows introduction of a range of functionality in side chains of amino acid, for example azides, alkynes, alkenes, and tetrazines [45, 46]. There are various approaches are available for unnatural amino acid incorporation within protein structure, however the most common strategy is based on expansion of the genetic code by co-expression of aminoacyl-tRNA synthetase (aaRS) and tRNA pairs from orthogonal species like *P. horikoshii* and *M. jannaschii* [47]. The ultimate result of this method is suppression of the amber stop codon and incorporation of the unnatural amino acid at nearly any site in the protein.

v) Coiled coil motifs: Incorporating coiled-coil peptide motifs either in protein cages or inside cargo molecules by genetic modification is a futuristic strategy for controlled cargo loading inside a protein cage. Structurally coiled-coil peptides are not true helices but are closely related. Coiled-coil assemblies are relatively simple, and sequence-to-structure relationships are available that permit their prediction and design. Peptide sequences which exhibit high affinity to specific motifs, metal, inorganic and polymer materials will be identified by several techniques such as phage display library and *de novo* design [48]. These peptides can be readily introduced on the interior surfaces of protein cage and exterior surface of the cargo molecule generally, through either genetic or chemical modification. Afterword proper titration with complementary peptides carrying cargo to the protein cage leads to controlled cargo loading inside the protein cage.

vi) By using novel genetic tag: This strategy is mainly used for cargo molecules (protein) only, an engineered cargo protein with an efficient genetic tag in general provides an excellent opportunity for cargo loading inside the protein cage. There are several protein tags available which are found suitable for cargo encapsulation inside protein cages for example polyhistidine tag was used in E2 protein cage for peptide coated gold nanoparticles encapsulation [49]. Beside polyhistidine tag, SpyTag/SpyCatcher system was one of the most common and robust ligation methods available for protein cage modification. The SpyTag/SpyCatcher-based technology was developed very recently and found widespread application in recombinant proteins including protein cages [50]. SpyTag is in general thirteen amino acids long and spontaneously reacts with the protein carrying SpyCatcher (12.3 kDa) to form a very strong covalent bond between the pairs. In general the protein cage is modified with spy-tag and the cargo is conjugated with spy-catcher [34]. The SpyTag/SpyCatcher-based ligation technique was successfully applied in MS2 bacteriophage capsid to encapsulate multiple enzymes [51].

12.3 Applications of Protein Cages in Biotechnology and Therapeutics

Although protein cages have many different applications (from chemical catalysis to nanoparticle generation), however, in therapeutic it is mainly used as a programmable synthetic container for targeted (drug/enzyme/therapeutic protein) delivery and vaccine development [8–11]. In the next section we are summarizing three major application areas of protein cage in therapeutics.

12.3.1 Protein Cage as Targeted Delivery Vehicles for Therapeutic Protein

In recent years protein based therapeutics has gained high popularity, mostly because of their high specificity and targeting capabilities compared to synthetic small molecule analogues [8, 31]. Furthermore it has been found that most of the protein-based therapeutics are biocompatible and they are available in solution form, so that they can be readily injected in patient [8]. Unquestionably, protein cargo encapsulation inside protein cage is powerful strategy to protect enzyme/therapeutic protein from harsh environmental condition (for e.g. proteases, nucleases, pH, chaotropic agent and temperature mediated degradation) and increase protein/enzyme shelf life (Figure 12.2) [8–11, 21]. Furthermore, it was found in many cases that encapsulated enzyme inside protein cage shows higher activity and protein cage acts as

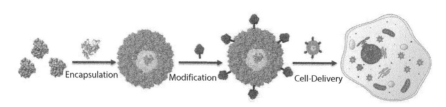

Figure 12.2 Schematic showing therapeutic protein encapsulation and further decoration of the protein cage for cell delivery. Far-right image depicts application of protein cage in cellular delivery.

a chaperone (due to protein crowding) compared to protein in same state present in solution [21]. However, like any other emerging technology, protein cage-based systems also have several limitations for example many protein cages have a tendency to aggregate, which leads to inactivation, also they are very difficult to be generate at industrial scale; In addition, most of the protein-based cages cause unwanted immune response. Furthermore, most of the protein cages show poor membrane permeability, and almost all protein-based drugs target receptors residing on the cell surfaces [52]. Among different protein-based cages, virus-like-particles (VLP) and ferritin are the two most widely used cages available for cell delivery application. Most common ways of cargo encapsulation inside any VLPs are following i) through passive diffusion ii) using electrostatic interaction and iii) direct genetic fusion of cargo to cage subunits [34]. Furthermore it has been found that it can be programmed to uptake cargo with different size/number or different cargo characteristics, e.g. surface charge, or affinity interaction via the internal modification on the protein cages [8–11]. In addition, VLP can be further modified externally to serve different purposes, such as fluorescence labeling for imaging, diminishing immune response by PEGylation or adding specific molecule (for example aptamer or nanobody) for target tissue delivery [4–11]. So far the best protein cage-based therapeutic (protein encapsulation and targeting) system were developed from VLPs, for example both P22 bacteriophage VLPs and CCMV-VLPs were found highly efficient in Cytochrome P450 (CYP, known for pro-drug activation) encapsulation and delivery [8, 9, 53]. CCMV-VLPs were also used for lysozyme encapsulation which further facilitates potential antibiotic use of lysozyme [54]. Similarly, P22 bacteriophage-based VLP was also used for several therapeutic protein encapsulation including α-galactosidase enzymes and glutathione (GSH) encapsulation [53, 55]. In addition, Steinmetz and co-workers recently develops tobacco mosaic virus (TMV)-based therapeutic cage by bio-conjugating thrombolytic enzymes streptokinase on TMV [56]. Apart from VLPs, ferritin was also found very effective in carrying therapeutic proteins (for example cytochrome C) in its cavity and further targeted delivery to the cancer cells [57]. Currently several VLP-based delivery platform are subjected to clinical trials.

12.3.2 Protein Cage-Based Encapsulation and Targeting of Anticancer Drugs

Apart from therapeutic protein/enzyme encapsulation, protein cage is also found very effective in targeted drug delivery application. Key features of any new drug delivery platform are centred on i) drug encapsulation and ii) its systemic release, and in both these categories the protein cage-based system provides excellent alternatives to existing ones. There are mainly four ways of drug encapsulation inside protein cages, (i) assembly/disassembly-based encapsulation, (ii) pore-based encapsulation, (iii) physical interaction-based encapsulation, and (iv) encapsulation by covalent modification [58]. It's already known from various studies that protein cages are highly effective in carrying anticancer drugs like doxorubicin, duanomycin, carbopaltin, and cisplatin, and they are also able to deliver these drugs to their target tissue with high payload [59]. In general intracellular delivery of highly hydrophobic drug such as paclitaxel is always difficult, mainly due to its insoluble nature and very high lipophilicity along with extremely poor pharmacokinetics, when administered in body via any known routes. As protein cages carry hydrophobic patches throughout its cavity so it could be an efficient alternative for super-hydrophobic drug loading and delivery. Interestingly, protein cages like ferritin, MS2 bacteriophage, adenovirus-based VLPs and recently developed synthetic cage (such as O3-33) already shows lot of promise in this field [8–11]. In order to achieve higher targeting ability, protein cages are routinely modified with cell targeting peptide or ligands (for e.g., folate) or aptamer (antibody equivalent of DNA) or nanobody (single-chain antibody) for precision targeting [4–11, 18].

12.3.3 Protein Cage-Based Immune-Therapy

In addition to protein-based cargo or drug delivery, protein cage nanostructure could be effectively used for immune-therapy and vaccine development. Irrespective of their structure (either icosahedral or filamentous) both geometry and multivaliancy of the protein cage contributes significantly towards immune cell activation [8]. In general protein cages interact with the immune system through toll-like receptors (for e.g. TLR9) activation. Furthermore, external decoration strategy with peptide antigen or oligodeoxynucleotides makes them one of the most powerful immune system activators. Recently, protein cage decorated with oligodeoxynucleotides (CpG) (known for dendritic cell activation) shows excellent immune response [60]. In a separate study protein cage (pyruvate dehydrogenase E2 particles) modified with melanoma-derived peptide gp100 and CpG activates strong T-cell response and as a result of this immunotherapy animal survival rate against melanoma increased several fold [61]. This type of strategy could be very efficiently used in future immunotherapy. In addition to promising immunotherapy application, protein cage-based vaccines are also gaining mass attention, recently VLP-based vaccine against Human papilloma virus such as GlaxoSmithKline Biologicals' Cervarix® or

Merck's Gardasi are readily available in market [8]. Beside this some other VLP-based vaccines are also under clinical trials. TVEC under the trade name Imlygic was recently approved for oncotherapy against melanoma [62]. Both immune-stimulatory properties and unique size of protein cage makes them ideal candidates for vaccine delivery.

12.4 Future Perspective

Enzyme replacement therapy (ERT) is a treatment procedure where deficient enzymes in the body are replaced by externally produced enzymes. This treatment is found very effective for several lysozyme storage disorder (LSD) disease including Gaucher disease, Fabry disease, Hunter disease and Pompe disease [63]. Unfortunately, this therapy is very expensive (presently costs over US$ 100 000/patient per year) [63]. In addition to cost, inability of the therapeutic enzyme to reach target sites or inability to cross blood–brain barrier and unwanted immune responses are the other major concerns associated with the present therapy [63]. Therefore, urgent clinical intervention is needed to enhance the therapeutic response of ERT by development of safe and efficient targeted delivery systems. Protein cages satisfied all the primary requirements to be an extremely useful enzyme carrier system, with targeting ability, as detailed out in the above sections. The idea of encapsulating highly clinically relevant enzymes inside a protein cage and targeting its delivery as a proof of concept is unique and has many promising applications in several diseases (Figure 12.2).

12.5 Conclusion

This idea that we can design nanoscale structures with targeted delivery ability that can be used in various applications including therapeutics may sound like science fiction, but these topics are already being researched across the world and progress is being made. Complex disease like cancer and neurodegenerative-diseases still remain major killers, and protein cage offers great hope for progress in diagnostics and treatment.

Acknowledgment

The authors are thankful to NIMR, New Delhi for providing us necessary research infrastructure. We would like to thank Ms. Sweta Ghosh for proof reading. This work is supported by Ramalingaswami Fellowship (BT/RLF/Re-entry/09/2019) by DBT. CT is the recipient of CSIR-NET-JRF fellowship. CT is supported by AcSIR for PhD (registration no. 10BB21A65011).

References

1. Liu, Y., Luo, D., Wang, T., Hierarchical structures of bone and bioinspired bone tissue engineering. *Small*, 12, 4611–4632, 2016.
2. Herzog, W., The multiple roles of titin in muscle contraction and force production. *Biophys. Rev.*, 10, 1187–1199, 2018.
3. Aumiller Jr, W.M., Uchida, M., Douglas, T., Protein cage assembly across multiple length scales. *Chem. Rev.*, 47, 3433–3469, 2018.
3. Luo, Q., Hou, C., Bai, Y., Wang, R., Liu, J., Protein assembly: Versatile approaches to construct highly ordered nanostructures. *Chem. Rev.*, 116, 13571–13632, 2016.
4. Flenniken, M.L., Uchida, M., Liepold, L.O., Kang, S., Young, M.J., Douglas, T., A library of protein cage architectures as nanomaterials. *Curr. Top. Microbiol. Immunol.*, 327, 71–93, 2009.
5. Edwardson, T.G.W., Levasseur, M.D., Tetter, S., Steinauer, A., Hori, M., Hilvert, D., Protein cages: From fundamentals to advanced applications. *Chem. Rev.*, 122, 9145–9197, 2022.
6. Li, Y. and Champion, J.A., Self-assembling nanocarriers from engineered proteins: Design, functionalization, and application for drug delivery. *Adv. Drug Delivery Rev.*, 189, 114462, 2022.
7. Zhang, Y., Ardejani, M.S., Orner, B.P., Design and applications of protein-cage-based nanomaterials. *Chem. Asian J.*, 11, 2814–2828, 2016.
8. Heddle, J.G., Chakraborti, S., Iwasaki, K., Natural and artificial protein cages: Design, structure and therapeutic applications. *Curr. Opin. Struct. Biol.*, 43, 148–155, 2017.
9. Steinmetz, N.F., Lim, S., Sainsbury, F., Protein cages and virus-like particles: From fundamental insight to biomimetic therapeutics. *Biomater. Sci.*, 8, 2771–2777, 2020.
10. Edwardson, T.G.W. and Hilvert, D., Virus-inspired function in engineered protein cages. *J. Am. Chem. Soc.*, 141, 9432–9443, 2019.
11. Schwarz, B., Uchida, M., Douglas, T., Biomedical and catalytic opportunities of virus-like particles in nanotechnology. *Adv. Virus Res.*, 97, 1–60, 2017.
12. Lavelle, L., Michela, J.P., Gingery, M., The disassembly, reassembly and stability of CCMV protein capsids. *J. Virol. Methods*, 146, 311–316, 2007.
13. Comellas-Aragones, M. *et al.*, A virus-based single-enzyme nanoreactor. *Nat. Nanotechnol.*, 2, 635–639, 2007.
14. Schoonen, L. *et al.*, Modular, bioorthogonal strategy for the controlled loading of cargo into a protein nanocage. *Bioconjug. Chem.*, 29, 1186–1193, 2018.

15. Hassani-Mehraban, A. et al., Feasibility of Cowpea chlorotic mottle virus-like particles as scaffold for epitope presentations. *BMC Biotechnol.*, 15, 80, 2015.
16. Rurup, W.F. et al., Predicting the loading of virus-like particles with fluorescent proteins. *Biomacromolecules*, 15, 558–563, 2014.
17. Wege, C. and Koch, C., From stars to stripes: RNA-directed shaping of plant viral protein templates-structural synthetic virology for smart biohybrid nanostructures. *Wiley Interdiscip. Rev. Nanomed. Nanobiotechnol.*, 12, e1591, 2020.
18. Jutz, G., van Rijn, P., Miranda, B.S., Böker, A., Ferritin: A versatile building block for bionanotechnology. *Chem. Rev.*, 115, 1653–1701, 2015.
19. Chakraborti, S. and Chakrabarti, P., Self-assembly of ferritin: Structure, biological function and potential applications in nanotechnology. *Adv. Exp. Med. Biol.*, 1174, 313–329, 2019.
20. Uchida, M., Kang, S., Reichhardt, C., Harlen, K., Douglas, T., The ferritin superfamily: Supramolecular templates for materials synthesis. *Biochim. Biophys. Acta*, 1800, 834–845, 2018.
21. Chakraborti, S. et al., Three-dimensional protein cage array capable of active enzyme capture and artificial chaperone activity. *Nano Lett.*, 19, 3918–3924, 2019.
22. Kumar, M. et al., A single residue can modulate nanocage assembly in salt dependent ferritin. *Nanoscale*, 13, 11932–11942, 2021.
23. Sim, S., Niwa, T., Taguchi, H., Aida, T., Supramolecular nanotube of chaperonin GroEL: Length control for cellular uptake using single-ring GroEL mutant as end-Capper. *J. Am. Chem. Soc.*, 138, 11152–11155, 2016.
24. Yamashita, I., Fabrication of a two-dimensional array of nano-particles using ferritin molecule. *Thin Solid Films*, 393, 12–18, 2001.
25. Okuda, M. et al., Self-organized inorganic nanoparticle arrays on protein lattices. *Nano Lett.*, 5, 991–993, 2005.
26. Kostiainen, M.A., Kasyutich, O., Cornelissen, J.J.L.M., Nolte, R.J.M., Self-assembly and optically triggered disassembly of hierarchical dendron-virus complexes. *Nat. Chem.*, 2, 394–399, 2010.
27. Liljestrom, V., Seitsonen, J., Kostiainen, M.A., Electrostatic self-assembly of soft matter nanoparticle cocrystals with tunable lattice parameters. *ACS Nano*, 9, 11278–11285, 2015.
28. Kostiainen, M.A. et al., Electrostatic assembly of binary nanoparticle superlattices using protein cages. *Nat. Nanotechnol.*, 8, 52–56, 2013.
29. Lai, Y.T., Cascio, D., Yeates, T.O., Structure of a 16-nm cage designed by using protein oligomers. *Science*, 336, 1129, 2012.
30. Stupka, I. and Heddle, J.G., Artificial protein cages – inspiration, construction, and observation. *Curr. Opin. Struct. Biol.*, 64, 66–73, 2020.
31. Naskalaska, A., Artificial protein cage delivers active protein cargos to the cell interior. *Biomacromolecules*, 22, 4146–4154, 2021.
32. Majsterkiewicz, K. et al., Artificial protein cage with unusual geometry and regularly embedded gold nanoparticles. *Nano Lett.*, 22, 3187–3195, 2022.

33. McConnell, S.A., Designed protein cages as scaffolds for building multi-enzyme materials. *ACS Synth. Biol.*, 9, 381–391, 2020.
34. Chakraborti., S., Lin, T.Y., Glatt, S., Heddle, J.G., Enzyme encapsulation by protein cages. *RSC Adv.*, 10, 13293–13301, 2020.
35. Edwardson, T.G.W., Mori, T., Hilvert, D., Rational engineering of a designed protein cage for siRNA delivery. *J. Am. Chem. Soc.*, 140, 10439–10442, 2018.
36. Lawrence, M.S., Phillips, K.J., Liu, D.R., Supercharging proteins can impart unusual resilience. *J. Am. Chem. Soc.*, 129, 10110–10112, 2007.
37. Azuma, Y., Zschoche, R., Tinzl, M., Hilvert, D., Quantitative packaging of active enzymes into a protein cage. *Angew. Chem. Int. Ed. Engl.*, 55, 1531–1534, 2016.
38. Beck, T., Tetter, S., Kunzle, M., Hilvert, D., Construction of Matryoshka-type structures from supercharged protein nanocages. *Angew. Chem. Int. Ed.*, 54, 937–940, 2015.
39. Kunzle, M., Eckert, T., Beck, T., Binary protein crystals for the assembly of inorganic nanoparticle superlattices. *J. Am. Chem. Soc.*, 138, 12731–12734, 2016.
40. Azuma, Y., Herger, M., Hilvert, D., Diversification of protein cage structure using circularly permuted subunits. *J. Am. Chem. Soc.*, 140, 558–561, 2018.
41. Jorda, J., Leibly, D.J., Thompson, M.C., Yeates, T.O., Structure of a novel 13 nm dodecahedral nanocage assembled from a redesigned bacterial microcompartment shell protein. *Chem. Commun.*, 52, 5041–5044, 2016.
42. Otten, R., How directed evolution reshapes the energy landscape in an enzyme to boost catalysis. *Science*, 370, 1442–1446, 2020.
43. Terasak, N., Azuma, Y., Hilvert, D., Laboratory evolution of virus-like nucleocapsids from nonviral protein cages. *Proc. Natl. Acad. Sci. U.S.A.*, 115, 5432–5437, 2018.
44. Wang, Y.H., Ferritin conjugates with multiple clickable amino acids encoded by C-terminal engineered pyrrolysyl-tRNA synthetase. *Front. Chem.*, 9, 779976, 2021.
45. Axup, J.Y. et al., Synthesis of site-specific antibody-drug conjugates using unnatural amino acids. *Proc. Natl. Acad. Sci. U.S.A.*, 109, 16101–16106, 2012.
46. Lang, K. and Chin, J.W., Cellular incorporation of unnatural amino acids and bioorthogonal labeling of proteins. *Chem. Rev.*, 114, 4764–4806, 2014.
47. Wals, K. and Ovaa, H., Unnatural amino acid incorporation in *E. coli*: Current and future applications in the design of therapeutic proteins. *Front. Chem.*, 2, 15, 2014.
48. Potekhin, S.A., *De novo* design of fibrils made of short alpha-helical coiled coil peptides. *Chem. Biol.*, 8, 1025, 2001.
49. Paramelle, D. et al., Specific internalisation of gold nanoparticles into engineered porous protein cages via affinity binding. *PLoS One*, 11, e0162848, 2016.
50. Keeble, A.H. and Howarth, A.H., Power to the protein: Enhancing and combining activities using the Spy toolbox. *Chem. Sci.*, 11, 7281–7291, 2020.

51. Giessen, T.W. and Silver, P.A., A catalytic nanoreactor based on *in-vivo* encapsulation of multiple enzymes in an engineered protein nanocompartment. *ChemBioChem*, 17, 1931–1935, 2016.
52. Bhaskar, S. and Lim, S., Engineering protein nanocages as carriers for biomedical applications. *NPG Asia Mater.*, 2017, 9, 371.
53. Wang, T. and Douglas, T., Protein nanocage architectures for the delivery of therapeutic proteins. *Curr. Opin. Colloid Interface Sci.*, 51, 101395, 2021.
54. Schoonen, L., Maassen, S., Nolte, R.J.M., van Hest, J.C.M., Stabilization of a virus-like particle and its application as a nanoreactor at physiological conditions. *Biomacromolecules*, 18, 3492–3497, 2017.
55. Patterson, D.P., LaFrance, B., Douglas, T., Rescuing recombinant proteins by sequestration into the P22 VLP. *Chem. Commun.*, 49, 10412–10414, 2013.
56. Pitek, A.S., Park, J., Wang, Y., Gao, H., Hu, H., Simon, D.I., Steinmetz, N.F., Delivery of thrombolytic therapy using rod-shaped plant viral nanoparticles decreases the risk of hemorrhage. *Nanoscale*, 10, 16547–16555, 2018.
57. Macone, A., Masciarelli, S., Palombarini, F., Quaglio, D., Boffi, A., Trabuco, M.C., Baiocco, P., Fazi, F., Bonamore, A., Ferritin nanovehicle for targeted delivery of cytochrome C to cancer cells. *Sci. Rep.*, 9, 11749, 2019.
58. Ramos, R., Bernard, J., Ganachaud, F., Miserez, A., Biomimetic and biopolymer-based enzyme encapsulation. *Small Sci.*, 2, 2100095, 2022.
59. Lee, N.K., Cho, S., Kim, I.S., Ferritin – a multifaceted protein scaffold for bio-therapeutics. *Exp. Mol. Med.*, 54, 1652–1657, 2022.
60. Shan, H., Dou, W., Zhang, Y., Qi, M., Targeted ferritin nanoparticle encapsulating CpG oligodeoxynucleotides induces tumor-associated macrophage M2 phenotype polarization into M1 phenotype and inhibits tumor growth. *Nanoscale*, 12, 22268–22280, 2020.
61. Molino, N.M., Neek, M., Tucker, J.A., Nelson, E.L., Wang, S.W., Viral-mimicking protein nanoparticle vaccine for eliciting anti-tumor responses. *Biomaterials*, 86, 83–91, 2016.
62. Ferrucci, P.F., Pala, L., Conforti, F., Cocorocchio, E., Talimogene laherparepvec (T-VEC): An intralesional cancer immunotherapy for advanced melanoma. *Cancer*, 13, 1383, 2021.
63. Safary, A., Khiavi, M.A., Mousavi, R., Barar, J., Rafi, M.A., Enzyme replacement therapies: What is the best option? *Bioimpacts*, 8, 153–157, 2018.

13
Green Nanostructures: Biomedical Applications and Toxicity Studies

Radhika Chaurasia[1], Omnarayan Agrawal[1], Rupesh[1], Shweta Bansal[2] and Monalisa Mukherjee[1]*

[1]*Amity Institute of Click Chemistry Research and Studies, Amity University, Uttar Pradesh, Noida, India*
[2]*Amity Institute of Applied Sciences, Amity University, Uttar Pradesh, Noida, India*

Abstract

In this modern era of nanotechnology, various engineered nanomaterials have been manufactured by a greener technique. Green Nanomaterial synthesis has evolved as an eco-friendly method of producing Nanomaterial with a wide range of biological, chemical, and physical features. It has emerged as a viable alternative, offering an economical, and energy-efficient method for producing a wide spectrum of nanomaterials. The creation of environment benign processes for producing nanostructures is a major aspect in the field of nanotechnology. Since the last two decades, nanoparticles have been constantly evaluated and utilized in a variety of industries. The physical and chemical processes involved for synthesis of nanomaterial require a lot of toxic chemicals that are harsh to the environment. The utilization of plant extracts, fungi, bacteria, and algae for the synthesis of nanoparticles has been acclaimed as a "green" technology. Metallic, non-metallic, and metal oxide nanoparticles have attracted a lot of attention due to their various properties, including their ability to combat with a variety of diseases and microbial infections. In this book chapter, we highlight numerous techniques for green synthesis of nanomaterial and the contribution of these nanoparticles for a variety of applications. The objective of this chapter is to provide a brief overview of the different green nanostructures used in the emerging field of Nanotechnology.

Keywords: Green nanostructures, biomedical application, nanomaterial, sustainable, toxicity, nanoparticle synthesis

---

Corresponding author: mmukherjee@amity.edu

Mousumi Sen and Monalisa Mukherjee (eds.) Bioinspired and Green Synthesis of Nanostructures: A Sustainable Approach, (307–324) © 2023 Scrivener Publishing LLC

13.1 Introduction

In pursuit of a sustainable and eco-friendly abode, research focusing on the green synthesis of materials has revolutionized the design, development and application of chemical products. Meticulous efforts for minimal waste products, synthesis of recyclable materials, and energy conservation have led to the research and discoveries of ingenious strategies. Currently, green nanoparticle (NP) synthesis is becoming extremely prevalent owing to its safety and suitability with living things [1]. Green synthesis represents a streamlined approach to manufacturing that supersedes conventional techniques, offering a superior alternative that is characterized by simplicity and efficiency. Therefore, the significance of green nanoparticles synthesis must be examined in terms of how it is produced, its quality, and potential biomedical application. The purpose of this chapter is to elucidate the potential of green synthesis in terms of its mechanism, interaction with biomolecules, and possible applications. This chapter endeavors to address queries such as, "How accurate is the green synthesis approach?" and "Why is green synthesis imperative?" Moreover, it aims to discuss the significance of green synthesis and enumerate its advantages and disadvantages. What are the advantages and drawbacks of green synthesis? As a result, fresh research findings should be used to examine and assess a platform for green synthesis as a research direction. Nanotechnology, which has greatly advanced in recent years, is one of the emerging field in science and technology that has developed most significantly, demonstrates unique intrinsic characteristics and has the ability to create novel structures, mechanisms, devices, and nanoplatforms with future applications in a variety of fields [2, 3]. Since decades metal and nonmetal based NPs such as iron, zinc, silver, selenium and gold nanoparticles have been employed as therapeutic agents [4]. Transition metal oxides, including CuO, TiO_2, Fe_3O_4, ZnO, and NiO NPs, have demonstrated successful applications as cutting-edge nanomaterial in the field of environment, healthcare, and energy sector. The high adsorption properties of these NPs have substantially increased their efficacy and potential for use [5]. According to research, biological methods known as "biosynthesis" or "green synthesis of NPs" is replacing conventional chemical and physical methods of synthesizing metallic and non-metallic nanoparticles [6]. Although traditional systems (chemical and physical) take less time to create vast quantities of nanoparticles, on the contrary to make them stable, they need lethal materials like stabilizing agents, which creates toxicity in the environment [7]. Green technology utilizing plants is emerging as an efficient, environment-friendly, cost effective owing to the availability of natural stabilizing agents present in the form of proteins provided by plant extract-mediated biogenesis of nanoparticles [8]. To diminish chemical toxicity in the environment, various metal oxide and non-metal oxide NPs are biologically synthesized through plant extract. With meticulous synthesis, this subtle method of modifying chemical synthesis enables for nanoparticle of different sizes and configurations. We believe that these green synthesis technique offers an innovative strategy for synthesis of nanomaterial which can be utilized in biomedical applications [9].

13.2 Moving Toward Green Nanostructures

Various traditional processes are adopted to produce nanoparticles however, these methods possessed toxicity. Use of hazardous chemical reagents for production of nanoparticles is a matter of concern owing to its endangering effect on the environment. Thus, biological approaches are emerging to fill the gap, achieving the goal of greener environment by adopting green synthesis using biological molecule, extracts derived from plant sources as it exhibit advantages over chemical methods [10]. In seeking new drugs, a growing number of individuals are shifting to natural products and extracts derived from plants. So, green synthesis, which utilizes extracts of microorganisms and natural ingredients made from plants, is superior than chemical methods.

13.3 Methods of Nanoparticle Synthesis

In the last 20 years, numerous physical, chemical, and biological pathways (Figure 13.1) have been developed to create nanoparticles with different

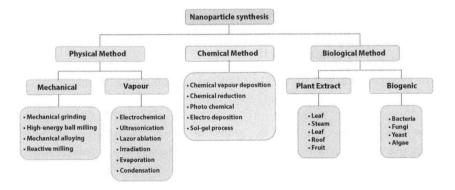

Figure 13.1 Classification of nanoparticle synthesis [66].

size range, forms, and compositions. This has been done to meet the increasing demand for both metallic and nonmetallic nanoparticles [11]. High-energy irradiation, laser ablation, and lithography are some of the physical techniques [12, 13]. The physical technique utilizes industrial machinery, operates at high pressure and temperature and require a lot of space to set up the machines. The popular methods for synthesizing nanomaterials in the chemical approach are sol-gel synthesis, co-precipitation, hydrothermal synthesis, chemical reduction, and photochemical reduction. The chemical method uses toxic materials that could harm both people and the environment [14].

13.4 Plant-Mediated Synthesis of Green Nanostructures

This section of the chapter focuses on the production of metal and metal oxide nanostructures from plant products. The potency for diverse plants to produce nanoparticles should be further investigated as a green route of nanoparticles fabrication [15]. Synthesis of nanoparticles through plant is a safer and simpler way that works at low temperatures and necessitates a small component from the environment [16]. Researchers are interested in plant-based nanoparticles because they are amiable to the environment. Furthermore, creating nanostructures from plants has other perks, such as safer solvents, using innocuous reagents, facilitating milder reaction conditions, feasible, and competent for utilization in medical, surgical, and pharmaceutical applications [17]. Furthermore, the criteria for making them, such as pressure, energy, temperature, and the materials they are made of, are very simple. The advantages of using plant extract to create nanostructures are they are easy to synthesise, safe to handle, and have variety of natural products that may help to reduce the amount of toxic materials in the nanoparticles.

13.4.1 Silver Nanoparticles

The green synthesis of silver NPs (AgNPs) exhibits a wide variety of applications, such as antimicrobial, anti-cancer, anti-inflammatory, and antiviral activity [18]. Moreover, due to its application in dentistry, AgNPs synthesis has gained considerable attention. The synthesis can be achieved from plant extracts or from microbes. AgNPs bear a protein capping which not only assist in internalization of the NPs for drug delivery but also resist aggregation of NPs as well. The protein capping encompasses a variety of functional groups extending the scope of functionalization of AgNPs with active drugs moieties. AgNPs synthesized from *Trichoderma harzianum* when compared with AgNPs without caps exhibited that capping and NPs work in synergism against plant pathogen *Sclerotinia sclerotiorum*. It has been observed that the size of the NPs plays an important part in their internalization.

Sr. no.	Source of AgNPs	Biomedical application	Size (nm)	Reference
1	*Budleja globosa* leaf extract	Antioxidant	16 nm	[19]
2	*Hagenia abyssinica* leaf extract	Antibacterial (Gram-positive and Gram-Negative), antioxidant, biosensor	22.2 nm	[20]
3	Leaf of *Syngonium. podophyllum*	Antifungal activity against *Candida albicans*	40 nm	[21]
4	*Aloe vera* leaf extract	Anti-glycating agent	30.5 nm	[22]
5	Leaf extract of *Cucumis prophetarum*	Anticancer, antibacterial, antioxidant	30–50 nm	[23]
6	*Azadirachta indica* leaf extract	Opto-electronics in medical devices	34 nm	[24]

13.4.2 Gold Nanoparticles

To prevent detrimental consequences in medical applications, a proposal has been designed for introducing metal or metal oxide NPs that do not utilize hazardous substances and is safer for the environment [25]. Use of biological systems such as microorganisms (like bacteria, fungi, and yeast) and plant extract or plant biomass could be an alternative to physical and chemical methods for creating metal NPs like Gold nanoparticles (AuNPs) [26, 27]. AuNPs becoming a new class of material with traits that set them apart from much more conventional types and convince significant promise for application in medical and biological research. Numerous studies on the antimicrobial activities and mutagenicity of gold nanoparticles have been reported over the last few decades [28]. Gold nanoparticles is one of the noble nanomaterials which has been explored much more. Numerous wet chemical practices, like citrate alleviation and the Brust-Schiffrin method, have been outlined as processes to create gold nanoparticles [29]. Regrettably, these ways utilize environmentally hazardous reagents, to avoid this, green methods have been utilized for making AuNPs which gained a lot of attention, in the crafting of gold nanoparticles, for example, plant extracts were used as reducing and capping agents. Plant extracts that are excellent at amassing toxins and avoiding them are the best choices for green nanoparticle manufacturing and environmental remediation.

Sr. no.	Source of AuNPs	Biomedical application	Size (nm)	Reference
1	Used extract of *Morinda citrifolia* roots	Anticancer and antidiabetic	12-38 nm	[30]
2	By using pomegranate fruit extract	As antibacterial agent	10-50 nm	[31]
3	Using rosa hybrida petal extract	In cancer therapy and imaging	10-11 nm	[32]
4	Synthesis of AuNPs by plants, such as Azadirachta indica leaf extract	It has strong antimicrobial properties	90-95 nm	[33]
5	Synthesized with *Z. officinale* extract at pH 7.4	For application in drug delivery, gene delivery or as a biosensor	5-15 nm	[34]
6	Synthesis of AuNPs using seed aqueous extract of *Abelmoschus esculentus*	As antifungal agent	45-75 nm	[35]

13.4.3 Zinc Oxide Nanoparticles

According to the given table, numerous studies have been conducted to create ZnO nanoparticles with a variety of morphologies using different plant extracts. These ZnO NPs have been used in a variety of applications, including antibacterial drugs, anti-cancer, antioxidants, anti-inflammatory, and antiviral activity and have attracted significant interest in the biomedical field. The synthesis can be achieved from plant extracts or from microbes. The integration of The NPs have been found to be significantly impacted by their sizes.

Sr. no.	Green nanostructures	Biomedical application	Size (nm)	Reference
1	ZnO NPs utilizing an aqueous plant extract from Cassia fistula	Antimicrobial, antioxidant, anti-inflammatory, and photodegradative activity	60–70 nm	[1]
2	ZnO NPs by using leaf of Azadirachta indica (L.)	Anticancer by inducing cancer cell apoptosis	18 nm	[36]
3	Synthesis of ZnO NPs from seaweed Sargassum myriocystum	Antibacterial activity and anti-biofilm actions against pathogens	36 nm	[37]
4	ZnO NPs are created from vitex negundo leaf and floral extract	Antibacterial behavior against pathogenic bacteria	38.17 nm	[38]
5	Synthesis of ZnO NPs by Aloe vera (Liliaceae)	Antibacterial and anti-biofilm activity	8–20 nm	[39]
6	Synthesis of ZnO NPs through Phyllanthus niruri	Antibacterial behavior against pathogenic bacteria	25.60 nm	[40]
7	Synthesis of ZnO NPs through Trifolium pratense	Antibacterial	60–70 nm	[41]

13.4.4 Selenium Nanoparticles

Selenium is a vital trace element essential for various enzymatic reactions in the human body. Analogous to other nanoparticles, Se NPs varies from the properties of the respective bulk materials. Several parameters such as the materials and methods utilized, the reaction time have a significant impact on the shape, size and other morphological characteristics of NPs [42]. Moreover, bioactive organic selenium can be synthesized by certain types of plants due to their ability to absorb selenium from soil, by fabricating selenium nanoparticles using bovine serum albumin, Kalimuthu *et al.* examined the effect of these particles on zebrafish embryonic cardiovascular functions. Very low dose of Se NPs is required to exhibit therapeutic activity for cardiovascular prevention. Antileukemic activity of Se NPs was recognized through citric acid–mediated green NPs stabilized by sodium alginate, exhibited strong antimicrobial activity against *P. aeruginosa*, *E. coli*, *S. aureus*, and *B. subtilis*. Additionally, the as-synthesized NPs exhibited excellent antioxidant property in conjugation with anticoagulant activity [43].

13.5 Microbe-Based Synthesis

Biogenic synthesis of nanoparticles aids in economical, biodegradable, and sustainable alternative to the traditional NPs synthesis. Variety of microorganisms can be utilized for the synthesis of nanoparticles having various shape and size. Synthesis of NPs from fungi, bacteria, yeast, and algae surmounts the use of hazardous chemicals. In comparison to other microbes, bacterial growth of NPs is preferred, owing to their ability to grow in a short time and ease of recovery of NPs. It has been noticed that NPs can be formed intracellularly and extracellularly, depending on the site of NP synthesis. During intracellular synthesis, the microbial cellular components, such as cell membrane, cell wall, cellular enzymes, and proteins, are involved. The process involves the incubation of microbial cells for growth in a suitable medium at an optimum temperature for a definite period. The cell broth is centrifuged to harvest the cell biomass, followed by incubation with the metal ion solution. The carboxyl group present in the microbial cell wall initiates electrostatic attraction of the metal ions, promoting passage of ions inside the cell. The intracellular conversion of these metal ions by enzymes, proteins, and cofactors results in the production of metallic NPs [44]. Multiple steps for the extraction of these NPs are performed, involving centrifugation and ultrasonication until the NPs are recovered, making the process sluggish and cumbersome [45].

For extracellular fabrication of NPs, the incubated cells are centrifuged to eliminate the cell mass. The enzymes reductase, required for the NP production, is present in supernatant. Once metal ion solution is added to the supernatant, the metal ions are transformed into metallic nanoparticles that can be characterized using a wide range of analytical methodologies. In comparison to intracellular NPs synthesis, the extracellular synthesis is time saving and convenient.

13.5.1 Bacteria-Mediated Synthesis of NPs

Numerous strains of bacteria have been shown to produce silver nanoparticles, irrespective of their taxonomic divisions [46]. If one looks closely, The majority of these isolates come from the soil or the ocean, and that They have participated in a number of metal biogeochemical cycles. These bacteria detoxify metals in their environment by reducing or precipitating inorganic ions or doing both into metal nanoclusters, as they are biologically adapted to performing it. Certain strains of *Rhodobacter sphaeroides*, *Shewanella algae*, *Lactobacillus* sp., and *Bacillus* sp. may independently produce metal nanoparticles such as gold, silver, titanium, platinum, and zinc. As a result, a single bacterium may produce nanoparticles made of many metals, and the ability of a particular bacterial strain to produce such particles depend on the physiological characteristics of the strain as well as the taxonomic group to which it belongs. The cold-adapted physiology of the strain, could be observed in a psychrotolerant bacterium, *M. psychrotolerans* at around 12°C however at approximately 22°C exhibited enhanced production rate [47]. *A. kerguelensis* supernatant failed to generate silver nanoparticles with the same temperature range as *P. antarctica*, indicating diverse features of different bacterial species in the cell-free culture supernatants that enable the production of silver nanoparticles [48]. The thermophilic bacteria *Geobacillus stearothermophilus* produces silver and gold nanoparticles. These extremeophilic bacteria produced monodispersed, spherical gold nanoparticles with a size range of 5 to 8 nm. Silver Nanoparticles were polydispersed, spherical in form, and ranged in size from 5 to 35 nm. The cysteine residues or free amine groups of the protein were crucial building blocks in the creation of nanoparticles. Additionally, seven distinct proteins with free amine or cysteine-containing molecules that facilitate the formation of nanoparticles have been found, ranging in size from 98 to 12 kDa [49]. The ZnO NPs synthesized by the conventional chemical methods require high production cost, generate harmful chemicals and generate hazardous waste, posing

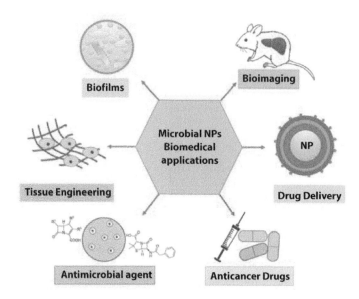

Figure 13.2 Biomedical applications of nanomaterial produced by microorganisms [67].

toxic effects on human and animal cells. ZnO NPs synthesised from zinc-tolerant probiotic *Lactobacillus plantarum* TA4 is economical and environment friendly. Multifarious NPs can be obtained by these nanofactories have various biomedical applications (Figure 13.2). The extremophilic bacterium *Ureibacillus thermosphaericus*, which was found in Ramsan geothermal hot springs in Mazandatan province, Iran, was able to synthesize extracellular silver nanoparticles at higher temperatures (60–80°C) and in a range of (0.001–0.1 M) concentration of $AgNO_3$. However, the maximum number of nanoparticles that could be produced at 0.01 M $AgNO_3$ at 80°C. Even at unfavorable temperatures, the biomolecules an extremophile produces at high temperatures, which may be crucial for the production of nanoparticles [50]. MR(muran)/CH(chitosan) nanoparticles are generated by the salt-tolerant bacteria *Halomonas maura* by the polyelectrolyte complexation of CH and MR solution under vigorous magnetic stirring [51].

13.5.2 Fungus-Mediated Synthesis of NPs

Utilization of fungus in the manufacture of plethora of nanoparticles has been the subject of further investigation in recent years. Owing to a host of benefits of fungi, it has been preferred for research purposes [52]. First, the advantage of employing fungus for the production of nanoparticles is that they are easier to handle than other microorganisms. Second, it can secrete a vast amount of enzymes which add-on to its benefits. However, one crucial challenge is the complexity to genetically manipulate some enzymes for overexpression in prokaryotes. A fungal specie *Fusarium oxysporum* has extensively been employed in the production of nanoparticles. Extracellular gold nanoparticles produced by this species are 20 to 40 nm in size, spherical or triangular in morphology, and highly dispersed that even after a month, there was no significant reunion [53].

The acidophilic fungus *Verticillium sp.* produces intracellular silver nanoparticles in the cytoplasm [54]. Astonishingly, Au–Ag alloy nanoparticles were extracted when *F. oxysporum* was used to create alloy nanoparticles, exhibiting for the first time fungus-mediated biosynthesis of Au-Ag alloy nanoparticles. When fungi were subjected to equimolar doses of $AgNO_3$ and $HAuCl_4$, extremely stable alloy nanoparticles between 8 and 14 nm in size were produced. It was shown that the fungus *F. oxysporum* secretes NADH-dependent enzymes, and it was proven that the composition of the nanoalloys affected the amount of NADH produced [55]. When *F. oxysporum* was treated with H_2PtCl_6, extracellular platinum nanoparticles with a size range of 5 to 30 nm were developed. These particles were stabilized by proteins [56]. *F. oxysporum* has also been used in the biosynthesis of spherical ZnS nanoparticles with an average size of 42 nm [57].

13.5.3 Actinomycete-Mediated Synthesis of NPs

Streptomyces viridogens strain [HM10], a new actinomycete, produces gold nanoparticles. According to reports, the strain uses an internal mechanism to create gold nanoparticles. Gold mycelium in the strain was revealed by electron microscopy (TEM) examination and X-ray diffraction analysis showed that their average size was 18 to 20 nm. The strain HM10 simultaneously displayed enhanced growth at $HAuCl_4$ concentrations of 1 and 10 mM. Additionally, the strain HM10 synthesised gold nanoparticles displayed significant antibacterial properties against *S. aureus* and *E. coli* [58]. In past few years, metallic nanoparticles have also been produced by the most extensively studied yeast, *Saccharomyces cerevisiae*. The biosynthesis of gold nanoparticles was carried out using 35 marine actinobacterial strains that were obtained from marine debris along the Busan coast in South Korea. The generation of extracellular gold nanoparticles was demonstrated by *Nocardiopsis sp.* MBRC-48 in that. FTIR, XRD, and UV-visible spectrum analyses have all been used for the characterization of the nanoparticles produced. FTIR analysis clearly illustrated that proteins, enzymes, and metabolites were responsible for the production of the nanoparticles. Elemental gold was found to be present, according to EDAX spectroscopy. The properties such as antibacterial, antioxidant, and cytotoxicity were assessed, and TEM examination indicated that they were spherical in shape and non-uniformly dispersed, with a size ranging from 7 to 15 nm [59].

13.6 Toxicity of Nanostructures

The recent concern for the synthesized nanoparticles is the indirect toxicological effect on human beings and the eco-system. The focus to synthesize green and environment friendly structures for biomedical purposes have resulted in green revolution. While synthesizing NPs, toxicity studies are a very crucial aspect. One of the most essential issues of nanoplatforms for biological applications is the biocompatibility of NPs [60]. Synthesis of nanostructures from natural sources and benign processing materials has led to a safe and clean habitat. However, characteristics of NPs include size, shape, chemical constituents, concentration, chemical reactivity, dispersion, and aggregation have a significant effect how nanoparticles interact with environment. As per research, nanoparticles having size 1 to 10 nm have higher cytotoxicity than their native counterparts. To interact with cellular components like nucleic acids, proteins, fatty acids, and carbohydrates, smaller size nanoparticles having larger specific surface area (SSA) is required. The smaller size of NPs presumably makes it feasible to penetrate the cell [61]. Moreover, in lieu of the internal material of the nanoparticle, surface coating and functionalization take part in the nanoparticle/environment contact. Due to the capacity of NPs to carry out highly precise functions, such as the targeted delivery of medications and imaging contrast agents (CAs) to tumor cells, the usage of NPs have increased significantly over the past decade, notably for the treatment of cancer [62]. Particularly, passive, and active targeting can both result in the buildup of nanocarriers in cancer cells. As a result of the enhanced permeability of tumor blood arteries in passive targeting, NPs gather close to the tumor site. Due to their unique physical characteristics, such as size, shape, and surface charge, NPs can passively accumulate to solid tumors or to the metastatic locations in the process known as the enhanced permeability retention (EPR) effect [63]. The utilization of nanoparticles for medical purposes will be thwarted by the existence of any hazardous surface toxic components. Thus, the therapeutic efficacy of nanoparticle-based medications will eventually be weighed against potential toxicity in medical use. According to studies selenium nanoparticles though are dietary supplements in higher concentration and long-term usage impart toxicity. The anti-microbial property of Ag and ZnO NPs are well established, nevertheless, higher concentration of these structures impose toxicity resulting in varied health issues [64, 65]. Hence, legitimate use of NPs is the best tool to overcome the toxicity-associated problems.

13.7 Conclusion

The green nanostructures were synthesized from plant extracts and biogenic agents have widespread biomedical applications. In comparison to the traditional NPs synthesis (physical and chemical methods), which utilizes harmful chemicals, the biological method of NPs synthesis has geared up due to its safe, biocompatible, sustainable, cost-effective nature. Intensive and extensive research is required to improve the eco-friendly properties and further reduce the toxicity of the green nanostructures. Metal-based and non-metal–based green nanostructures can be synergistically used with carbon based green nanostructures to unleash new possibilities in nanoscience and green chemistry.

References

1. Suresh, D., Nethravathi, P.C., Rajanaika, H., Nagabhushana, H., Sharma, S.C., Green synthesis of multifunctional zinc oxide (ZnO) nanoparticles using Cassia fistula plant extract and their photodegradative, antioxidant and antibacterial activities. *Mater Sci. Semicond. Process.*, 31, 446–54, 2015.
2. Kalpana, V.N. and Devi Rajeswari, V., A review on green synthesis, biomedical applications, and toxicity studies of ZnO NPs. *Bioinorg. Chem. Appl.*, 2, 1–12, 2018
3. Mirzaei, H. and Darroudi, M., Zinc oxide nanoparticles: Biological synthesis and biomedical applications. *Ceram. Int.*, 43, 1, 907–14, 2017.
4. Folorunso, A., Akintelu, S., Oyebamiji, A.K., Ajayi, S., Abiola, B., Abdusalam, I. *et al.*, Biosynthesis, characterization and antimicrobial activity of gold nanoparticles from leaf extracts of Annona muricata. *J. Nanostruct. Chem.*, 9, 2, 111–7, 2019.
5. Pakzad, K., Alinezhad, H., Nasrollahzadeh, M., Green synthesis of Ni@ Fe3O4 and CuO nanoparticles using Euphorbia maculata extract as photocatalysts

for the degradation of organic pollutants under UV-irradiation. *Ceram. Int.*, 45, 14, 17173–82, 2019.
6. Pugazhendhi, A., Kumar, S.S., Manikandan, M., Saravanan, M., Photocatalytic properties and antimicrobial efficacy of Fe doped CuO nanoparticles against the pathogenic bacteria and fungi. *Microb. Pathog.*, 122, 84–9, 2018.
7. Salam, H.A., Sivaraj, R., Venckatesh, R., Green synthesis and characterization of zinc oxide nanoparticles from Ocimum basilicum L. var. purpurascens Benth.-Lamiaceae leaf extract. *Mater. Lett.*, 131, 16–8, 2014.
8. Cai, W. and Chen, X., Nanoplatforms for targeted molecular imaging in living subjects. *Small*, 3, 11, 1840–54, 2007.
9. Zhang, Y., Nayak T, R., Hong, H., Cai, W., Biomedical applications of zinc oxide nanomaterials. *Curr. Mol. Med.*, 13, 10, 1633–45, 2013.
10. Parveen, K., Banse, V., Ledwani, L., Green synthesis of nanoparticles: Their advantages and disadvantages, in: *AIP Conference Proceedings*, AIP Publishing LLC, p. 20048, 2016.
11. Soni, M., Mehta, P., Soni, A., Goswami, G.K., Green nanoparticles: Synthesis and applications. *IOSR J. Biotechnol. Biochem.*, 4, 3, 78–83, 2018.
12. Mafuné, F., Kohno, J., Takeda, Y., Kondow, T., Sawabe, H., Formation of gold nanoparticles by laser ablation in aqueous solution of surfactant. *J. Phys. Chem. B*, 105, 22, 5114–20, 2001.
13. Krishnia, L., Thakur, P., Thakur, A., Synthesis of nanoparticles by physical route, in: *Synthesis and Applications of Nanoparticles*, A. Thakur, P. Thakur, S.M.P. Khurana (Eds.), pp. 45–59, Springer Nature Singapore, Singapore, 2022, Available from: https://doi.org/10.1007/978-981-16-6819-7_3.
14. Mughal, S.S. and Hassan, S.M., Comparative study of AgO nanoparticles synthesize via biological, chemical and physical methods: A review. *Am. J. Mater. Synth. Process.*, 7, 2, 15–28, 2022.
15. Chung, I.-M., Park, I., Seung-Hyun, K., Thiruvengadam, M., Rajakumar, G., Plant-mediated synthesis of silver nanoparticles: Their characteristic properties and therapeutic applications. *Nanoscale Res. Lett.* [Internet], 11, 1, 40, 2016, Available from: https://doi.org/10.1186/s11671-016-1257-4.
16. Goodsell, D.S., *Bionanotechnology: Lessons from nature*, John Wiley & Sons, Hoboken, New Jersey, 2004.
17. Abdel-Halim, E.S., El-Rafie, M.H., Al-Deyab, S.S., Polyacrylamide/guar gum graft copolymer for preparation of silver nanoparticles. *Carbohydr. Polym.*, 85, 3, 692–7, 2011.
18. Wypij, M., Jędrzejewski, T., Trzcińska-Wencel, J., Ostrowski, M., Rai, M., Golińska, P., Green synthesized silver nanoparticles: Antibacterial and anticancer activities, biocompatibility, and analyses of surface-attached proteins. *Front. Microbiol.*, 12, 632505, 2021.
19. Carmona, E.R., Benito, N., Plaza, T., Recio-Sánchez, G., Green synthesis of silver nanoparticles by using leaf extracts from the endemic Buddleja globosa hope. *Green Chem. Lett. Rev.*, 10, 4, 250–6, 2017.

20. Melkamu, W.W. and Bitew, L.T., Green synthesis of silver nanoparticles using Hagenia abyssinica (Bruce) JF Gmel plant leaf extract and their antibacterial and anti-oxidant activities. *Heliyon*, 7, 11, e08459, 2021.
21. Yasir, M., Singh, J., Tripathi, M.K., Singh, P., Shrivastava, R., Green synthesis of silver nanoparticles using leaf extract of common arrowhead houseplant and its anticandidal activity. *Pharmacogn. Mag.*, 13, Suppl 4, S840, 2017.
22. Ashraf, J.M., Ansari, M.A., Khan, H.M., Alzohairy, M.A., Choi, I., Green synthesis of silver nanoparticles and characterization of their inhibitory effects on AGEs formation using biophysical techniques. *Sci. Rep.*, 6, 1, 1–10, 2016.
23. Hemlata, Meena, P.R., Singh, A.P., Tejavath, K.K., Biosynthesis of silver nanoparticles using Cucumis prophetarum aqueous leaf extract and their antibacterial and antiproliferative activity against cancer cell lines. *ACS Omega*, 5, 10, 5520–8, 2020.
24. Ahmed, S., Saifullah, Ahmad, M., Swami, B.L., Ikram, S., Green synthesis of silver nanoparticles using Azadirachta indica aqueous leaf extract. *J. Radiat. Res. Appl. Sci.*, 9, 1, 1–7, 2016.
25. Kanchi, S., Kumar, G., Lo, A.-Y., Tseng, C.-M., Chen, S.-K., Lin, C.-Y. et al., Exploitation of de-oiled jatropha waste for gold nanoparticles synthesis: A green approach. *Arab. J. Chem.*, 11, 2, 247–55, 2018.
26. Aromal, S.A. and Philip, D., Green synthesis of gold nanoparticles using Trigonella foenum-graecum and its size-dependent catalytic activity. *Spectrochim. Acta Part A Mol. Biomol. Spectrosc.*, 97, 1–5, 2012.
27. Sankar, R., Karthik, A., Prabu, A., Karthik, S., Shivashangari, K.S., Ravikumar, V., Origanum vulgare mediated biosynthesis of silver nanoparticles for its antibacterial and anticancer activity. *Colloids Surf. B Biointerfaces* [Internet], 108, 80–4, 2013, Available from: https://www.sciencedirect.com/science/article/pii/S0927776513001628.
28. Tao, C., Antimicrobial activity and toxicity of gold nanoparticles: Research progress, challenges and prospects. *Lett. Appl. Microbiol.*, 67, 6, 537–43, 2018.
29. Slepička, P., Slepičková Kasálková, N., Siegel, J., Kolská, Z., Švorčík, V., Methods of gold and silver nanoparticles preparation. *Mater. (Basel)*, 13, 1, 1, 2019.
30. Suman, T.Y., Rajasree, S.R.R., Ramkumar, R., Rajthilak, C., Perumal, P., The green synthesis of gold nanoparticles using an aqueous root extract of Morinda citrifolia L. *Spectrochim. Acta Part A Mol. Biomol. Spectrosc.*, 118, 11–6, 2014.
31. Basavegowda, N., Sobczak-Kupiec, A., Fenn, R.I., Dinakar, S., Bioreduction of chloroaurate ions using fruit extract Punica granatum (pomegranate) for synthesis of highly stable gold nanoparticles and assessment of its antibacterial activity. *Micro Nano Lett.*, 8, 8, 400–4, 2013.
32. Noruzi, M., Zare, D., Khoshnevisan, K., Davoodi, D., Rapid green synthesis of gold nanoparticles using Rosa hybrida petal extract at room temperature. *Spectrochim. Acta Part A Mol. Biomol. Spectrosc.* [Internet], 79, 5, 1461–5,

2011, Available from: https://www.sciencedirect.com/science/article/pii/S1386142511003337.
33. Anuradha, J., Abbasi, T., Abbasi, S.A., Green'synthesis of gold nanoparticles with aqueous extracts of neem (Azadirachta indica). *Res. J. Biotechnol.*, 5, 1, 75–9, 2010.
34. Kumar, K.P., Paul, W., Sharma, C.P., Green synthesis of gold nanoparticles with Zingiber officinale extract: Characterization and blood compatibility. *Process Biochem.* [Internet], 46, 10, 2007–13, 2011, Available from: https://www.sciencedirect.com/science/article/pii/S1359511311002510.
35. Jayaseelan, C., Ramkumar, R., Rahuman, A.A., Perumal, P., Green synthesis of gold nanoparticles using seed aqueous extract of Abelmoschus esculentus and its antifungal activity. *Ind. Crops Prod.*, 45, 423–9, 2013.
36. Elumalai, K. and Velmurugan, S., Green synthesis, characterization and antimicrobial activities of zinc oxide nanoparticles from the leaf extract of Azadirachta indica (L.). *Appl. Surf. Sci.*, 345, 329–36, 2015.
37. Agarwal, H., Venkat Kumar, S., Rajeshkumar, S., A review on green synthesis of zinc oxide nanoparticles – An eco-friendly approach. *Resour. Technol.* [Internet], 3, 4, 406–13, 2017, Available from: https://www.sciencedirect.com/science/article/pii/S2405653717300283.
38. Sundrarajan, M., Ambika, S., Bharathi, K., Plant-extract mediated synthesis of ZnO nanoparticles using Pongamia pinnata and their activity against pathogenic bacteria. *Adv. Powder Technol.*, 26, 5, 1294–9, 2015.
39. Qian, Y., Yao, J., Russel, M., Chen, K., Wang, X., Characterization of green synthesized nano-formulation (ZnO–A. vera) and their antibacterial activity against pathogens. *Environ. Toxicol. Pharmacol.*, 39, 2, 736–46, 2015.
40. Anbuvannan, M., Ramesh, M., Viruthagiri, G., Shanmugam, N., Kannadasan, N., Synthesis, characterization and photocatalytic activity of ZnO nanoparticles prepared by biological method. *Spectrochim. Acta Part A Mol. Biomol. Spectrosc.*, 143, 304–8, 2015.
41. Dobrucka, R. and Długaszewska, J., Biosynthesis and antibacterial activity of ZnO nanoparticles using Trifolium pratense flower extract. *Saudi J. Biol. Sci.*, 23, 4, 517–23, 2016.
42. Chaudhary, S., Umar, A., Mehta, S.K., Selenium nanomaterials: An overview of recent developments in synthesis, properties and potential applications. *Prog. Mater. Sci.*, 83, 270–329, 2016.
43. Alhawiti, A.S., Citric acid-mediated green synthesis of selenium nanoparticles: Antioxidant, antimicrobial, and anticoagulant potential applications. *Biomass Convers. Biorefinery*, 12, 5, 1–10, 2022.
44. Lahiri, D., Nag, M., Sheikh, H.I., Sarkar, T., Edinur, H.A., Pati, S. *et al.*, Microbiologically-synthesized nanoparticles and their role in silencing the biofilm signaling cascade. *Front. Microbiol.*, 12, 636588, 2021.
45. Mohd Yusof, H., Rahman, A., Mohamad, R., Zaidan, U.H., Samsudin, A.A., Biosynthesis of zinc oxide nanoparticles by cell-biomass and supernatant

of Lactobacillus plantarum TA4 and its antibacterial and biocompatibility properties. *Sci. Rep.*, 10, 1, 1–13, 2020.
46. Mohanpuria, P., Rana, N.K., Yadav, S.K., Biosynthesis of nanoparticles: Technological concepts and future applications. *J. Nanopart. Res.*, 10, 3, 507–17, 2008.
47. Parikh, R.Y., Ramanathan, R., Coloe, P.J., Bhargava, S.K., Patole, M.S., Shouche, Y.S. et al., Genus-wide physicochemical evidence of extracellular crystalline silver nanoparticles biosynthesis by Morganella spp. *PLoS One*, 6, 6, e21401, 2011.
48. Parak, W.J., Boudreau, R., Le Gros, M., Gerion, D., Zanchet, D., Micheel, C.M. et al., Cell motility and metastatic potential studies based on quantum dot imaging of phagokinetic tracks. *Adv. Mater.*, 14, 12, 882–5, 2002.
49. Fayaz, A.M., Girilal, M., Rahman, M., Venkatesan, R., Kalaichelvan, P.T., Biosynthesis of silver and gold nanoparticles using thermophilic bacterium Geobacillus stearothermophilus. *Process Biochem.*, 46, 10, 1958–62, 2011.
50. Juibari, M.M., Abbasalizadeh, S., Jouzani, G.S., Noruzi, M., Intensified biosynthesis of silver nanoparticles using a native extremophilic Ureibacillus thermosphaericus strain. *Mater. Lett.*, 65, 6, 1014–7, 2011.
51. Raveendran, S., Palaninathan, V., Nagaoka, Y., Fukuda, T., Iwai, S., Higashi, T. et al., Extremophilic polysaccharide nanoparticles for cancer nanotherapy and evaluation of antioxidant properties. *Int. J. Biol. Macromol.*, 76, 310–9, 2015.
52. Dhillon, G.S., Brar, S.K., Kaur, S., Verma, M., Green approach for nanoparticle biosynthesis by fungi: Current trends and applications. *Crit. Rev. Biotechnol.*, 32, 1, 49–73, 2012.
53. Mukherjee, P., Senapati, S., Mandal, D., Ahmad, A., Khan, M.I., Kumar, R. et al., Extracellular synthesis of gold nanoparticles by the fungus Fusarium oxysporum. *ChemBioChem*, 3, 5, 461–3, 2002.
54. Sastry, M., Ahmad, A., Khan, M.I., Kumar, R., Biosynthesis of metal nanoparticles using fungi and actinomycete. *Curr. Sci.*, 85, 2, 162–70, 2003.
55. Ahmad, A., Senapati, S., Khan, M.I., Kumar, R., Sastry, M., Extra-/intracellular biosynthesis of gold nanoparticles by an alkalotolerant fungus, Trichothecium sp. *J. Biomed. Nanotechnol.*, 1, 1, 47–53, 2005.
56. Vickers, N.J., Animal communication: When I'm calling you, will you answer too? *Curr. Biol.*, 27, 14, R713–5, 2017.
57. Mirzadeh, S., Darezereshki, E., Bakhtiari, F., Fazaelipoor, M.H., Hosseini, M.R., Characterization of zinc sulfide (ZnS) nanoparticles biosynthesized by Fusarium oxysporum. *Mater. Sci. Semicond. Process.*, 16, 2, 374–8, 2013.
58. Balagurunathan, R., Radhakrishnan, M., Rajendran, R.B., Velmurugan, D., Biosynthesis of gold nanoparticles by actinomycete Streptomyces viridogens strain HM10, *Indian J. Biochem. Biophys.*, 8, 5, 331–5, Oct 2011.
59. Manivasagan, P., Alam, M.S., Kang, K.-H., Kwak, M., Kim, S.-K., Extracellular synthesis of gold bionanoparticles by Nocardiopsis sp. and evaluation of its

antimicrobial, antioxidant and cytotoxic activities. *Bioprocess Biosyst. Eng.*, 38, 6, 1167–77, 2015.
60. Elmowafy, E.M., Tiboni, M., Soliman, M.E., Biocompatibility, biodegradation and biomedical applications of poly (lactic acid)/poly (lactic-co-glycolic acid) micro and nanoparticles. *J. Pharm. Investig.*, 49, 4, 347–80, 2019.
61. Huang, Y.-W., Cambre, M., Lee, H.-J., The toxicity of nanoparticles depends on multiple molecular and physicochemical mechanisms. *Int. J. Mol. Sci.*, 18, 12, 2702, 2017.
62. Panebianco, F., Climent, M., Malvindi, M.A., Pompa, P.P., Bonetti, P., Nicassio, F., Delivery of biologically active miR-34a in normal and cancer mammary epithelial cells by synthetic nanoparticles. *Nanomed. Nanotechnol. Biol. Med.*, 19, 95–105, 2019.
63. Maeda, H., Toward a full understanding of the EPR effect in primary and metastatic tumors as well as issues related to its heterogeneity. *Adv. Drug Delivery Rev.*, 91, 3–6, 2015.
64. Prabhu, S. and Poulose, E.K., Silver nanoparticles: Mechanism of antimicrobial action, synthesis, medical applications, and toxicity effects. *Int. Nano Lett.*, 2, 1, 1–10, 2012.
65. Sirelkhatim, A., Mahmud, S., Seeni, A., Kaus, N.H.M., Ann, L.C., Bakhori, S.K.M. *et al.*, Review on zinc oxide nanoparticles: Antibacterial activity and toxicity mechanism. *Nano-Micro Lett.*, 7, 3, 219–42, 2015.
66. Boldoo, T., Ham, J., Kim, E., Cho, H. Review of the photothermal energy conversion performance of nanofluids, their applications, and recent advances. *Energies*, 13, 21, 5748, 2020.
67. Ghosh, S., Ahmad, R., Zeyaullah, M., Khare, S. K. Microbial nano-factories: Synthesis and biomedical applications. *Front. Chem.*, 9, 626834, 2021.

14
Future Challenges for Designing Industry-Relevant Bioinspired Materials

Warren Rosario[1] and Nidhi Chauhan[2]*

[1]*School of Engineering (SoE), University of Petroleum and Energy Studies (UPES), Bidholi, Dehradun, India*
[2]*School of Health Sciences & Technology (SoHST), University of Petroleum and Energy Studies (UPES), Bidholi, Dehradun, India*

Abstract

Ever since its genesis, life has successfully been able to spread throughout the planet and has survived even in the harshest of conditions. It has done so through the process of "Evolution by natural selection." This ability of nature to perfect designs and materials over time has caught the eye of scientists all over the world and led to the development of materials and systems inspired from nature. This concept is called bioinspiration whose core tenet is the belief that, the solution to practically all our issues can be found in nature. A cursory examination of prior civilizations reveals that we have long known the benefits of biologically inspired materials, despite being uninformed of their specific workings and mechanisms. Now, with the help of advanced characterization and analysis tools, we are finally able to crack open nature's treasure chest and use this unprecedented wealth of knowledge to engineer materials that can tackle problems that have tormented us. Using nanotechnology, scientists have developed bioinspired materials that are far superior to the traditional materials in use today. Superior properties, non-toxic nature, and other qualities, however, are insufficient in today's capitalist environment to make these materials mainstream. These advanced bioinspired materials must overcome additional obstacles before they can be embraced by the general public. These obstacles can be broadly divided into two categories: a) difficulty in developing feasible synthesis methods; and b) challenges in ensuring the quality, stability, and sustainability of these materials. We will examine the applications of various bioinspired materials in this chapter, as well as the different hurdles that they must overcome in

---

Corresponding author: nidhi.chauhan@ddn.upes.ac.in

Mousumi Sen and Monalisa Mukherjee (eds.) Bioinspired and Green Synthesis of Nanostructures: A Sustainable Approach, (325–352) © 2023 Scrivener Publishing LLC

the coming future to be viable alternatives to existing materials. Nature is an ideal scientist who has spent millions of years developing and refining biological materials and systems with lots of time and resources, all of which are luxuries we as humans with finite lifespans cannot afford. We must be exceedingly judicious with our time and resources, concentrating our efforts on producing materials that can meet our requirements without sacrificing too much. Bioinspired materials have a bright future ahead of them, and once all of the production and material issues are resolved, they can greatly help humanity. Nature is a never-ending source of information and inspiration that will inspire us to come up with wonderful ideas and promote innovation.

Keywords: Bioinspired materials, nanotechnology, optics, energy, medical, biomimic, industrial nanomaterials

14.1 Introduction

About 3.7 billion years ago the earliest forms of life emerged on the earth and ever since then they have been engaged in the process of evolution [1, 2]. This process of continuous evolution has enabled living organisms to develop functions that are optimally suited to not only survive but thrive in their respective ecosystems [2]. They have developed body parts made of unique materials that show remarkable optical, mechanical and other properties that increase their chance of survival ensuring procreation. A few examples of this are the ants that can lift object weighing many times their own weight, some bird feathers that can repel water and dirt and many more [3, 4]. We also have butterflies with beautifully colored wings which are actually due to the wing's morphology that helps scatters light to get said colours [5].

Such abundance of materials with unique abilities found in nature have always fascinated humans since long back. And using this as an inspiration we have tried to mimic these materials and their mechanisms to solve our problems. This process of using materials and methods inspired from nature is called as bioinspiration. Even today scientists are always looking towards nature when trying to come up with answers to the problems that we are facing. This used to be a challenging task, but the use of modern technology has really helped us in understanding the mechanisms behind many of the biological materials and increased our scope to observe and adopt even more biological phenomenon [6]. The application of nanotechnology has enabled us to develop bioinspired materials using unique structures which can result in desired properties [7]. With our increasing awareness regarding the scarcity of resources and the resulting pollution there is a growing push towards the development of more bioinspired materials with better sustainability. As a result, bioinspired materials are growing in popularity which make studying them, their properties, and their fabrication techniques extremely important. There are many bioinspired materials that have been developed, showing great promise in solving many of our problems. But on the road to mass production, there are still some obstacles that we have yet to overcome. In this chapter we have attempted to shine a spotlight on different promising bioinspired materials and the many challenges we face in designing them as per industrial standards.

14.2 Bioinspired Materials

Manmade materials that are made such that they emulate the properties of biological materials such as structural, mechanical and so on, are called as bioinspired materials [8]. Natural materials have developed over the years to possess immensely beneficial properties like greater mechanical strength while being very lightweight, strong adhesion, water and dust resistance and thermal insulation just to name a few [9–14]. As a result, they have the curiosity and attention of world's majority materials scientists. In the past few decades many researchers have spent time studying biological materials and tried to come up with ways to synthetically recreate them for our advantage. Synthesis of materials inspired from nature that can emulate certain properties is at the forefront of materials science research. Nature has become the muse of many scientists for creating innovative materials with great utility [15].

14.3 Applications of Bioinspired Materials and Their Industrial Relevance

The use of nanotechnology related fabrication and modern characterization techniques has enabled us to precisely understand biological materials, their structures, and chemical composition. And this has helped researchers in designing, fabricating, and optimizing the bioinspired materials properties. Mechanical modeling of biological materials is a new tool that has garnered the attention of many scientists [16]. The use of novel nanomaterials with bioinspired structures and mechanisms has enabled us to develop innovative technologies. And these technologies have enabled us to apply these bioinspired materials in a huge variety of fields [17]. To get a more concrete idea of what these materials are and how relevant they are to the industry we shall have an in depth look at their application in a couple of fields mentioned in the following sections and in the later section, discuss the challenges that need to be faced in the future for their adoption by industries.

14.4 Bioinspired Materials in Optics

Nature is like an art gallery displaying a range of beautiful colors produces by living organisms that have evolved body parts with unique structures and materials that can manipulate light to this effect. They have developed these optical abilities for escaping predators by using camouflage, attracting partners and even prey, and for many other tasks. These optical feats are generally achieved via pigments, bioluminescence and structural coloring which has by far generated the most buzz among many scientists [17–19]. These structures are called as photonic crystals which scatter light to produce majestic colors. This has thus inspired the development of novel micro/nanostructures in the construction of optical devices and for coloring requirements. Structurally generated colors offer excellent brightness, saturation, color change based on incidence angle and vivid iridescence over pigment generated color which tends to fade away over time.

14.4.1 Applications in Optics

Structures for color generation are mostly fabricated using the atomic layer deposition (ALD) technique. Inspired by the *Papilo* butterfly wing scales, a perception of the color green was generated by the deposition of TiO_2 (n=2.5) and Al_2O_3 (n=1.5) on PS (Polystyrene) films by ALD (n is the refractive index) [20]. Structures that cause scattering or reflection of light for color generation are classified based on their dimensions as 3D, 2D, 1D, and 0D. For 3D, we have the scales of the white beetle *Cyphochilus* spp. that create the perception of a brilliant white color [21], which has inspired the fabrication of a material with improved performance and with room available for more improvements (Figure 14.1A) [22, 23]. Then there are the 2D disordered stacks on the scales of fishes like Japanese koi and lookdown made of guanine crystal sheets and cytoplasm that produced a silvery glitter useful for camouflage (Figure 14.1B) [24]. For the case of 1D scatterers, we have been using fibers on fabrics and other products to generated vivid light scattered colors. Inspired by the triangular hairs of the silver ant, a triangular fiber with a bigger diameter is produced that can reduce heating (Figure 14.1C). But there aren't enough studies comparing them to determine which is the better one and as such more research is required before the adoption of a commercial fiber technology [25–27]. There are also the flat cone-shaped hairs of *Calothyrza margaritifera* another type of white beetle, which are full of spheres of about 550 nm size which scatter light to get a beautiful white color [28]. 0D scatterers made from transparent ceramics are used in low-cost white paint and even TiO_2 nanoparticles are sometimes used to get white color in fibers. As a result, the latest research is focused on studying their combined effect so as to mimic it.

Another major optical application of bioinspired materials is as antireflectors. They help in the optimum utilization of the solar radiation incident on earth, which is of great importance. Antireflectors surfaces (ARS) inspired by moth eyes are used in imaging and other optoelectronic devices as they limit the interference of the penetrating light due to their smaller size (sub-wavelength range) and ordered pattern. Moth eye inspired ARS materials help improve light transmission in transparent substrates like quartz and glass, which is extremely useful in lasers, and they can also improve the absorption light. ARS are prominently used in devices such as light sensors, solar cells and displays where absorption of incident light is necessary for optimum performance [29–32]. On the other hand, there are also butterfly wing scales with complex hierarchical structures that are beneficial in photothermal, broad band light absorption and photocatalysis applications (Figure 14.1D). Although due to the delicate nature of these structures, replicating them has proven to be very hard [33, 34]. There are still a large variety of organisms in nature that have demonstrated antireflective properties and through further research we can successfully imitate their structures as well for many more optical applications [35–38].

14.4.2 Bioinspired Materials in Energy

With the ever-increasing pace of human growth, there is a simultaneous increase in the need for more energy. With this there is a constant search for new materials for energy generation, conversion and storage that can

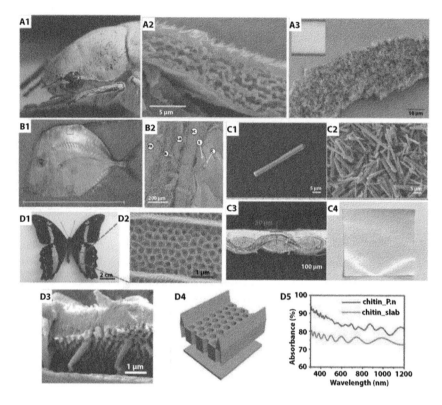

Figure 14.1 (A)(1) shows scales of *Cyphochilus* that are ~220 μm in length and 60 μm in width, (2) shows a highly magnified image of those scales obtained via scanning electron microscopy (SEM) and (3) shows a SEM image of porous white cellulose acetate film which is a material inspired by these *Cyphochilus* scales. (A) Reproduced with permission [23]. Copyright 2019, Nature Publishing Group. (B)(1) Image of the lookdown, Selene vomer and (2) The SEM image shows the epidermis (e) from the horizontal section in the mid-flank region along with the scales (s), stratum argenteum (sa) and many other interesting parts. (B) Reproduced with permission [24]. Copyright 2015, Royal Society. (C) Shows SEM images of (1) single ZnO microrod (MR), (2) ZnO MRs, (3) cross-sectional view of the T-ZM (ZnO microrod/PDMS) fabric and (4) a photograph of the prepared T-ZM fabric. (C) Reproduced with permission [27]. Copyright 2019, Elsevier. (D) Images relating to *Papilio nireus* butterfly and its inspired material (1) digital photograph, (2) top view SEM image of the wing, (3) sectional view SEM image of the wing, (4) simulated structure and (5) simulative optical absorption vs wavelength curves for the butterfly wing and the simulated structure. (D) Reproduced with permission [34]. Copyright 2016, Nature Publishing Group.

keep up with the increasing demand. There is also a requirement for these materials for to be sustainable for the environment so as to not cause additional problems.

14.4.3 Applications in Energy

Bioinspired materials developed with the assistance of nanotechnology have provided us with viable solutions to our problems in the field of energy [39]. Taking cues from the natural world we have been able to design and develop materials that have shown to enhance the performance of energy materials used in industry. The electric eel is a creature in the wild with a fascinating ability to generate electricity. Inspired by it we have developed a transparent, soft and flexible power generator that has an open circuit voltage of 110 V per cell. This power source is made up of polyacrylamide hydrogel with ion selective hydrogel membranes in a series that repeat similar to those in the eel [40]. There are also examples of other power generators which draw inspiration from hummingbird wings (Figure 14.2A) and many others [41]. Another important aspect for the energy industry is the conversion part and since, all the energy sources available on earth originated from the sun, solar energy conversion is of extreme significance. As a result, many bioinspired solar harvesting technologies are sought after. There is a special category of materials that mimic structural and other properties from natural materials called as gel materials. Their unique 3D structures offer a large surface area with many pores and reaction sites giving them properties that are extremely beneficial in energy conversion, storage and other energy related applications [42].

In the previous section we have seen the bioinspired optical materials which helped in the designing of solar cells which enhanced absorption, but apart from solar cells there are also other technologies for harnessing solar energy. Hydrogen evolution reaction (HER) and oxygen evolution reaction are a few examples of electrochemical reactions for energy conversion. Catalysts play a crucial role in the generation of electrochemical energy. In nature enzymes act as catalysts in specific instances of light harvesting, electron transfer, etc. Enzymes are cost-effective, efficiently use metal ions, highly selective and operate under moderate pressure and temperature. Using enzymes as catalysts has the potential to greatly escalate the energy conversion ability of electrochemical reaction and with them being much less harmful to the environment there is much hope with their application in the industry [43]. Materials inspired by the structure of venation systems and other qualities of leaves are used in the making of better solar cells and other types of displays. These materials

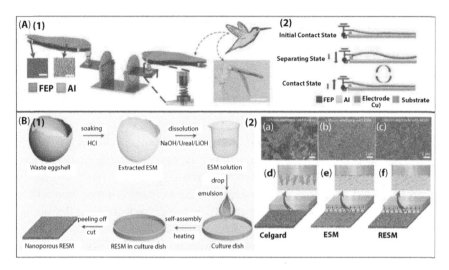

Figure 14.2 (A) Illustrates the schematic and experimental structure of hummingbird inspired triboelectric Nanogenerator (H-TENG), (1) 3D model of H-TENG wind harvester along with real hummingbird and photograph of real H-TENG, and (2) Working mechanism of the flag inside the wing as a second mode to harvest the mechanical motion resulting from the wind which mimics the hummingbird's flapping motion. (A) Reproduced with permission [41]. Copyright 2017, Nature Publishing Group. (B) Illustrations for regenerated eggshell membrane (RESM); (1) Schematic diagram for the preparation process of RESM and (2) SEM images of Li electrode after long-term cycling under 5.0 C in LiFePO4 cells (a) commercial Celgard, (b) ESM and (c) RESM separators respectively; schematic representation of dendrite growth on electrodes in presence of (d) commercial Celgard, (e) ESM and (f) RESM separators respectively. (B) Reproduced with permission [46]. Copyright 2018, Elsevier.

have great flexibility, strength, best current transport and improved electrochemical properties [44, 45]. Lithium-ion batteries are widely used for all kinds of energy storage needs due to their high energy density, but they suffer a noticeable drop in storage capacity due to the formation of dendrites. This problem is solved quiet uniquely by using regenerated eggshell membrane (RESM) inspired by the eggshell membranes available in nature as a nanoporous separators, which have shown to significantly reduce the formation of dendrites, Figure 14.2B shows the synthesis process and test results for Li-ion batteries with RESM [46]. $CoMoO_x$ hexagonal nanostructure having superior mechanical strength with interconnected channels and pores is another example of nature inspired (honeycomb) material which has helped in the development of high-performance lithium storage devices [47]. There were also, NiO hollow microspheres made using yeast as a nature derived template.

These hollow microspheres of NiO enhanced the working of lithium-ion batteries by increasing discharge capacity and showed better cycle performance when also coated with carbon [48]. These and many other examples have definitely proved the significance of using bioinspired materials for applications in energy industry.

14.4.4 Bioinspired Materials in Medicine

The advancement in nanoscience has also made possible for the development of materials inspired from nature and natural processes which have found applications in the medical industry as well [49–52]. Using nanomaterials such as graphene, fullerenes, CNTs, polymers, self-assembly molecules, etc. scientists are trying to revolutionize the fields of drug delivery, therapy, wound healing and many other in the medical domain [53]. Majority of the biological systems are made of simple minerals and biomacromolecules, but their structures and properties are extremely complex. For this reason, it becomes extremely essential to use bioinspired materials in biomedical applications [54, 55]. The following sub-sections show the application of various bioinspired materials to signify their relevance to the industry of medicine.

14.5 Applications in Medicine

A fascinating phenomenon found in nature is the self-assembly of biological macromolecules. With this as a base inspiration, scientists have successfully developed an innovative range of polymers that show self-assembly and are able to form 2D and 3D structures [56]. Polymersomes are analogous to liposomes but with higher physical and chemical stability. A basic overview of the similarities and differences between liposomes and polymersomes can be seen in Figure 14.3A. They can be used to make nanoreactors and artificial organelles, which then can be used for drug delivery applications [57–60]. Polymer vesicles that respond to stimuli are also developed by integrating block copolymers having stimuli-responsive chemical groups within the structure. These polymer vesicles then either start leaking or disassemble due to certain stimuli and precisely deliver the cargo [61, 62]. The stimuli in such cases can be from pH, temperature, light, redox reactions and so on. These polymers can also be developed to have intrinsic stimuli by loading it with biomolecules like proteins, DNA,

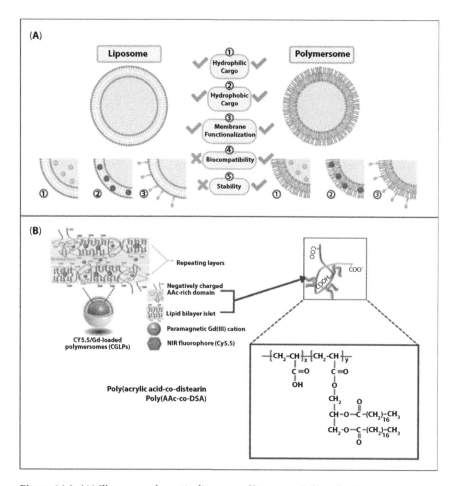

Figure 14.3 (A) Illustrates schematic diagrams of liposome (left) and polymersome (right), along with examples of their ability (1) to load hydrophilic cargo, (2) to load hydrophobic cargo and (3) for membrane functionalization; along with the advantages of polymersome over liposomes for (4) biocompatibility and (5) stability. (B) Shows the schematic representation of polymersomes used for diagnosis that can embed Cy5.5, a hydrophobic NIR fluorophore and the paramagnetic probe Gd(III) simultaneously within vesicular walls. (B) Reproduced with permission [65]. Copyright 2015, Elsevier.

enzymes, etc. that are naturally responsive [63]. Due to stability and non-intrusive behaviour in the blood stream, polymer vesicles are also used to mimic simple cells [64]. Diagnostic polymersomes containing dye with emission in near-infrared region and paramagnetic cations of gadolinium (Ga(III)) were administered in mice, as can be seen in Figure 14.3B. And the scientists were successful in imaging the dual-labeled tumor regions [65].

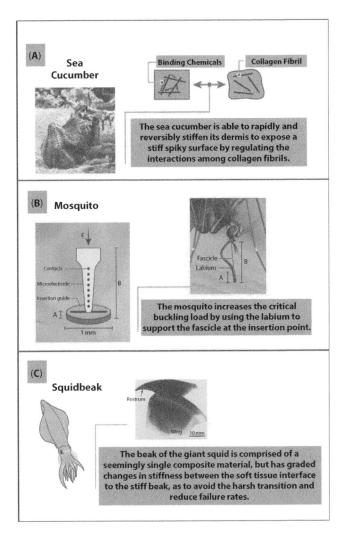

Figure 14.4 This figure displays different solutions for electrodes for neural implants inspired by a few animals; (A) Sea cucumbers use chemicals that binds the collagen fibrils which stiffens its dermis, (B) The female mosquito's labium to brace and guide its fascicle through the host's skin is used to develop insertion guide to reduce buckling of flexible electrodes, and (C) Squid's beak is made up of a singular composite material that can undergo transitions from soft to hard. (A) Reproduced with permission [76]. Copyright 2018, Elsevier. (B) Reproduced with permission [79]. Copyright 2018, Nature Publishing Group. (C) Reproduced with permission [78]. Copyright 2013, American Chemical Society Publications.

Safe storage of biological samples and organs for transplantation is a great challenge faced by the medical community. The present industrial methods are not enough to meet our demands. As a result, scientists have been looking to nature for answers. Bioinspired materials and methods have the potential to solve the cryopreservation problem by mimicking the cold tolerance biomechanisms used by animals. And along with the assistance of modern technology, the cryopreservation approaches developed are hypothermic preservation (inspired by rodents, marsupials, bats, etc.) [66], supercooling preservation (inspired by mollusks, reptiles, fishes, etc.) [67, 68], freezing preservation (inspired by insects, amphibians & reptiles) [69] and vitrification (inspired by invertebrates & plants) [70]. Protein cages are synthetic, multifunctional and highly symmetrical structures inspired by proteins and primarily used in biomedical applications. They have three types of interfaces namely, subunit interface, interior and exterior interfaces which can be functionalized to be used as sequestration, biomineralization, storage and delivery of nucleic acids and in some recent studies also for drug storage and delivery [71, 72]. Proteins along with peptides has also been the inspiration behind peptide nanotubes, hollow tubes which are used in tissue engineering [73–75]. Neural implant is a domain that is at the frontier of the medical industry. In order to develop optimal microelectrode materials that are soft but can also stiffen during insertion scientists like always went to nature and have found a few amazing solutions [76]. Polymer systems inspired by sea cucumber (changes stiffness based on chemical triggers) [77], squid's beak (gradual change from soft to stiff) [78], mosquito (microneedles inserted by using an insertion guide) [79] and many other animals are used for making these microelectrodes and other injection mechanisms. The following Figure 14.4 shows the various animals in nature that have inspired scientists to develop solutions in neural engineering.

14.6 Future Challenges for Industrial Relevance

Tremendous efforts by scientists and researchers have helped in the realization of many bioinspired materials that can mimic natural materials [80]. The synthesis of these bioinspired materials is only made possible due to advances in nanotechnology and our understanding of natural phenomenon. Despite this our present understanding of nature and its very many mysteries, is still peripheral which means a scrupulous investigation is warranted [81]. Materials in nature are formed naturally in milder conditions, even though they have complex structure. But the preparation of

Table 14.1 Bioinspired material properties along with corresponding sources and the future challenges they face classified on the basis of the three industries mentioned earlier.

S. no.	Industry of application	Bioinspiration source	Bioinspired material property	Future challenges	Ref.
1	Optical	Papilio palinurus butterfly wing	Perception of the colour green	• Synthesis of colour producing and ARS micro/nanostructure requires expertise and cutting-edge equipments • Most bioinspired materials are still in early stages of research	[20]
2		Cyphochilus beetle scales	Perception of the colour white		[23]
3		Lookdown fish scales	Polarization and broadband reflectance used for camouflage		[24]
4		Saharan silver ant *Cataglyphis bombycina* hairs	Optical properties resulting in silver colour and heat shielding		[25, 27]
5		3D nanostructures of butterfly wings	Solar cells with enhanced absorption in the far to near infrared region		[34]
6		Black colored regions on male peacock spiders	Anti-reflective surfaces that can reflect less than 1% of the incident light		[36]

(Continued)

Table 14.1 Bioinspired material properties along with corresponding sources and the future challenges they face classified on the basis of the three industries mentioned earlier. (*Continued*)

S. no.	Industry of application	Bioinspiration source	Bioinspired material property	Future challenges	Ref.
7		Velvet black scales of the West African Gaboon viper	Black colour due to nanostructures capable of super absorption in the UV and near IR range independent of viewing angle		[38]
8	Energy	The electric organ of electric eel	Biocompatible electric power source	• Expensive and complex fabrication process	[40]
9		Wings of the hummingbird	Unique flutter mechanics and shape adaptive qualities to develop nanogenerator	• Need for sustainable sources for raw materials	[41]
10		Enzymes	Catalysts for electrochemical energy generation	• Low energy output due to early research stage	[43]
11		Structure of leaf venation system	Inspired nanomaterials capable of enhancing solar cell performance		[44]
12		Leaf structures	Inspired nanomaterials capable of photocatalytic H2 generation and purification		[45]

(*Continued*)

Table 14.1 Bioinspired material properties along with corresponding sources and the future challenges they face classified on the basis of the three industries mentioned earlier. (*Continued*)

S. no.	Industry of application	Bioinspiration source	Bioinspired material property	Future challenges	Ref.
13		Eggshell membrane	Provide mechanical and thermal stability along with preventing dendrite formation in Li-ion batteries		[46]
14		Honeycomb structures	Inspired nanomaterials show significant enhancement in Li-ion storage capacity		[47]
15	Medical	Liposomes	Inspired materials called as polymersomes show both hydrophilic and hydrophobic properties along with membrane functionalization, biocompatibility and stability	• Polymersomes struggle with bulk hydration, functionalization in some cases decreases stability, have low circulation time in blood, lack of understanding regarding the effect of other body parameters due to limited *in vivo* testing	[58, 59]

(*Continued*)

Table 14.1 Bioinspired material properties along with corresponding sources and the future challenges they face classified on the basis of the three industries mentioned earlier. (*Continued*)

S. no.	Industry of application	Bioinspiration source	Bioinspired material property	Future challenges	Ref.
16		Hibernation in rodents, marsupials, some birds like speckled mousebird, etc.	Understanding the mechanism of hypothermic preservation	• Limited knowledge regarding cold-tolerant species and the cold adaptation mechanism • Need for further research and testing in practical conditions. • Affordable and large-scale synthesis methods	[66]
17		Mollusks, reptiles, fishes, etc.	Understanding the mechanism of supercooling preservation		[67, 68]
18		Insects, amphibians and reptiles	Understanding the mechanism of freezing preservation		[69]
19		Invertebrates and plants	Understanding the mechanism of vitrification		[70]
20		Proteins	Protein cages capable of sequestration, biomineralization, storage and delivery of nucleic acids		[71, 72]
21		Sea cucumber	Ability to rapidly and reversibly stiffen		[77]
22		Squid's beak	Gradually change hardness		[78]
23		Mosquito	Mechanism to support and guide microneedles		[79]

bioinspired materials has proved to be extremely difficult when trying to maintain precise control on structural aspects [82]. As a result, we need to develop processes and methods that can enable us to have precise control on the structural parameters, along with physical and chemical properties so as to get desired properties in the prepared materials [83]. The field of bioinspired materials is where many disciplines come together. Hence, for meaningful advancement and practical industrial applications an interdisciplinary approach to research is critical [84]. The data presented in Table 14.1 provides a brief overview of the different natural sources of inspiration, the properties of the bioinspired materials made or in development and the challenges they face to become relevant in the industry all classified on the basis of their applications in the optical, energy and medical industries.

To understand the challenges before us in more detail, we explore various issues as per the previously mentioned industries.

14.7 Optics-Specific Challenges

As seen in the previous sub-section, bioinspired materials have a wide range of applications in the field of optics. The structures responsible for colour generation are mostly micro and nanostructures that need to be synthesized with great care using complex methods. Apart from ALD method, some other synthesis methods for bioinspired optical materials are layer-by-layer assembly, spin-coating, sol-gel, focused ion-beam CVD and a few other alterations of these methods [85]. As the structure is what helps in the manipulation of light, it has to be made with the least number of defects and a uniform quality needs to be maintained. Hence, due to these synthesis methods that require expertise and high-end equipment, the fabrication cost of these materials increases which is a big reason for the hesitation of industries to commit. The 3D structural materials that show enhanced scattering as compared to the 2D materials are very rare and still in their initial research stages [86]. Usually structural and pigmentary colours are studied in isolation but there are cases where both mechanisms interact to produce colours. These colours are possible to get only via both mechanisms and in few cases, it is possible via a singular mechanism but with great difficulty. As a result, more research is required on this subject to achieve more complex and durable colours due to structural and pigmentation [87]. In ARS too, the synthesis methods require expertise, expensive equipment and are also tedious. To get the desired antireflective properties the materials have to be precisely synthesized while taking into account various factors like aspect ratio and period of ARS [88]. Optical microstructures like those inspired by the wings of butterfly are extremely complex and delicate which makes manufacturing them very expensive and thus limits their practical adoption. Despite all this, the fact still remains that the utilization of more bioinspired materials in the field of optics will result in phenomenal advancements [81].

14.8 Energy-Specific Challenges

Going into the future, the adoption of bioinspired materials in the energy industry is of extreme importance. Despite all the amazing results produced by bioinspired materials in energy applications, there is still a long way to go. The technology for designing and development of materials that can accurately imitate natural materials and processes is still in its early stages. The successful integration of these materials in our present technology still requires a lot of research and a detailed understanding of natural phenomenon. For electrochemical energy conversion reactions, the procurement of raw materials and other chemicals needs to be from economical and sustainable sources. And this also goes for solar based process that need to be scaled to higher levels for industrial usage [43]. There is also the need to develop synthesis methods that don't use expensive equipment, hard to get precursors and are relatively straightforward in execution. Only when these requirements are fulfilled, there will be actual chance of getting practical bioinspired materials on an industrial scale.

14.9 Medicine-Specific Challenges

Despite the major strides taken by the medical industry with bioinspired materials, there is still a lot to be desired. With the biological macromolecule inspired polymers it is difficult to get bulk hydration as they have a large hydrophobic fraction (~65 wt%) [89]. There are potential workarounds for this problem, but they sometimes distort the structure of the polymers [90]. Polymers that are modified with biomolecules for responsive stimulation are unable to maintain long-term stability and their structures degrade [63, 91, 92]. Even in drug delivery applications of polymersome, the blood circulation times need to be increased by optimizing the carrier system to efficiently target the tumor cells [65]. In most studies there is no *in vivo* testing of these polymer nanosystems. Also, these studies lack information regarding the effect of varying molecular parameters which is the main reason they have been proposed as models and not as alternatives. More research is required to perfectly understand the interactions between biological molecules and self-assembling polymers [93]. Our present knowledge of cold-tolerant species is still limited and as such we need more insight into their cold adaption mechanisms. To achieve this, we need more advanced 3D temperature sensors that are non-invasive and operate in real time along with genomic and proteomic tools for screening. For the practical utilization of cryopreservation technologies on the industrial scale more research on the various anti-cold mechanisms in nature is of paramount importance [94]. Whenever biological substances like enzymes are used in bioinspired materials, they tend to either increase the cost or make the process complicated and also introduce the problem of a shorter lifespan [84]. Apart from these, there are many other reasons, including manufacturing difficulties, which have hindered the large-scale industrial application of such bioinspired materials and have even limited their medical uses.

14.10 Conclusion

Nature has demonstrated a fascinating range of materials and mechanisms that have been fine-tuned by millions of years of evolution and adaptation. These natural materials hold the key to many of our questions and problems. With nature as our muse and modern nanotechnology as our tools, mankind has developed materials and technologies that have the potential to revolutionize the world. With our growing problems of pollution, we are in a desperate bid to develop materials and products that are sustainable with costing us too much. Bioinspired materials have given us hope and a promise for a better future. But the road to Elysium is not an easy one and many obstacles stand in our way. In this chapter, we have seen the industrial relevance of bioinspired materials in relative detail by focusing on the fields of optics, energy and medicine. Even though we see how beneficial these materials are in the mentioned field, this does not mean that their application is limited within the said fields. Bioinspired materials have also found use-cases in sensing, construction, adhesive manufacturing, communication, thermoregulation and many other fields [87]. We then also went through the various roadblocks that need to be crossed for an industry level application of these materials. We lack insight into the many enigmas surrounding natural processes. In which case more research, especially multidisciplinary research seems like the only path forward. Along with this, there is also a requirement of more advance technologies and methods to study and fabricate these materials in a cost-effective way.

When nature and science join forces, no problem seem big enough and with continuous innovation more and more bioinspired materials will gain widespread use. Till then it is imperative that we don't lose sight of the big picture. Nature is extremely beautiful and a treasure trove of possibilities, so we must be mindful of it and strive to preserve and protect it for the future.

References

1. Nutman, A.P., Bennett, V.C., Friend, C.R.L., van Kranendonk, M.J., Chivas, A.R., Rapid emergence of life shown by discovery of 3,700-million-year-old microbial structures., *Nature*, 537, 535–538, 2016. https://doi.org/10.1038/nature19355.
2. Oldroyd, D.R., Charles Darwin's theory of evolution: A review of our present understanding. *Biol. Philos.*, 1, 133–168, 1986, https://doi.org/10.1007/BF00142899.
3. Naskar, K. and Raut, S.K., Food–carrying strategy of the ants Pheidole Roberti. *Int. J. Tech. Res. Appl.*, 3, 55–58, 2015.
4. Genzer, J. and Marmur, A., Biological and synthetic self-cleaning surfaces. *MRS Bull.*, 33, 742–746, 2008, https://doi.org/10.1557/mrs2008.159.
5. Jiang, X., Shi, T., Zuo, H., Yang, X., Wu, W., Liao, G., Investigation on color variation of Morpho butterfly wings hierarchical structure based on PCA. *Sci. China Technol. Sci.*, 55, 16–21, 2012, https://doi.org/10.1007/s11431-011-4528-4.
6. Wang, Y., Naleway, S.E., Wang, B., Biological and bioinspired materials: Structure leading to functional and mechanical performance. *Bioact. Mater.*, 5, 745–757, 2020, https://doi.org/10.1016/j.bioactmat.2020.06.003.
7. Mishra, S., Sharma, S., Javed, M.N., Pottoo, F.H., Barkat, M.A., Harshita, M.S., Alam, Amir, M., Sarafroz, M., Bioinspired nanocomposites: Applications in disease diagnosis and treatment. *Pharm. Nanotechnol.*, 7, 206–219, 2019, https://doi.org/10.2174/2211738507666190425121509.
8. McKittrick, J., Chen, P.-Y., Tombolato, L., Novitskaya, E.E., Trim, M.W., Hirata, G.A., Olevsky, E.A., Horstemeyer, M.F., Meyers, M.A., Energy absorbent natural materials and bioinspired design strategies: A review. *Mater. Sci. Eng.: C*, 30, 331–342, 2010, https://doi.org/10.1016/j.msec.2010.01.011.
9. Lewis, R.V., Spider silk: Ancient ideas for new biomaterials. *Chem. Rev.*, 106, 3762–3774, 2006, https://doi.org/10.1021/cr010194g.

10. McConney, M.E., Anderson, K.D., Brott, L.L., Naik, R.R., Tsukruk, V.V., Bioinspired material approaches to sensing. *Adv. Funct. Mater.*, 19, 2527–2544, 2009, https://doi.org/10.1002/adfm.200900606.
11. Lintz, E.S. and Scheibel, T.R., Dragline, egg stalk and byssus: A comparison of outstanding protein fibers and their potential for developing new materials. *Adv. Funct. Mater.*, 23, 4467–4482, 2013, https://doi.org/10.1002/adfm.201300589.
12. Liu, M., Zheng, Y., Zhai, J., Jiang, L., Bioinspired super-antiwetting interfaces with special liquid–solid adhesion. *Acc. Chem. Res.*, 43, 368–377, 2010, https://doi.org/10.1021/ar900205g.
13. Nishimoto, S. and Bhushan, B., Bioinspired self-cleaning surfaces with superhydrophobicity, superoleophobicity, and superhydrophilicity. *RSC Adv.*, 3, 671–690, 2013, https://doi.org/10.1039/C2RA21260A.
14. Bahners, T., Schlosser, U., Gutmann, R., Schollmeyer, E., Textile solar light collectors based on models for polar bear hair. *Sol. Energy Mater. Sol. Cells*, 92, 1661–1667, 2008, https://doi.org/10.1016/j.solmat.2008.07.023.
15. Liu, K. and Jiang, L., Bio-inspired design of multiscale structures for function integration. *Nano Today*, 6, 155–175, 2011, https://doi.org/10.1016/j.nantod.2011.02.002.
16. Wu, M.S., Strategies and challenges for the mechanical modeling of biological and bio-inspired materials. *Mater. Sci. Eng.: C*, 31, 1209–1220, 2011, https://doi.org/10.1016/j.msec.2010.11.012.
17. Vukusic, P. and Sambles, J.R., Photonic structures in biology. *Nature*, 424, 852–855, 2003, https://doi.org/10.1038/nature01941.
18. Vukusic, P., Sambles, J.R., Lawrence, C.R., Colour mixing in wing scales of a butterfly. *Nature*, 404, 457–457, 2000, https://doi.org/10.1038/35006561.
19. Teyssier, J., Saenko, S.V., van der Marel, D., Milinkovitch, M.C., Photonic crystals cause active colour change in chameleons. *Nat. Commun.*, 6, 6368, 2015, https://doi.org/10.1038/ncomms7368.
20. Crne, M., Sharma, V., Blair, J., Park, J.O., Summers, C.J., Srinivasarao, M., Biomimicry of optical microstructures of Papilio palinurus. *Europhys. Lett.*, 93, 14001, 2011.
21. Vukusic, P., Hallam, B., Noyes, J., Brilliant whiteness in ultrathin beetle scales. *Sci. (1979)*, 315, 348–348, 2007, https://doi.org/10.1126/science.1134666.
22. Kadolph, S.J. and Marcketti, S., *Textiles*, 12th ed., Pearson, Boston, 2016.
23. Burg, S.L., Washington, A., Coles, D.M., Bianco, A., McLoughlin, D., Mykhaylyk, O.O., Villanova, J., Dennison, A.J.C., Hill, C.J., Vukusic, P., Doak, S., Martin, S.J., Hutchings, M., Parnell, S.R., Vasilev, C., Clarke, N., Ryan, A.J., Furnass, W., Croucher, M., Dalgliesh, R.M., Prevost, S., Dattani, R., Parker, A., Jones, R.A.L., Fairclough, J.P.A., Parnell, A.J., Liquid–liquid phase separation morphologies in ultra-white beetle scales and a synthetic equivalent. *Commun. Chem.*, 2, 100, 2019, https://doi.org/10.1038/s42004-019-0202-8.

24. Zhao, S., Brady, P.C., Gao, M., Etheredge, R.I., Kattawar, G.W., Cummings, M.E., Broadband and polarization reflectors in the lookdown, *Selene vomer*. *J. R. Soc. Interface*, 12, 20141390, 2015, https://doi.org/10.1098/rsif.2014.1390.
25. Willot, Q., Simonis, P., Vigneron, J.-P., Aron, S., Total internal reflection accounts for the bright color of the saharan silver ant. *PLoS One*, 11, e0152325, 2016, https://doi.org/10.1371/journal.pone.0152325.
26. Shi, N.N., Tsai, C.-C., Camino, F., Bernard, G.D., Yu, N., Wehner, R., Keeping cool: Enhanced optical reflection and radiative heat dissipation in Saharan silver ants. *Sci.* (1979), 349, 298–301, 2015, https://doi.org/10.1126/science.aab3564.
27. Wang, Y., Shang, S., Chiu, K., Jiang, S., Mimicking Saharan silver ant's hair: A bionic solar heat shielding architextile with hexagonal ZnO microrods coating. *Mater. Lett.*, 261, 127013, 2020, https://doi.org/10.1016/j.matlet.2019.127013.
28. Lafait, J., Andraud, C., Berthier, S., Boulenguez, J., Callet, P., Dumazet, S., Rassart, M., Vigneron, J.-P., Modeling the vivid white color of the beetle Calothyrza margaritifera. *Mater. Sci. Eng.: B*, 169, 16–22, 2010, https://doi.org/10.1016/j.mseb.2009.12.026.
29. Zhu, J., Yu, Z., Burkhard, G.F., Hsu, C.-M., Connor, S.T., Xu, Y., Wang, Q., McGehee, M., Fan, S., Cui, Y., Optical absorption enhancement in amorphous silicon nanowire and nanocone arrays. *Nano Lett.*, 9, 279–282, 2009, https://doi.org/10.1021/nl802886y.
30. Zhu, J., Hsu, C.-M., Yu, Z., Fan, S., Cui, Y., Nanodome solar cells with efficient light management and self-cleaning. *Nano Lett.*, 10, 1979–1984, 2010, https://doi.org/10.1021/nl9034237.
31. Lee, C., Bae, S.Y., Mobasser, S., Manohara, H., A novel silicon nanotips antireflection surface for the micro sun sensor. *Nano Lett.*, 5, 2438–2442, 2005, https://doi.org/10.1021/nl0517161.
32. Xu, H., Liu, L., Teng, F., Lu, N., Emission enhancement of fluorescent molecules by antireflective arrays. *Research*, 2019, 1–8, 2019, https://doi.org/10.34133/2019/3495841.
33. Dou, S., Xu, H., Zhao, J., Zhang, K., Li, N., Lin, Y., Pan, L., Li, Y., Bioinspired microstructured materials for optical and thermal regulation. *Adv. Mater.*, 33, 2000697, 2021, https://doi.org/10.1002/adma.202000697.
34. Yan, R., Chen, M., Zhou, H., Liu, T., Tang, X., Zhang, K., Zhu, H., Ye, J., Zhang, D., Fan, T., Bio-inspired plasmonic nanoarchitectured hybrid system towards enhanced far red-to-near infrared solar photocatalysis. *Sci. Rep.*, 6, 20001, 2016, https://doi.org/10.1038/srep20001.
35. Lee, J.-W., Dai, Z., Han, T.-H., Choi, C., Chang, S.-Y., Lee, S.-J., de Marco, N., Zhao, H., Sun, P., Huang, Y., Yang, Y., 2D perovskite stabilized phase-pure formamidinium perovskite solar cells. *Nat. Commun.*, 9, 3021, 2018, https://doi.org/10.1038/s41467-018-05454-4.
36. McCoy, D.E., McCoy, V.E., Mandsberg, N.K., Shneidman, A.V., Aizenberg, J., Prum, R.O., Haig, D., Structurally assisted super black in colourful

peacock spiders. *Proc. R. Soc. B: Biol. Sci.*, 286, 20190589, 2019, https://doi.org/10.1098/rspb.2019.0589.
37. McCoy, D.E. and Prum, R.O., Convergent evolution of super black plumage near bright color in 15 bird families. *J. Exp. Biol.*, 222, 208140, 2019, https://doi.org/10.1242/jeb.208140.
38. Spinner, M., Kovalev, A., Gorb, S.N., Westhoff, G., Snake velvet black: Hierarchical micro- and nanostructure enhances dark colouration in Bitis rhinoceros. *Sci. Rep.*, 3, 1846, 2013, https://doi.org/10.1038/srep01846.
39. Thekkekara, L.V., and Gu, M., Large-scale waterproof and stretchable textile-integrated laser-printed graphene energy storages. *Sci. Rep.*, 9, 11822, 2019, https://doi.org/10.1038/s41598-019-48320-z.
40. Schroeder, T.B.H., Guha, A., Lamoureux, A., VanRenterghem, G., Sept, D., Shtein, M., Yang, J., Mayer, M., An electric-eel-inspired soft power source from stacked hydrogels. *Nature*, 552, 214–218, 2017, https://doi.org/10.1038/nature24670.
41. Ahmed, A., Hassan, I., Song, P., Gamaleldin, M., Radhi, A., Panwar, N., Tjin, S.C., Desoky, A.Y., Sinton, D., Yong, K.-T., Zu, J., Self-adaptive bioinspired hummingbird-wing stimulated triboelectric nanogenerators. *Sci. Rep.*, 7, 17143, 2017, https://doi.org/10.1038/s41598-017-17453-4.
42. Luo, L., Zhao, P., Yang, H., Liu, B., Zhang, J.-G., Cui, Y., Yu, G., Zhang, S., Wang, C.-M., Surface coating constraint induced self-discharging of silicon nanoparticles as anodes for lithium ion batteries. *Nano Lett.*, 15, 7016–7022, 2015, https://doi.org/10.1021/acs.nanolett.5b03047.
43. Artero, V., Bioinspired catalytic materials for energy-relevant conversions. *Nat. Energy*, 2, 17131, 2017, https://doi.org/10.1038/nenergy.2017.131.
44. Han, B., Huang, Y., Li, R., Peng, Q., Luo, J., Pei, K., Herczynski, A., Kempa, K., Ren, Z., Gao, J., Bio-inspired networks for optoelectronic applications. *Nat. Commun.*, 5, 5674, 2014, https://doi.org/10.1038/ncomms6674.
45. Fu, J., Zhu, B., You, W., Jaroniec, M., Yu, J., A flexible bio-inspired H2-production photocatalyst. *Appl. Catal. B: Environ.*, 220, 148–160, 2018, https://doi.org/10.1016/j.apcatb.2017.08.034.
46. Ma, L., Chen, R., Hu, Y., Zhang, W., Zhu, G., Zhao, P., Chen, T., Wang, C., Yan, W., Wang, Y., Wang, L., Tie, Z., Liu, J., Jin, Z., Nanoporous and lyophilic battery separator from regenerated eggshell membrane with effective suppression of dendritic lithium growth. *Energy Storage Mater.*, 14, 258–266, 2018, https://doi.org/10.1016/j.ensm.2018.04.016.
47. Mei, J., Liao, T., Spratt, H., Ayoko, G.A., Zhao, X.S., Sun, Z., Honeycomb-inspired heterogeneous bimetallic Co–Mo oxide nanoarchitectures for high-rate electrochemical lithium storage. *Small Methods*, 3, 1900055, 2019, https://doi.org/10.1002/smtd.201900055.
48. Tian, J., Shao, Q., Dong, X., Zheng, J., Pan, D., Zhang, X., Cao, H., Hao, L., Liu, J., Mai, X., Guo, Z., Bio-template synthesized NiO/C hollow microspheres with enhanced Li-ion battery electrochemical performance. *Electrochim. Acta*, 261, 236–245, 2018, https://doi.org/10.1016/j.electacta.2017.12.094.

49. Aricò, A.S., Bruce, P., Scrosati, B., Tarascon, J.-M., van Schalkwijk, W., Nanostructured materials for advanced energy conversion and storage devices. *Nat. Mater.*, 4, 366–377, 2005, https://doi.org/10.1038/nmat1368.
50. Davis, S.S., Biomédical applications of nanotechnology — Implications for drug targeting and gene therapy. *Trends Biotechnol.*, 15, 217–224, 1997, https://doi.org/10.1016/S0167-7799(97)01036-6.
51. Walcarius, A., Minteer, S.D., Wang, J., Lin, Y., Merkoçi, A., Nanomaterials for bio-functionalized electrodes: Recent trends. *J. Mater. Chem. B*, 1, 4878, 2013, https://doi.org/10.1039/c3tb20881h.
52. Du, D., Yang, Y., Lin, Y., Graphene-based materials for biosensing and bioimaging. *MRS Bull.*, 37, 1290–1296, 2012, https://doi.org/10.1557/mrs.2012.209.
53. Andre, R., Tahir, M.N., Natalio, F., Tremel, W., Bioinspired synthesis of multifunctional inorganic and bio-organic hybrid materials. *FEBS J.*, 279, 1737–1749, 2012, https://doi.org/10.1111/j.1742-4658.2012.08584.x.
54. Sanchez, C., Arribart, H., Giraud Guille, M.M., Biomimetism and bioinspiration as tools for the design of innovative materials and systems. *Nat. Mater.*, 4, 277–288, 2005, https://doi.org/10.1038/nmat1339.
55. Barthelat, F., Biomimetics for next generation materials. *Philos. Trans. R. Soc. A: Math. Phys. Eng. Sci.*, 365, 2907–2919, 2007, https://doi.org/10.1098/rsta.2007.0006.
56. Huber, M.C., Schreiber, A., von Olshausen, P., Varga, B.R., Kretz, O., Joch, B., Barnert, S., Schubert, R., Eimer, S., Kele, P., Schiller, S.M., Designer amphiphilic proteins as building blocks for the intracellular formation of organelle-like compartments. *Nat. Mater.*, 14, 125–132, 2015, https://doi.org/10.1038/nmat4118.
57. Palivan, C.G., Fischer-Onaca, O., Delcea, M., Itel, F., Meier, W., Protein-polymer nanoreactors for medical applications. *Chem. Soc. Rev.*, 41, 2800–2823, 2012, https://doi.org/10.1039/C1CS15240H.
58. Messager, L., Gaitzsch, J., Chierico, L., Battaglia, G., Novel aspects of encapsulation and delivery using polymersomes. *Curr. Opin. Pharmacol.*, 18, 104–111, 2014, https://doi.org/10.1016/j.coph.2014.09.017.
59. Tanner, P., Baumann, P., Enea, R., Onaca, O., Palivan, C., Meier, W., Polymeric vesicles: From drug carriers to nanoreactors and artificial organelles. *Acc. Chem. Res.*, 44, 1039–1049, 2011, https://doi.org/10.1021/ar200036k.
60. Onaca-Fischer, O., Liu, J., Inglin, M., Palivan, C.G., Polymeric nanocarriers and nanoreactors: A survey of possible therapeutic applications. *Curr. Pharm. Des.*, 18, 2622–2643, 2012, https://doi.org/10.2174/138161212800492822.
61. Qiao, Z.-Y., Cheng, J., Ji, R., Du, F.-S., Liang, D.-H., Ji, S.-P., Li, Z.-C., Biocompatible acid-labile polymersomes from PEO-b-PVA derived amphiphilic block copolymers. *RSC Adv.*, 3, 24345, 2013, https://doi.org/10.1039/c3ra42824a.
62. Ahmed, F. and Discher, D.E., Self-porating polymersomes of PEG–PLA and PEG–PCL: Hydrolysis-triggered controlled release vesicles. *J. Controlled Release*, 96, 37–53, 2004, https://doi.org/10.1016/j.jconrel.2003.12.021.

63. Akbarzadeh, A., Rezaei-Sadabady, R., Davaran, S., Joo, S.W., Zarghami, N., Hanifehpour, Y., Samiei, M., Kouhi, M., Nejati-Koshki, K., Liposome: Classification, preparation, and applications. *Nanoscale Res. Lett.*, 8, 102, 2013, https://doi.org/10.1186/1556-276X-8-102.
64. Marguet, M., Bonduelle, C., Lecommandoux, S., Multicompartmentalized polymeric systems: Towards biomimetic cellular structure and function. *Chem. Soc. Rev.*, 42, 512–529, 2013, https://doi.org/10.1039/C2CS35312A.
65. Huang, W.-C., Chen, Y.-C., Hsu, Y.-H., Hsieh, W.-Y., Chiu, H.-C., Development of a diagnostic polymersome system for potential imaging delivery. *Colloids Surf. B: Biointerfaces*, 128, 67–76, 2015, https://doi.org/10.1016/j.colsurfb.2015.02.008.
66. Geiser, F., Metabolic rate and body temperature reduction during hibernation and daily torpor. *Annu. Rev. Physiol.*, 66, 239–274, 2004, https://doi.org/10.1146/annurev.physiol.66.032102.115105.
67. Miya, T., Gon, O., Mwale, M., Christina Cheng, C.-H., The effect of habitat temperature on serum antifreeze glycoprotein (AFGP) activity in Notothenia rossii (Pisces: Nototheniidae) in the Southern Ocean. *Polar Biol.*, 37, 367–373, 2014, https://doi.org/10.1007/s00300-013-1437-y.
68. Richter, M.M., Williams, C.T., Lee, T.N., Tøien, Ø., Florant, G.L., Barnes, B.M., Buck, C.L., Thermogenic capacity at subzero temperatures: How low can a hibernator go? *Physiol. Biochem. Zool.*, 88, 81–89, 2015, https://doi.org/10.1086/679591.
69. Storey, K.B. and Storey, J.M., Molecular biology of freezing tolerance. *Compr. Physiol.*, Wiley, 3, 1283–1308, 2013, https://doi.org/10.1002/cphy.c130007.
70. Sformo, T., Walters, K., Jeannet, K., Wowk, B., Fahy, G.M., Barnes, B.M., Duman, J.G., Deep supercooling, vitrification and limited survival to −100°C in the Alaskan beetle *Cucujus clavipes puniceus* (Coleoptera: Cucujidae) larvae. *J. Exp. Biol.*, 213, 502–509, 2010, https://doi.org/10.1242/jeb.035758.
71. Uchida, M., Klem, M.T., Allen, M., Suci, P., Flenniken, M., Gillitzer, E., Varpness, Z., Liepold, L.O., Young, M., Douglas, T., Biological containers: Protein cages as multifunctional nanoplatforms. *Adv. Mater.*, 19, 1025–1042, 2007, https://doi.org/10.1002/adma.200601168.
72. Suci, P.A., Kang, S., Young, M., Douglas, T., A streptavidin–protein cage janus particle for polarized targeting and modular functionalization. *J. Am. Chem. Soc.*, 131, 9164–9165, 2009, https://doi.org/10.1021/ja9035187.
73. Yu, X., Wang, J., Du, Y., Ma, Z., He, W., Nanotechnology for tissue engineering applications. *J. Nanomater.*, 2011, 1–2, 2011, https://doi.org/10.1155/2011/506574.
74. Hartgerink, J.D., Granja, J.R., Milligan, R.A., Ghadiri, M.R.; Self-assembling peptide nanotubes. *J. Am. Chem. Soc.*, 118, 43–50, 1996, https://doi.org/10.1021/ja953070s.
75. Scanlon, S. and Aggeli, A., Self-assembling peptide nanotubes. *Nano Today*, 3, 22–30, 2008, https://doi.org/10.1016/S1748-0132(08)70041-0.

76. Shoffstall, A.J. and Capadona, J.R., Bioinspired materials and systems for neural interfacing. *Curr. Opin. Biomed. Eng.*, 6, 110–119, 2018, https://doi.org/10.1016/j.cobme.2018.05.002.
77. Capadona, J.R., Shanmuganathan, K., Tyler, D.J., Rowan, S.J., Weder, C., Stimuli-responsive polymer nanocomposites inspired by the sea cucumber dermis. *Sci. (1979)*, 319, 1370–1374, 2008, https://doi.org/10.1126/science.1153307.
78. Fox, J.D., Capadona, J.R., Marasco, P.D., Rowan, S.J., Bioinspired water-enhanced mechanical gradient nanocomposite films that mimic the architecture and properties of the squid beak. *J. Am. Chem. Soc.*, 135, 5167–5174, 2013, https://doi.org/10.1021/ja4002713.
79. Shoffstall, A.J., Srinivasan, S., Willis, M., Stiller, A.M., Ecker, M., Voit, W.E., Pancrazio, J.J., Capadona, J.R., A mosquito inspired strategy to implant microprobes into the brain. *Sci. Rep.*, 8, 122, 2018, https://doi.org/10.1038/s41598-017-18522-4.
80. Kaur, J., Rajkhowa, R., Afrin, T., Tsuzuki, T., Wang, X., Facts and myths of antibacterial properties of silk. *Biopolymers*, 101, 237–245, 2014, https://doi.org/10.1002/bip.22323.
81. Dou, S., Xu, H., Zhao, J., Zhang, K., Li, N., Lin, Y., Pan, L., Li, Y., Bioinspired microstructured materials for optical and thermal regulation. *Adv. Mater.*, 33, 2000697, 2021, https://doi.org/10.1002/adma.202000697.
82. Zhao, N., Wang, Z., Cai, C., Shen, H., Liang, F., Wang, D., Wang, C., Zhu, T., Guo, J., Wang, Y., Liu, X., Duan, C., Wang, H., Mao, Y., Jia, X., Dong, H., Zhang, X., Xu, J., Bioinspired materials: From low to high dimensional structure. *Adv. Mater.*, 26, 6994–7017, 2014, https://doi.org/10.1002/adma.201401718.
83. Shi, Y., Zhang, J., Pan, L., Shi, Y., Yu, G., Energy gels: A bio-inspired material platform for advanced energy applications. *Nano Today*, 11, 738–762, 2016, https://doi.org/10.1016/j.nantod.2016.10.002.
84. Bhattacharya, P., Du, D., Lin, Y., Bioinspired nanoscale materials for biomedical and energy applications. *J. R. Soc. Interface*, 11, 20131067, 2014, https://doi.org/10.1098/rsif.2013.1067.
85. Zhang, C., Mcadams, D.A., Grunlan, J.C., Nano/micro-manufacturing of bioinspired materials: A review of methods to mimic natural structures. *Adv. Mater.*, 28, 6292–6321, 2016, https://doi.org/10.1002/adma.201505555.
86. Mandal, J., Fu, Y., Overvig, A.C., Jia, M., Sun, K., Shi, N.N., Zhou, H., Xiao, X., Yu, N., Yang, Y., Hierarchically porous polymer coatings for highly efficient passive daytime radiative cooling. *Sci. (1979)*, 362, 315–319, 2018, https://doi.org/10.1126/science.aat9513.
87. Schroeder, T.B.H., Houghtaling, J., Wilts, B.D., Mayer, M., It's not a bug, it's a feature: Functional materials in insects. *Adv. Mater.*, 30, 1705322, 2018, https://doi.org/10.1002/adma.201705322.

88. Boden, S.A. and Bagnall, D.M., Tunable reflection minima of nanostructured antireflective surfaces. *Appl. Phys. Lett.*, 93, 133108, 2008, https://doi.org/10.1063/1.2993231.
89. Discher, D.E. and Ahmed, F., Polymersomes. *Annu. Rev. Biomed. Eng.*, 8, 323–341, 2006, https://doi.org/10.1146/annurev.bioeng.8.061505.095838.
90. Cui, J., van Koeverden, M.P., Müllner, M., Kempe, K., Caruso, F., Emerging methods for the fabrication of polymer capsules. *Adv. Colloid Interface Sci.*, 207, 14–31, 2014, https://doi.org/10.1016/j.cis.2013.10.012.
91. Chemin, M., Brun, P.-M., Lecommandoux, S., Sandre, O., le Meins, J.-F., Hybrid polymer/lipid vesicles: Fine control of the lipid and polymer distribution in the binary membrane. *Soft Matter*, 8, 2867, 2012, https://doi.org/10.1039/c2sm07188f.
92. Bozzuto, G. and Molinari, A., Liposomes as nanomedical devices. *Int. J. Nanomed.*, 10, 975, 2015, https://doi.org/10.2147/IJN.S68861.
93. Palivan, C.G., Goers, R., Najer, A., Zhang, X., Car, A., Meier, W., Bioinspired polymer vesicles and membranes for biological and medical applications. *Chem. Soc. Rev.*, 45, 377–411, 2016, https://doi.org/10.1039/C5CS00569H.
94. Dou, M., Lu, C., Rao, W., Bioinspired materials and technology for advanced cryopreservation. *Trends Biotechnol.*, 40, 93–106, 2022, https://doi.org/10.1016/j.tibtech.2021.06.004.

15

Biomimetic and Bioinspired Nanostructures: Recent Developments and Applications

Sreemoyee Chakraborty[1], Debabrata Bera[1], Lakshmishri Roy[2] and Chandan Kumar Ghosh[3]*

[1]*Department of Food Technology and Biochemical Engineering, Jadavpur University, Kolkata, West Bengal, India*
[2]*Department of Food Technology, Techno Main Salt Lake, Kolkata, West Bengal, India*
[3]*School of Materials Science & Nanotechnology, Jadavpur University, Kolkata, West Bengal, India*

Abstract

This chapter focuses on the recent developments and novel applications of bio-inspired and biomimetic nanomaterials as functionally advanced biomolecules which have a huge prospect in research, development and engineering industries. Biomimetics (biomimicry) is the development of novel biomaterials which not only mimic the composition of natural systems but also copy their structure, morphology, and functionality. These bioinspired materials generally have their origins in nature and are designed by studying and imitating the remarkable biological processes of organisms and pathways of occurrence of different natural phenomena. In general, bio-inspired structures are synthesized by the following systems: (i) elementary biomimetics, (ii) high-level biomimetics, (iii) intelligent biomimetics, (iv) living-organism, and (v) bioinspired macromolecular arrangements. The core technology of bio-inspiration is built upon deciphering how biological materials are constructed and understanding the interactions that cause their unique properties.

Many of the functional details of biotic processes and structures belong to the nanorange and hence can be perfectly twinned and understood in detail using the knowledge of nanomaterials and nanotechnology. The interphase between biotechnology and nanotechnology comprises a hitherto unexplored/barely

---

*Corresponding author: chandan.kghosh@jadavpuruniversity.in

Mousumi Sen and Monalisa Mukherjee (eds.) Bioinspired and Green Synthesis of Nanostructures: A Sustainable Approach, (353–404) © 2023 Scrivener Publishing LLC

explored area of modern science with huge potential applications in medical, pharmaceutical, biotechnology and food engineering fields. Advances in design of nanostructured materials can make way for us to restructure systems and create new materials by modifications at the (sub)molecular level, creating new functionalities.

Current researches in this are concerned with the absorption and distribution of nanoparticles in the micro and macro-organisms, development of highly sensitive sensors, methods to counteract pollution, designing smart nano materials, and others. The potential of biomimetic approaches on a nano scale encompasses recent breakthroughs in biomedical, biomechanics, and cell-free enzymatic manufacture of nanoparticles. Bioinspired strategies for synthesizing bottom-up nanocomposites by using biomolecules for potential applications in colorimetry, electrochemistry, surface plasmon resonance, Raman scattering, and microbalance sensors have also been explored.

Other medical and biomedical applications include development of biomimetic synthetic calcium phosphate for supplementing the strength of bones and teeth, therapeutic drug delivery systems including biomimetic liposomes, hydrogels, micelles, and other misc nanostructures. Brain-machine interfaces utilizing biomimetic neuronal modelling have shown to greatly improve the performance and control of prosthetics. This genre of nanobioengineering is still in its initiation stages and has ground-breaking scope of application in material science and engineering, food engineering, agriculture, biomedical, and pharmaceutical industries involving development of programmable nanomaterials.

Keywords: Biomimetic, bioinspired, nanomaterial, biotechnology, nanotechnology, bio-interfaces

15.1 Introduction

In all things of nature, there is something of the marvellous – Aristotle

For centuries, humans have tried to harness this "marvel" of nature for designing a broad range of materials, structures, models, systems, and processes. In the course of evolution, which encompasses a few billion years, nature has found solutions to almost every possible challenge, scientific technical or otherwise, faced by humanity. It is the use of these solutions to address different problems in engineering sciences that came to be known as biomimetics. By examining how nature resolves diverse problems, from the mechanical challenges in tree branches and animal bones to the development of colours on butterfly wings, we may learn about natural problem-solving strategies that are often resource-efficient and have been perfected over millions of years [1-3].

This branch of bioengineering is a combination of two different classes of biotechnological phenomenon—bioinspiration (bionics) and biomimetics (biomimicry). Bioinspiration entails creating high-end engineering blueprints at different scales inspired by nature's design principles to address some of humanity's most pressing problems. Biomimicry, on the other hand, is a direct replication of techniques or processes designed by nature in nanoscale, microscale, or macroscale in different fields of research, such as robotics, bioelectronics, medicine, core engineering, and energy generation.

Nature's "solution manual" is quite comprehensive considering the wide range of functions that natural systems have evolved to perform. Advanced engineering techniques, such as light-harvesting, charge transfer, molecular complementarity, structural sustenance, signal transduction, actuation, gating, self-assembly, autopoiesis, replication, sensing, catalysis, trafficking or any combinations of these functions have always been abundant in nature and have inspired researchers to develop sustainable biosystems replicating these processes. The logical design of nanostructured materials and engineering systems is inspired by the relevant aspects in living systems (Figure 15.1), such as the water collecting ability of *Stenocara gracilipes* beetles, the strength and dexterity of spider silk; the self-cleaning ability of lotus leaves; the antifogging function of mosquitoes' eyes; the superior walking ability of water striders; the adhesivity of mussels in marine environments; the anisotropic wetting and pigmentation in butterflies' wings; the anti-reflection capabilities of cicadas' wings; among others.

The biomimetic method confers functional improvements to newly developed systems and processes such greater biocompatibility, biodistribution, and targeting characteristics. By leveraging biological molecules, animals, and cells—or drawing inspiration from them, novel microstructures and nanostructures are now rapidly progressing toward development of perfect bioprocess-based systems.

Nanotechnology garners the capacity to control atoms at atomic scales between one and several nanometres (generally less than 100 nm) to comprehend, develop, and employ material structures, mechanisms with radically new characteristics and functionalities deriving from their unique structures. From their initial level of organization, all biological systems have their most fundamental characteristics and functions specified at the nanoscale. Therefore, the main goal of nanotechnology in biological systems is to use low-energy linkages to hierarchically build molecules into objects and vice versa. Under the two sides already mentioned, nanotechnology offers tools and platforms for the analysis and

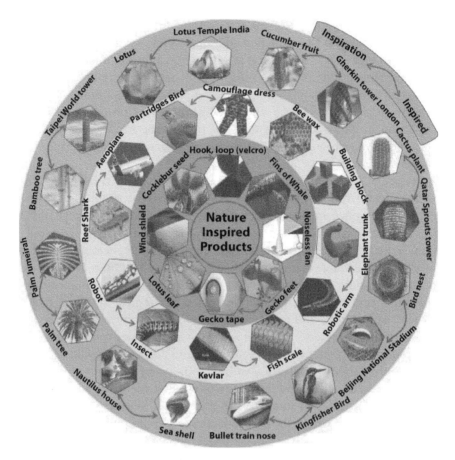

Figure 15.1 Commercially available nature-inspired products [4]. Reproduction does not require permission.

modification of biological processes, while biology serves as the origin of motivation for developing novel devices and systems integrated from the nanoscale [4].

In several engineering fields, biomimetic and bioinspired microstructures and nanostructures are incredibly significant. The potential uses of these microstructures and nanostructures, which we encounter every day and which were inspired by nature itself, are multitudinous. The use of carbon nanostructures, polymer nanocomposites, hybrid scaffolds and protein nanostructures in biomimetic and bioinspired systems in fields of engineering, medicine, biotechnology, food processing, agriculture, bioremediation, electronics, robotics, etc. is immense. These bioinspired and biomimetic structures have an advantage over synthetically created micro- and nanostructures because of their great biocompatibility. These microstructures and nanostructures provide a viable answer for many unsolved bioengineering issues if more is learned about them and the accompanying difficulties are overcome. Biomimetic delivery systems have the potential to enhance the stability and *in vivo* targeting, as well as better biocompatibility biodistribution and targeting properties than synthetic NPs.

15.2 Designing Bioinspired and Bioimitating Structures and Pathways

Even though the terms "biomimicry" and "bioinspired" are relatively new progress of human civilization has always been influenced by nature. Earliest recorded evidence of bioinspired and bio-mimicked structures dates back to ancient Egypt where elite bodyguards wore crocodile skins as protective gears against weapons and nobles of Chinese Shang Dynasty wore breastplates made from turtle shells. Moreover, over the centuries, several attempts to emulate the flight of birds and insects have been made, but actual flying bionics were first properly documented by Leonardo da Vinci (AD 1496) who was especially attuned to the natural world. The extensive study of birds and several wing designs for the initial flying machines, particularly those based on bat wings, which the artist referred to as an impenetrable membrane, can be found in da Vinci's journals. Not simply birds and animals, but even a little dry fruit or seed (burr) served as an inspiration for the Velcro fabric fastening system. To control light for structural colour, optical index matching, or light polarization, cuticular nanostructures seen in bioluminescent fireflies and jellyfishes have been identified. These discoveries have led to the designing of LED lens with increased light transmission in the visible spectrum comparable to a smooth surface lens.

Biomimetics can involve any of the following steps as a process:

- Identify flaws or concerns with materials, then look for natural materials or creatures displaying a similar problem.
- Embrace inspiration through watching and researching natural phenomena and natural materials systems.

- Examine biomimetic simulations and examine their optimization in the context of natural phenomena and natural materials systems.

One significant area of biomimetics is the synthesis of materials by biomimetic techniques; among the different materials, the synthesis of nanomaterials through biomimetic methods is the most dynamic and promising area. Here, functional biomimetic synthesis and process biomimetic synthesis are separated into two categories. By using a variety of substances and techniques, functional biomimetic synthesis attempts to imitate certain characteristics of natural materials, structures, and systems. For instance, artificial bones made of diverse materials (bioceramics, biopolymers/composites, etc.) have been created using various techniques and are anticipated to simultaneously have extra durability, good biocompatibility, and impact strength. Process-based biomimetics is a type of synthesis technique that aims to create different desirable nanostructures by imitating the arrangement, orientations, or procedures of biosystems. Examples include, numerous distinct nanosuperstructures (satellite structures, [9] dendrimer, [9] pyramids, [10] cubes, [11] two-dimensional nanoparticle arrays, [12] AuNP tubes, [13] etc.) created *in vitro* by simulating the protein manufacturing process. It is clear from the explanation above that there are substantial differences between functional biomimetics and process biomimetics. The former's goal is to duplicate the "properties," while the latter's goal is to emulate the "methods" [14].

In general, biomimetic systems are developed by any one of the following strategies:

(i) Basic biomass templates for biomimetic synthesis;
(ii) Advanced biomimetic synthesis using integrated soft and rigid membranes;
(iii) Interactive biomimetic synthesis using carriers and liquid membranes;
(iv) Plants and microorganisms used in living-organism biomimetic synthesis;
(v) Bioinspired synthesis by rearrangement of bio-macromolecule.

The development of a basic knowledge of biologically inspired materials, the application of this rudimentary understanding to engineering challenges, and the creation of materials with improved attributes organized according to hierarchically are key steps for designing bioinspired materials. The first essential step in realizing this idea is elucidating structure and property mechanisms and developing well-strategic theories, which is referred to as bio-material science [5–8]. This opens the door for the following exciting phase of developing new sophisticated matter by supplying crucial knowledge with as-yet-untapped techniques from natural designs.

The creation of bioinspired materials often includes actions that expose the design principles found in nature. These principles serve as the foundation for the creation of innovative structured materials that are created to solve specific technical issues. This process begins with identifying the distinctive, intriguing properties of biological substances or systems, and then via scientific study, the basic mechanisms behind these properties are systematized or postulated to meet the requirements of pertinent engineering applications. The following stage uses these organic principles to develop and create materials with specific mechanical and/or functional qualities and improved performance. Technology transfer from laboratory to pilot or mass industrial production and the need for highly sophisticated fabrication instruments are two challenges in this endeavour, but as this chapter demonstrates, bioinspired materials have offered many valuable perception in fostering novel ideas in a variety of fields, including pharmaceutical, biomedical, marine, aerospace, agriculture, heavy engineering, transportation, and even housing. Additionally, this process encourages a more peaceful coexistence between human culture and the natural environment by allowing us to review our understanding of biology.

An intriguing field in material innovation has been the design of novel materials using naturally honed biological components and processes. The functional and structural categories of these bioinspired materials receive the majority of attention, and both depend on their hierarchical structures. While bio-functionalized materials focuses on designs that offer enhanced wettability, bioactivity, and stimuli-responsiveness, bionic materials features designs that offer superior tensile strength and durability with light weight [15].

15.3 Nanobiomimicry—Confluence of Nanotechnology and Bioengineering

Living things and natural phenomena have particular traits that enable them to coexist in peace with their surroundings. Understanding these natural phenomena helps us create cutting-edge technologies to explore uncharted regions. Numerous environmentally friendly solutions to challenges encountered in daily life have been developed using biomimicry and nanotechnology. In this segment, we examine how the principles of biomimicry, nanotechnology, and their interaction can be applied in different scientific and technological scenarios. The science of nanotechnology examines how to create materials, systems, and technologies with atomic accuracy. The natural world serves as an inspiration for many nanotechnology advancements. Even while research on the nanoscale is frequently seen as a component of the future, it really forms the foundation for the materials and systems that make up both our living and non-living worlds. Nanoscience is present in many living things, including butterflies with varied hues, certain insects that shine at night, and geckos who can defy gravity by walking on walls or ceilings. The development of display technologies was influenced by nanobiostructures, such as the embedded interference patterns on butterfly wings, the hydrophobic self-cleaning properties of the lotus led to the synthesis of nanomaterials possessing similar properties, and spiders provided the inspiration for the most significant inventions, such as optical devices and sensors.

Exceptional answers to difficult issues can be found in the environment in the form of tiny, nanoscale structures with a high degree of similarity. Researchers now have new scientific methods at their disposal to thoroughly examine the relationships between natural form and function. Nanobiomimicry, which refers to the nanoscale duplicating of living things in materials, structures, or processes as found in nature, is an effect of nanotechnology on the biomimicry trend [16]. Memory polymers, for example, which can recall their structure at a certain temperature after stretching or expanding, are made using nanotechnology. Other substances can alter color and transparency, store information, or use sensors to convert sound and light between one another. Engineering design may include bioinspiration into a variety of technologies at many scales, including nano, micro, and macro [17, 18]. Biological imitation of nanoscale structures and processes is called nanobiomimicry.

Bacteria, viruses, diatoms, and biomolecules are just a few examples of the wide range of nanoscale materials that nature offers as potential templates for designing of new materials. With the aid of nanoparticles and nanocomposites, nanobiomimetic technology can produce existing biomaterials in greater quantities, combine biomolecules and materials with inorganic ones to create hybrid systems, mimic biology at the molecular level by synthesizing structures that resemble biological structures, and create new materials with novel properties. In several domains, including tissue regeneration, chemistry, physics, biology, medicine, engineering, robotics, and many biomaterial technologies, the nanobiomimetic technology is employed. According to various functional models, such as assembly, self-healing, sensing and responding, self-cleaning, water collection, solar transformations, material recycling, and energy conservation, this branch of nanotechnology that manipulates metals to mimic nature generally follows these principles [19]. When compared to more traditional lithographic approaches, important components of nanodevices including nanowires, quantum dots, and nanotubes have been created through the study of nanobiomimicry in an effective and straightforward manner. Then, several of these biologically derived structures evolved into photovoltaic, sensor, filtration, insulation, and medical device applications. Nanobiomimetics is a highly interdisciplinary area that calls for cooperation between experts in biology, engineering, physics, material science, nanotechnology, and other related fields. Science has been able to create microscopic biological reproductions because of the burgeoning area of nanotechnology in the last century.

15.4 Biofunctionalization of Inorganic Nanoparticles

An essential step in creating biomimetic and bioinspired nanostructures is biofunctionalization (Figure 15.2), which entails altering the physico-chemical characteristics of a material's surface to enable broad spectrum applications with a focus on biology (such as a prosthetic fitting, an implant or additive applied to a live person) to enhance the biological interactions with the organism [20]. This innovative idea enables the amplification of surface qualities to provide advantageous effects including higher encapsulated component digestibility and bioaccessibility, as well as quicker biological impact response times, among other advantages [21, 22].

In this regard, nanoparticles have been used in a variety of industries, including the agricultural, pharmaceutical, biotechnological, food, and bio-chemical sectors, among others, for the nanoencapsulation of substances or live cells. However, the goal of the planned nanostructure is taken into consideration while analyzing polymer techniques or matrices [23].

15.4.1 Strategies to Develop Biofunctionalized Nanoparticles

Bio-functionalization is the process of combining chemical features of two or more compounds to enhance their activity by enhancing the action of one of these substances. One technique for bio-functionalization combines a chemical with low water solubility with one with high water solubility in stoichiometric proportions to ensure the right interaction between functional groups that can act as anchors [24, 25]. These formulations enable the employment of cutting-edge technologies like electrospray, which produces fibers with a specific molecular arrangement and enables a variety of applications, including controlled-release systems for chemicals with biological activity and packaging materials [26]. In addition to the preceding, functional groups are used as anchors for the attachment of biological substances to certain polymeric monomers to take use of their capabilities [27]. The desired conclusion is always ensured during biofunctionalizations, which are carried out utilizing both physical and chemical means, by accounting for the chemical makeup of the substances. To guarantee the stability of the components after biofunctionalization, the system must be kept at a low humidity content. Biofunctionalization requires careful consideration of low-energy interactions including electrostatic interactions and Van der Waals forces. To ensure its continuing usage in *in vitro* and *in vivo* research, lyophilization, spray drying, or other physical techniques may be performed [28–30].

The development of nanoparticle production techniques has been defined by the necessity to synthesis safe and functional chemicals, streamline operations, and significantly enhance encapsulation performance. A more regulated and targeted release is possible with a biofunctionalized surface. As a result, there are several uses for nanoparticles, and their future growth is quite promising [31]. Since the particle size for human use may be greater than 100 nm and frequently necessitates the use of non-toxic biocompatible polymers, the surface chemistry and shape of the nanomaterial are important elements to take into consideration in biofunctionalization [32].

15.4.2 Fate of Biofunctionalized Nanoparticles

The highly complex fluids known as physiological media, which include salts, sugars, lipids, proteins, amino acids, and enzymes, have the ability to destabilise nanoparticles and lead to their assemblage through Van der Waals interactions. A successful biomedical use thus requires the NPs' colloidal stability in the biological fluid. In an ideal scenario, the NPs would also be capable of navigating biological barriers and avoiding immune system detection, which would prevent them from being used as intended since they would build up unspecifically in the liver and spleen. The body's principal immunological reaction to foreign substances like NPs is the absorption of serum proteins, or opsonins, onto their surface. NPs that are freely moving throughout the body are recognized by phagocytes as a result of the opsonization, and they are then removed from the circulation [33]. To create robust nanoparticles with outstanding colloidal stability, preserve the inorganic core's physicochemical characteristics, and specify how NPs interact with their biological environment, a suitable coating is required.

Additionally, coating offers a versatile surface chemistry for future biomolecule functionalization and allows for the tailoring of the NP's outer layer for certain biological interactions.

The second step in creating bioactive NPs is selecting the best biofunctionalization strategy after covering the inorganic core and making sure the NPs have the necessary chemical moieties and colloidal stability. This is not an easy procedure to perform given the large variety of nanomaterials and biomolecules that are now available. Without standardized methods or procedures for routine functionalization, every distinct circumstance (between a nanomaterial and a biomolecule) requires careful consideration [34].

15.4.3 Biofunctionalization Nanoparticles with Different Organic Compounds

With a focus on the relevance of their appropriate (bio)functionalization to guarantee their consistency in biological media and to maintain the metabolic activity of the attached biomolecules, this section's main goal is to provide a concise summary of the most recent advancements in chemical modification of smart nanomaterials for bioapplications.

15.4.3.1 *Carbohydrates*

Compared to more complicated biomolecules, the functionalization of NPs with carbohydrates is comparatively simple. The density and distribution of carbohydrates on NP surfaces are particularly crucial since they can influence how well NPs identify molecules. Low carbohydrate density can result in unwanted protein absorption and other non-specific biological interactions of the NPs, whereas excessive surface coverage can interfere with recognition processes owing to steric hindrance [35].

15.4.3.2 *Nucleic Acid*

For over two decades, biomolecules based on nucleic acids have been a crucial component of bionanotechnology. They have made it easier to build a variety of materials with special features, including sensors [36], imaging

[37], and delivery systems [38, 39]. The controlled assembly of nanoscale materials has substantially improved thanks to DNA's predictable and programmable self-assembling characteristics, which are recognized by their complementary chemical motifs [40, 41].

15.4.3.3 Peptides

Peptides are shorter than proteins in length because they comprise 50 amino acids or less and are short chain amino acid monomers joined by amide bonds. Peptides have the ability to stabilise and biofunctionalize NPs and can be obtained naturally or made artificially. Wang *et al.* show that it is simple to create several functional peptide stabilized AuNPs in a single surface coating process and that the functions may be targeted specifically on a microarray. Researchers created a simple method for producing stable AuNPs that have both single and dual biological functions. The particles display the distinctive biological cargo identification characteristics without any signs of particle conglomeration and non-specific binding [42]. The length, hydrophobicity, and charge of peptide ligands all contribute to the stability they provide, which can occasionally lead even to enhanced stability.

15.4.3.4 DNA

One application for the controlled functionalization of NPs with DNA is the development of DNA-based nanoconstructs as sensor tools. NP-based DNA sensors can recognize infections and genetic diseases by binding to certain disease-related DNA strands and creating a quantitative signal to show the amount of DNA present. This approach makes extensive use of Au NPs that have been modified with primer sequences that bind to the desired DNA.

15.4.3.5 Antibody

The immobilization of antibody (Ab) on nanostructured materials is required for the majority of immune-based test nanosystems, including biosensors, antibody arrays, and cellular targeting systems. Since it is the most abundant in human blood, immunoglobulin G (IgG), one of the five major types of antibodies, is widely used to biofunctionalize nanomaterials. Given their high 3D complexity as proteins with a molecular weight of 180 kDa and four polypeptide chains, general issues for the conjugation of Ab to nanomaterials must first be recognized. At the ends of the Y's arms are the two identical antigen-binding sites, known as Fab fragments, and the four protein chains are formed into a unique "Y-shaped" geometry (antigen binding fragment). It is because of where the antigen binding sites are that antigen binding is influenced by the final molecular orientation of the Ab on a nanomaterial's surface.

15.4.3.6 Enzyme

Due to their significance in biotechnology and biomedicine, enzymes are frequently utilized in biofunctionalization due to their ease of handling, simplicity in separating from the reacting mixture and reusing, as well as their inexpensive price. The immobilization in NPs frequently lowers the enzymes' diffusional constraints and/or improves their catalytic activity. Enzyme electrodes are a prime area of interest in the study on AuNPs-based biosensors. Glucose biosensors are one such instance [43]. In a different research, N-phosphonomethyl iminodiacetic acid was added to magnetic nanoparticles (MNP) to immobilise urease. As a result, the carboxyl groups that urease had been adsorbed to by the carbodi- imide process were added to the surface coating [44].

15.4.3.7 Stability of Biofunctionalized Nanoparticles

To functionalize NPs with biomolecules, a number of challenges must be addressed. One of the most difficult tasks is to keep the NPs stable in solution while conjugation is occurring. Furthermore, there are no accepted techniques for NP functionalization due to the wide variety of distinct NPs and biomolecules that have been reported thus far. As a result, the selection of a coupling method is influenced by a number of factors, including the stability of the NP, the functional groups, and the circumstances surrounding the bioconjugation (temperature, pH, solvent type, surfactant structure, ionic strength, and the biomolecule to attach). The ability to adjust the orientation may also be necessary to maintain the conjugated biomolecule's activity after joining the NP [45].

15.4.3.8 Applications of Biofunctionalized Nanoparticles

With the help of biofunctionalized nanoparticles, it is now feasible to increase the biological activity of several bioactive substances. The most important factors affecting a nanomaterial's toxicity are its shape, size, and place of origin. The biological effect, however, might be hampered by the physical-chemical properties that the nanomaterial acquires throughout the biofunctionalization process. These "toxic properties" can be found by characterization, and they can be verified through *in vitro* testing. Science has a challenge in designing safe nanoparticles with the right concentrations of a phytochemical or extract while taking into consideration variables like size and structure. Prior to completing an *in vitro* biological assessment, it is critical to define the biofunctionalized materials' biological activity (autoinflammatory, antioxidant, antiproliferative, etc.) to determine their potential toxicity [32].

The largest supply of hydrocarbons to the environment is still wastewater and sludge samples gathered from various oilfield companies. Even though these businesses employ a number of conventional techniques to get rid of these unpleasant pollutants, they are still present, in either high or low amounts, in the sample that has been treated. The low/high molecular weight monocyclic and polycyclic aromatic hydrocarbons, for instance, may not be completely detoxified by some techniques like incineration and stabilization (entrapping the contaminants in a durable, mechanically strong material while making them less toxic). Because this issue uses so little energy and requires so little investment, it necessitates certain improvements or developments in the contemporary technologies in conjunction with functionalized nanomaterials. Through the use of certain chemical reactive agents, monocyclic and polycyclic aromatic hydrocarbon molecules are oxidized to produce CO_2 and water in chemical methods to oxidation therapy. The most current advanced oxidation methods utilized to treat monocyclic and polycyclic aromatic hydrocarbon molecules are supercritical water oxidation (SCWO), wet air oxidation (WAO) and photocatalytic oxidation (PO). In WAO, oxygen is used as an oxidant at high pressure and temperature to break down harmful organic molecules into CO_2, H_2O, and other products. The catalytic ability of functionalized nanoparticles for the catalytic oxidation of oily wastes has attracted a lot of interest. The petroleum industry's oily sludge was treated using catalytic moist air oxidation. Recent studies indicate that biofunctionalized nanomaterials for the environment can speed up the biodegradation of redox-sensitive pollutants [46]:

(i) It has cumulative impacts on biological and chemical processes, and it makes use of metabolic niches such as cathodic depolarization and bioremediation.
(ii) the efficiency is increased by the initial high contaminant concentration, and the desirable reactions are preferred at high concentrations;

(iii) change toxins into harmless final products that are quickly biodegradable or less hazardous versions of the pollutants.

15.5 Multifarious Applications of Biomimicked/Bioinspired Novel Nanomaterials

Although many nanomaterials have been created intentionally by copying the nanostructures found in nature, this chapter will highlight on some of the most significant materials with the most promise for future applications. This section focuses on how advances in the field of biomimetics have been made possible by nanoscience and nanotechnology for a variety of applications. It is clear how crucial it was to carefully examine the nanostructures seen in nature to properly recreate them in the lab. It is crucial to stress that if we continue to observe nature carefully and comprehend how it functions, it will never cease to astound us. The parts that follow focus on some of the cutting-edge uses of these bioinspired materials and how they're shaping the future.

15.5.1 Implementation of Nanobiomimicry for Sustainable Development

Given that they don't produce trash over a prolonged period of time, natural processes are sustainable. Using nature as a model, biomimicry has been successful in discovering novel strategies for attaining sustainability. Although the fundamentals of nanotechnology and biomimicry are not new, we must update and expand the classification of nanobiomimicry potential to address sustainable engineering if we are to build sustainable engineering solutions based on this technology. Nanobiomimicry is hardly a style of building or indeed a completed piece of design. The designer can address problems with design functions, such as flexibility, adaptability, the ability to be strong under tension, wind resistance, sound isolation, cooling, and heating, etc., by looking for a local ecosystem or species of animals in nature. This is referred to as the design process, or method of identifying solutions. Nanobiomimicry has long been a source of inspiration for artists, scientists, observers, and philosophers, as we've previously described. However, one of the more recent applications of nanobiomimicry is in organic architecture. To create architecture that is in tune with nature, organic architecture takes a design strategy that draws from natural principles and refers to local sites and cultural links. A structure that an organic architect creates is modelled on an organism visually or aesthetically, yet it is structured or has a purpose of providing sustainability. As biology does in nature, biomimicry influences architecture on several levels. These levels may be categorized into three groups: (1) form, (2) process, and (3) ecosystem. Numerous nanomaterials are currently in existence, and by incorporating nanoparticles into conventional materials, it is possible to create superior nanocomposites with brand-new multifunctional features. Additionally, it is anticipated that these materials will significantly enhance their strength, functionality, and longevity [47].

Modules were used in the construction of the Manuel Gea Gonzalez (Hospital) project in Mexico. This structure was influenced by two distinct species, a marine sponge and lotus leaves, rather than just one (Figure 15.2). Sea sponges are beneficial to the survival of other reef organisms by making carbon biologically available. They also process organic elements like carbon, nitrogen, and phosphorus as well as filter water and separate bacteria. Additionally, the reef is shielded by sea sponges from sharp changes in temperature, light, and nutrient density. Lotus leaves have a great water repellent and self-cleaning ability due to their high hydrophobicity and unique surface structure (Figure 15.3) [48]. Nano TiO2, a nanomaterial with anti-pollution and smog-eating capabilities that helps break down pollution particles into less harmful particles, was used to coat the building. It can also inhibit the formation of microorganisms

Figure 15.2 Bioinspired architectural construction of Manuel Gea Gonzalez (Hospital), Mexico.

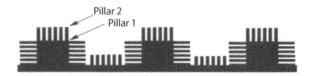

Figure 15.3 Biomimetic design inspired from the double nanostructured pillars on the surface of a lotus leaf [48]. Reproduction does not require any permission.

and biofilms. Additionally, the external paint has water resistant and self-cleaning qualities.

Nanotechnology offers the potential to create materials that match those found in nature but have better mechanical qualities, are more durable, and are more sustainable. Engineers were able to study and replicate creatures thanks to advances in nanotechnology [49]. To generate smart building materials, inorganic nanoparticles for solar energy storage and conversion, and sustainability akin to that of living things, nanoengineers and designers are attempting to develop unique nanomaterials and structures that interact with the ecosystem [50]. With the use of technological advancements, architects may create structures that are naturally cooler by studying the laws of nature and mimicking their physical characteristics. The "Waterloo International Terminal," created by Nicolas Grimshaw & Partners, serves as an illustration at this level of biomimicry of form and function, using glass panels to mimic a pangolin's unique scales (Figure 15.4).

Often referred to as "animal architecture," structures that mimic biological organisms provide architects successful instances of sustainable systems (Zari, 2014, p. 6). The most intricate degree of biomimicry is system mimicry. It's crucial to keep in mind that nothing in nature lives in a vacuum. By creating a sustainable system that depends on various businesses leveraging one another, this concept could be applied to architecture. It may begin out with a microlevel and work its way up to something larger, like ecofriendly skyscrapers. An architectural design for the Garden by the Bay in Singapore (Figure 15.5) provides two cooled conservatories, the Flower Dome (cool dry biomimicry) and Cloud Forest (cool moist biomimicry), and is intended to be powered by solar-powered mega "Super trees" (with the aid of using nanomaterials and devices) [51]. This degree of biomimicry can involve copying living things using cutting-edge architectural techniques or materials to promote sustainability.

Figure 15.4 Waterloo International Terminal—biomimicry of form and function.

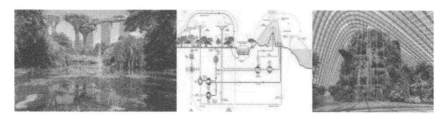

Figure 15.5 Garden by the Bay, Singapore.

When impacted by biomimetic processes, biomimicry is enhanced at this level by mimicking material organization patterns that are present in many nature inspired green constructions and play a significant part in their structural efficiency. In addition to other functions, it might be utilized for self-cleaning, self-healing, detecting and reacting, water collection, and solar changing. Additionally, it offers the possibility to create materials that are systemic, recyclable, or energy-efficient, among other things. The previous idea of mimicking nature in "Organic Architecture" of the 20th century, which was linked to visual imitation, is being moved by these new nanogoods to further expanded degrees of contact with nature.

15.5.2 Bioinspired Nanomaterials for Biomedical and Therapeutic Applications

As a category of user-friendly biomaterials, bio-inspired nanomaterials have proven to be potent in addressing diverse biomedical problems by serving as adaptable instruments for bioimaging, biosensing, antibacterial treatment, biocatalysis, and biotherapy [52–55]. The superficial plane of the functional nanomaterials can be subjected to a variety of chemical modifications or biological interactions, enabling it to recognise a particular spot in complicated surroundings both *in vitro* and *in vivo*. Surface engineering offers a different way to make specificity of biofunctional nanomaterials, whereas many natural systems demand a very particular lock-and-key strategy for intermolecular interactions [56, 57]. One essential requirement for guaranteeing the biosafety of nanomaterials is the attachment of biological molecules, such as proteins, amino acids, enzymes, nucleic acids, and antibodies, at the surface of nanomaterials. An advanced nanomaterial's key characteristics include despondant toxicity, excellent biocompatibility, and distinctive responses to stimuli including magnetism, light, electricity, and which broaden its bio-applications [58–60].

Recent studies on the interplay of nanomaterials with natural systems have mostly concentrated on their therapeutic uses and on identifying any potential negative consequences [61, 62]. The first approach takes use of nanoparticles' size resemblance to common cell components like proteins, which enables them to enter cellular partitions both mechanically and by energy-driven endocytosis. The form, size, surface features, and chemical makeup of the nanoparticles, as well as their chemical content, heavily influence how they enter cells. Researchers have created a wide range of biological uses for nanomaterials, motivated by nature's amazing degree of intelligence. Significant advancements in biomimicry and biomaterial composites have been made over the past few decades, including the creation of smart robotics, nanomaterials for tissue engineering and orthopedic implants, and a new class of materials that mimic the adaptability and self-regulation of living things [63].

Biological templates are support systems that serve as vessels, such as viral capsids. These vessels, in particular, may be used to produce novel materials, as well as act as carriers for drugs, catalysts, DNA testing, and immunoassays [64]. A new functional nanomaterial was recently created by scientists utilizing biological macromolecular assemblies as templates [65].

Protein cages are highly balanced, synthetic, multifunctional protein constructions with three unique interfaces: the inside, the exterior, and the contact between subunits. Recently, our lab and others have undertaken substantial study on these protein architectures [66, 67]. These subunits may be altered chemically and genetically, giving rise to a single cage with several functions and a variety of uses in fields ranging from biomedicine to electronics. The pioneer of protein cages employed in the manufacture of inorganic nanoparticles were ferritins, a class of proteins that have as their main function the storage and sequestration of iron (Fe) [68]. The self-assembly of these protein structures was the subject of several research. By creating protein cages, it is feasible to illustrate the crucial role that conserved hydrogen-bonding interactions play in establishing geometrical specificity for cage construction. The structure and stability of the isolated ferritin monomer, as well as the homogeneous chemical change of the cage interior under physiological circumstances, may also be understood fundamentally. The main uses of protein cages include the creation of anode materials for lithium-ion batteries, storage of nucleic acids, biomineralization, transport and delivery of nucleic acids between various chemical environments, and sequestration, to name a few. By genetically inserting silver-binding peptides into the ferritin cage and then chemically converting the silver ions associated with the peptides to silver nanoparticles *in situ*, ferritin-encapsulated silver nanoparticles were created [69–71].

It is now possible to construct molecules with remarkable physicochemical and mechanical characteristics because to recent advancements in macromolecular research. Dendrimers are a subclass of these macromolecules that are highly branched, symmetrical, and monodisperse. They have a well-defined number of end groups that may be functionalized to exhibit certain functionalities [72]. The development of innovative applications for the transport of genes and medications has been facilitated by the remarkable flexibility in dendritic synthesis, the rich physico-chemistry, and self-assembly. These dendrimers include poly(amidoamine) (PAMAM) dendrimers, which imitate globular proteins by having a topology similar to that of biomacromolecules. Dendrimers are more durable than proteins for pharmaceutical and biomedical applications, yet there are some slight but significant distinctions between the two. For instance, globular proteins are sensitive to destruction by temperature, pH, light of UV and visible since they are basically folded structures made of linear polypeptides. The protein surfaces are more diverse and the interiors are packed closely together, with a combination of hydrophilic and phobic zones. Dendrimers, at the same time, are spherical structures that are covalently attached, and their homogenous surfaces and distinct interiors provide a structural integrity for exclusive and reliable biological activities. Apoferritin can be utilized for a variety of purposes due to its capacity to deconstruct and restructure under specific pH control to load the medicinal chemical daunomycin. Insoluble and soluble pharmaceuticals can be enclosed for delivery thanks to the fusion of a changeable internal and external surface and penetrable hydrophilic and phobic channels via the cage. PAMAM dendrimers, like apoferritins, are particularly appealing for producing monodispersed metallic nanoparticles. The first copper nanoparticles were created in a PAMAM dendrimer scaffold by Crook *et al.* [73]. Since then, a variety of metal nanoparticles enclosed in a dendrimer have been created by several author groups [74].

Wang *et al.* [75] recently employed dendrimer-encapsulated gold nanoparticles as anti-cancer medication carriers. When compared to free anti-cancer medicines, these dendrimer-encapsulated pharmaceuticals demonstrated significantly lower cytotoxicity (Figure 15.6). Gold nanoparticles coated in dendrimers were also used by Jeong *et al.* [76] to covalently sequester a electrochemical carcinoembryonic monoclonal antigen for extremely sensitive immunosensing. The use of nanoparticles encased in dendrimers as protein mimics has also been effective, and they display catalytic activity similar to that of catalase, an enzyme that helps healthy cells get rid of excess reactive oxygen species. Due to similarities

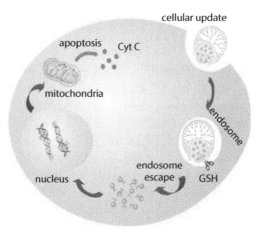

Figure 15.6 Schematic diagram demonstrating the release of anti-cancer medicines from dendrimer-encapsulated gold nanoparticles when intracellular glutathione (GSH) is present (DEGNPs; yellow ellipsoid within branched dendrimer) [75]. Copyright © 2013, American Chemical Society. Reprinted (adapted) with permission from ACS Publications.

in size, globular form, and dendritic effect, the generation 9 PAMAM dendrimer used here offers significant advantages for the creation of synthetic enzymes. Dendrimers and protein cages have both been used as contrast agents in MR imaging with strong water proton relaxivity [77].

Recently, protein-based nanomedicine platforms, constructed from spontaneously assembled amino acid subunits of the same protein or a mix of proteins, have drawn the greatest interest in the development of drug delivery systems because of their biocompatibility, biodegradability, and low toxicity [78]. Drugs can be stored inside a protein cage and then administered just to cells with the desired effects. Prospective synthetic cargos can be kept in protein cages because, once the nucleic acids from the viruses have been extracted, protein cages become the structural shells of viruses. Additionally, their consistent cage sizes prevent the nanoparticles from aggregating and permit the loading of relatively uniform concentrations of medicines [78, 79].

The applicability of using the mineralized HFn protein cage as a cage for cancer cell-specific targeting *in vitro* was demonstrated by Uchida *et al.* [80]. To mineralize the protein cage, Fe(II) was introduced into the interior of the protein cage as iron oxide (Fe_2O_3) at a stoichiometric ratio of 3000 Fe per cage. A mutant variant known as RGD4C-Fn was created by genetically altering the ligand peptide RGD4C and attaching it to the 24 subunits on the external side of the HFn. The mineralized RGD4C-Fn mutant has the potential to be used in medication administration and can function as an imaging agent that targets cancer cells specifically. This preservation of the cage's structural integrity combined with surface alteration due to protein mutations gives a method for changing other comparable protein cage structures. PbS quantum dots (QDs) enclosed in HSAF (AFt-Pbs) have recently been shown by Bradshaw et al. [81] to have dose-dependent anti-tumor action against two colorectal cancer (CRC) cell lines. When CRC cells were exposed to Aft-PbS *in vitro*, ROS were produced, which caused apoptosis and a reduction in cell growth. To create radioisotope-immobilized protein cages for prospective use in nuclear medicine, apo-ferritin can also be employed as a template [82, 83]. The radioactive ions are shielded and stabilized by apoferritin in the core/shell nanoparticles as opposed to directly loading radioactive ions like $^{177}Lu^{3+}$ or $^{90}Y^{3+}$ to a chelating agent. It has been demonstrated that encapsulating curcumin in the apoferritin cavity greatly improves its bio-stability and absorptibility while preserving its pharmaceutical benefits.

To create very monodisperse nanoparticles for protein assays, protein cages can be employed as templates. The protein cage plays several roles in these processes, such as I creating a confined environment that is optimal for the production of highly monodisperse nanoparticles, (ii) inhibiting the aggregation of the produced nanoparticles, and (iii) frequently initiating a mineralization response [84]. Apoferritin has frequently been employed as a protein cage to create size-restricted bioinorganic nanocomposites because of the distinctive form of its cavity [85–88]. This function has been achieved by both bacterial multi-enzyme complexes and viral cages [89]. In some regions of their inner cavity, these systems exhibit high charge concentrations that can act as mineralization nucleation sites. Monodispersed single crystal nanoparticles may be created inside cages and capsids using these techniques. Additionally, the protein shell itself may be changed to provide the nanoparticles various capabilities.

A biological technique for creating functionalized nanoparticles for efficient usage as label agents in a bioaffinity test was provided by Jaaskelainen et al. [90]. A rapid, affordable, and environmentally responsible technique of creating nanoparticles with intrinsic antigen binding activity makes the system desirable, especially for large-scale applications.

Since the tissues and organs of the human body are made up of nano-structured proteins, nanomaterials are an excellent choice as scaffolds for tissue engineering. The natural assembly of proteins and peptides produce solid nanofibrils or nanotubes (NTs) that served as inspiration for the creation of protein-like chain structures known as peptide NTs which have varied applications in tissue engineering [91–93]. Zhang et al. [94] and Cui et al. [95] focused on the self-assembly of peptides and proteins in the construction of nanostructured biological materials. When employed for biological sensing, biological NTs have the benefit of being biocompatible. Short detection periods, great stability, and significant current densities were demonstrated using thiol-modified peptide NTs mounted on gold electrode surfaces coated with enzymes for the detection of glucose and ethanol [96].

Biomolecules including nucleic acids, amino acids, and sugars are important bioorganic materials that can be used in place of synthetic polymers because of their general biocompatibility and supramolecular features. These newly created materials and systems have a wide range of uses, including sensors, diagnostic tools, tissue engineering, and medication delivery. As a result, advancements in biomedical engineering with biomimetic capabilities, such as robotic nanodevices, organs-on-chips, and tissue engineering, have been made possible by the creation of bio-inspired nanomaterials. Additionally, recent developments in biotechnology have made it possible to create and directly isolate nanomaterials from biological sources, such as cellular membranes, organelles, and exosomes, as well as to combine them with synthetic nanomaterials. These materials have special properties that will help the biomedical field advance. A nanofiber mat with two bioactive components and a biomimetic matrix structure was made by Han *et al.* [97].

The nanofiber mat demonstrated excellent mechanical properties and a porous structure suited for cell growth and migration by combining homogenous blending with electrospinning technology. This material has a great deal of promise for use as bone healing materials.

It appears to be a viable substitute for a soft tissue adhesive given that the new injectable hydrogel's assessed adhesion ability was better than that of the conventional Porcine Fibrin Sealant Kit.

Target-triggered nanomaterials have been suggested as potential delivery systems with potential benefits including targeted administration of anticancer drugs, reduced cytotoxicity, and enhanced therapeutic efficacy. Zhou *et al.* mixed the dimeric prodrug of paclitaxel (diP) with the redox sensitive star-shaped polymeric prodrug (PSSP) to create an intracellular drug release mechanism (diP@PSSP) for tumor cells [99]. The polymeric prodrug of diP@PSSP micelles, which similarly showed excellent therapeutic efficiency in red blood cells, was readily received by HeLa cells. When combined with photothermal treatment (PTT) and chemodynamic therapy (CDT), polypyrrole (PPy)-modified Fe_3O_4 nanoparticles (PPy@ Fe_3O_4 NPs) have been shown to inhibit the development and spread of non-small cell lung cancer [100]. The results of the *in vitro* and *in vivo* tests showed that PPy@Fe_3O_4 NPs were effective as NIR-sensitive chemodynamic/

photothermal cancer therapy agents that could lower MMP2, MMP9, and MMP13 levels. It offers a cutting-edge method for treating non-small cell lung cancer. Another illustration is the pH-sensitive NIR-II photothermal hydrogel developed by Xia et al. for combination cancer treatment. [101]. It included glucose oxidase (GOx) and had a photothermal activity at room temperature. The hydrogels were produced using the second near-infrared (NIRII) photothermal agent, a solution of alginate having pH-sensitive charge transfer nanoparticles (CTNs), and GOx. Hydrogel's starvation treatment, which involved eating glucose, caused the tumour microenvironment to become more acidic and downregulated the production of HSP90 while also exhausting the tumour cells. The suggested hydrogel was able to entirely prevent lung metastasis and breast cancer in mouse model and limit the development of subcutaneously implanted tumours by combining modest NIR-II PTT and fasting treatment. The logical planning, development, and operation of NIR photoactivatable agents were briefly discussed by Yu et al. in their summary of current developments in NIR photoactivatable immunomodulatory nanoparticles for combinational cancer therapy [102]. According to their research, this technique may hold great potential for the clinical applications of serious illnesses like cancer, infectious diseases, and autoimmune disorders. Recent advancements in biomimetic nanomaterials in ferroptosis-related cancer nanomedicine were highlighted by Zhang et al. Numerous nanoinducers that cause ferroptosis have unanticipated drawbacks, such as short circulation times, immunological exposure, and inefficient tumor targeting [103]. Due to their distinct physicochemical characteristics, biomimetic nanomaterials could be able to offer fresh approaches to these constraints.

15.5.3 Nanomaterial-Based Biosensors for Environmental Monitoring

Numerous difficulties in previous decades have prompted a pressing need to guarantee global food safety. The practice of increasing food manufacturing has presented many difficulties for agricultural ecosystems, including the persistence of residual pesticide particles, the accumulation of toxic elemental chemicals, contamination with heavy metals, all of which have had a negative impact on the agricultural environment. Numerous health problems, including illnesses of the nervous system, metabolic issues, infertility, unsettling of cellular biological processes, and respiratory and immune disorders, result from the ingestion of such harmful components through agricultural goods. Considering the 220,000 documented yearly mortality brought on by the hazardous effects of residual pesticides will help you understand the urgency of monitoring the agri based ecosystems. The current methods used to monitor agroecosystems rely on methods like HPLC, GC-MS, etc., which have quite a few disadvantages, including being expensive, time-consuming, and requiring complicated equipment and specialized skilled operators (Figure 15.7).

The science of nanotechnology has advanced significantly over the last few decades, and this advancement has greatly aided the creation of inexpensive, fast reacting, and economically plausible bio and nanosensors for identifying various contaminants in natural agroecosystems that have the advantage of being safe for human health. A biosensor is a tool that detects biochemical processes by sending signals proportional to the amount of target analyte in the reaction. A system referred to as a "nanosensor" has at least one nanostructure employed in its design to detect chemicals, change in light intensity, temperature, gases, biological agents, electric fields, etc. (Figure 15.8). The quick development of bio and nanosensors for detection of various analytes, ranging from the detection of multiple pesticides, nucleic acid, metal ions, amino acids, to the detection of whole microbes, has been made possible by the advent of this nanotechnology [104]. An interesting new technology called nanomaterial enabled sensors allows for the precise detection of environmental pollutants at the nanomolar to picomolar levels [105–109]. The ability of these sensors to quickly and easily detect contaminants in the field without the use of expensive lab equipment is what has people interested in them.

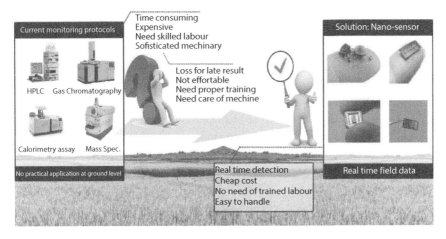

Figure 15.7 Bioinspired nanosensors for real time field data monitoring [104] Reproduction does not require any permission.

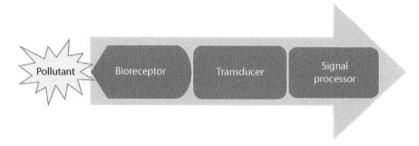

Figure 15.8 Biosensor operation [154]. Reproduction does not require any permission.

15.5.3.1 Nanosensor Design

Nanosensors are elemental, tiny devices designed to detect a specific chemical, biological component, or environmental condition. In comparison to their macroscale analogs, these sensors detect at a level that is significantly lower, are portable, and cost-effective. The following guidelines govern how a typical nanosensor device functions [104]:

1. Sample preparation: Sample could be a homogeneous or intricate mixture of solid, liquid, or gas phases. Agro based sample preparation is highly complicated process often due to the presence of contaminants. Specific chemicals, functional groupings of molecules, or living things are present in the sample that the sensors can target. The analyte, or target molecule or organism, can be a molecule (such as pesticides, colors, toxic elements, vitamins, hormones, antibiotics, etc.), a biomolecule (such as hormones, enzymes, DNA/RNA, allergens, etc.), an ion (such as metals, halogens, volatile compounds, or surfactants), a gas or vapor (such as carbon dioxide, oxygen, water vapors, etc) (pH, relative humidity, temperature, light, etc.)
2. Recognition: Specific substances can identify the analytes present in the sample. A biosensor is a tool that detects biological or chemical systems by generating signals proportional to the amount of target analyte in the reaction. The use of a biosensor, a device that produces signals proportional to the concentration of an analyte in the reaction, allows the detection of biological or chemical processes. These molecules may detect target molecules at nanoscales

and picometer scales by simulating various natural phenomena to produce a positive or negative detection result.
3. Signal transduction: Using various signal transduction techniques, these simple devices have been divided into several different categories, including optical, colorimetric, electrochemical, piezoelectric, conductometric, pyroelectric, electronic, and gravimetric transducers. Detected measurements are transformed into transmittable signals, which are then further processed to provide the data.

Therefore, they are made up of three parts: a signal processor, a transducer, which relays the presence of the analyte, and one or more nanomaterials that act as a particular recognition element (bioreceptor). Although each nanosensor may be classified using these three categories, these parts are not always different entities inside a sensor. The detection of numerous analytes simultaneously by a single sensor is known as multiplex detection (Figure 15.9).

First, a class is chosen, and then a particular pollutant of interest is chosen (i). The probe is then developed after determining the amount of analytes the sensor will be able to detect (ii). A signal transduction and one nanomaterial in combination make up the two fundamental components of a nanoprobe, which may additionally have a recognition component (iii). The final decision is made about the sensor deployment format (iv) (Figure 15.9).

Quantum dots (QDs), carbonaceous nanoparticles, and metal nanomaterials are the most common nanomaterials utilized in nanobiosensors because they can match advanced design specifications for sensors such downsizing, mobility, and quick signal response times. Selectivity is a crucial component of a good biosensor design. A wide range of recognition components, such as antibodies [110–114], aptamers [115–120], enzymes [121], and functional proteins [122], have been used in the creation of nanosensors. Optical, electrochemical, and magnetic signal transduction are the three key techniques of signal transduction used in nanosensors. Optical methods, especially colorimetric sensors can detect signals in both UV and visible spectrum, electrochemical sensing techniques with high specificity and the ability to be easily simplified and miniaturized, and magnetic transduction techniques with little background signal are all examples of optical techniques. A varied range of analytes can be identified by nanomaterial-based sensors – pesticides (organophosphates, neonicotinoids, triazine), metals (mercury, lead, chromium, cadmium) and pathogens [123].

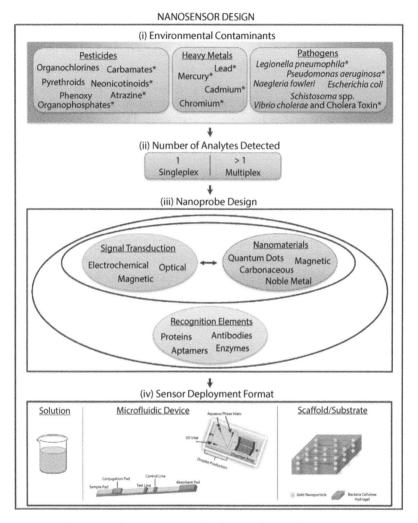

Figure 15.9 Nanosensor design schematic [123]. Reproduction does not require any permission.

15.5.3.2 Operation of a Biomimetic Sensor

Environmental pollution monitoring calls for quick, dependable, economical, and tiny instruments. Biomimetic sensors are those that employ biomimetic components as a biorecognition system, such as molecularly imprinted polymers, combinatorial ligands, and biomimetic catalysts. The most recent developments in biosensing tools for pollution detection include enzyme, antibody, aptamer, whole cell, DNA-based, and biomimetic sensors. Designing biosensors makes use of a number of detecting concepts, including conductometry, amperometry, luminescence, etc. They differ from each other in terms of design, profitability, sensitivity, and quickness. Depending on the sensitive element, each one has a unique selectivity and detection limit. Due to their benefits over conventional ones, mimetic biosensors are steadily receiving interest from researchers and consumers. The creation of a biomimetic sensor requires the careful and cost-effective integration of biotechnology and analytical methods [124–126]. Biosensors are used particularly for assessing ecological concerns. In these situations, biosensors are crucial for completing the particular chemical investigations [127, 128].

The two most crucial parts of the detecting process, the transducer and the bioreceptor, are used to categorize biosensors. Biosensors can be classified into three groups depending on the type of bioreceptor: the microbial group, which includes microbial biosensors, the biocatalytic group, which includes enzymatic biosensors, and the bioaffinity group, which includes immunosensors, aptasensors, and genosensors [129, 130]. Based on the physicochemical properties and operation of the transducer, biosensors are categorized into electrochemical (conductometric, impedimetric, amperometric, and potentiometric biosensors), optical (fiber-optic, Raman spectroscopy-based, surface plasmon resonance, and FTIR-based biosensors), and mass-based categories (magnetoelectric and piezoelectric biosensors) [131].

15.5.3.3 *Applications in Environmental Monitoring*

The foundation of biomimetics was set years ago, despite the fact that the nomenclature may appear fresh. Its guiding premise is to discover solutions that closely resemble the workings of a natural system, particularly when it comes to an organism's structure or particular interactions with the environment. The produced goods are effective and suitably tailored to actual surroundings [132].

The development of biomimetic sensors began with an analysis of the fundamental concepts behind the related enzymatic biosensors. High sensibility, selectivity, sensitivity, and ease of operation were to be maintained, but some drawbacks were to be lessened. The difficulties that need to be resolved generally stem from the unique properties of each enzyme, such as problems with inactivation or expensive purification and standardization procedures. In these situations, study was done to identify long-term fixes for developing imitation systems. Metal conjugates, molecularly imprinted polymers, nanozymes, and nanotbes are among the bases for some of the models that have been created [133].

The field of biomimetic sensors has made tremendous advancements during the last several years. Initially, bi- or uni-dimensional structures were used to build biomimetic sensors. Triangular assemblies were thereafter frequently employed, and the findings showed superior performances, sometimes even surpassing those of natural models [133]. The first step in creating precise instruments is to identify the ideal ligand for the targeted analyte. The recognition systems' choice of peptides has an impact on the sensor's affinity [134]. Simulation and computer modeling [135] are two processes that raise the effectiveness of these gadgets.

Environmental contaminants detection using biomimetic sensors is a field that is currently evolving. Modified nanoparticles [136–138], metal chalcogenides nanocrystals attached with various microorganisms [139], technological upgradation of classical imprinted electrodes [140], and nanozymes for phenol removal [141] are just a few of the promising directions that research has opened for the development of nanobiosensors.

For example, colorimetric and spectrophotometric detection of heavy metals in normal water and river water [142–144], potentiometric detection of chemicals in river water, drinking water, industrial wastewater, sewage water, and air [135, 145–149], amperometric, potentiometric, and fluorometric/colorimetric detection of soil tap water [150–153], and potentiometric detection of toxins in water [140, 154] are some examples.

Along with these biomimetic sensors, there is another category of sensors called "bioinspired" that draws its inspiration from biological processes, such as photosynthesis and fluorescence. It is important to concentrate on biosensors that take advantage of photosynthesis among those that use plant tissues as biomediators. The biosensors known as photobiobiosensors are capable of identifying specific damaging and dangerous contaminants that are often present in both surface and groundwater. Using the energy from sunshine, the process of photosynthesis transforms gases like carbon dioxide into organic molecules, principally sugars. Triazines, phenylureas, diazines, and heavy metals are among a number of contaminants that are important to the ecosystem and are particularly sensitive to photosynthesis. The following chemical categories apply to these substances: (i) phenyl carbamides or ureas or arylureas, (ii) triazines, (iii) diazine, (iv) phenols [155]. Naessens *et al.* developed a biosensor for the detection of harmful substances in 2000 [156] based on the alteration of *Chlorella vulgaris* chlorophyll fluorescence. The development of biosensing tools for the detection of photosynthetic herbicides by Marty *et al.* 1995's [157] team made use of the chloroplast and thylakoids membranes from a variety of species. They also used enzymes and antibodies as alternative recognition components for different pesticide, insecticide, and organophosphorus chemical substances. Atrazine, diuron, ioxynil, terbuthylazine, prometryn, and linuron were only a few of the herbicides that Tibuzzi et al. recorded in 2007 [158] using an optical biosensor built on the Chlamydomonas reinhardtii, a green photosynthetic alga. Scognamiglio et al. built a multi-biomediator fluorescence biosensor in 2009 [159] employing a brand-new adaptable portable device. For instance, Giardi et al. 2005 [160] described a gas chromatography-mass spectrometry-based fluorescence multi-biosensor that relied on the activity of thylakoids from microflora to estimate the presence of various contaminants on actual test samples from the Tiber river, the Acqua Marcia, the Valle del Sorbo, and the Po river (GC-MS).

Although there are many different applications for biosensors, those developed for environmental analysis hold great promise for systems that will eventually monitor the environment. Although routine analysis on innumerable samples that needs to be screened each day still requires high tech instruments like HPLC or GC-MS, biosensors represent a useful, straightforward, and quick way to significantly reduce the number of tests. The sensitivity and selectivity requirements for biomimetic sensors should be satisfied when they are utilized with complex, erratic environmental samples with variable compositions. [161].

15.5.4 Biomimetic Nanostructure for Advancement of Agriculture and Bioprocess Engineering

Along with industry and the production of electricity and heat, agriculture is one of the three main human activities that are responsible for planetary boundary breaches. It produces nearly a third of all greenhouse gas emissions and has a significant impact on the contamination of soil, freshwater, coastal ecosystems, and food chains in general. Most agricultural problems, including soil erosion and dwindling (essential) biodiversity, are the result of improper application of already available technologies. Toxic and entropic agro-technologies tend to require strategies that replicate the operation of naturally existing ecosystems that are pertinent to agriculture. Applications of agrotechnology that are eco-therapeutic and sustainable must move away from the chemical model of agrosystems and toward the ecological system-design model of agriculture.

Nanotechnology and nanomaterials have become more widely used in agriculture during the past ten years to better care for plants, boost crop output, defend plants from viruses and pests, and enhance the storage of agricultural goods. Nanotechnology is used to create nanoparticles for more effective delivery of a variety of conventional agrochemicals, including micronutrients and pesticides, to retain the ability of plants to grow and mature healthily.

Given the increasing reliance on insecticides and pesticides and the concurrent awareness of the impact of dumping these chemicals on the environment, there is a pressing need for agrotechnology to transition to more environmentally friendly technological systems that can be implemented on farms. After making notable contributions, using nanotechnology to agriculture is now a tried-and-true way to advance sustainability. High temperatures [162], pressure [163], acidification [164], and toxic metals [165] are often used in the production of nanomaterials as well as in their application.

Due to numerous benefits including a sustainable pathway, a variety of plant metabolites, cutting-edge extraction and separation techniques, the creation of a comprehensive phytoconstituent database, and instrumental support, the approach to nanosynthesis has drastically changed over the last two decades from traditional methods to plant-based protocols.

Precision farming has become quite popular and is seen as a solution to many issues in farming thanks to the introduction and development of bio-inspired nanomaterials. Nanomaterials are anticipated to play an integral role in a number of fields, including soil moisture regulation, fight against insecticide resistance, nanoparticle-mediated gene or DNA transfer in plants, food preservation and processing, recycling agricultural waste, reducing use of pesticides, plant breeding, and smart packaging. The potential for producing these nanomaterials in a more environmentally friendly manner has given farming operations a boost [165].

The basic goal of fertilizers is to keep the mineral content in balance while increasing production. If these mineral fertilizers are used excessively, soil fertility may suffer significantly. Greenly synthesized nanofertilizer formulations have been shown to efficiently defend against pests while also supplying a balanced supply of vital minerals with minimal human involvement. Guo *et al.* [166] have recently evaluated the many types of nanomaterials that are frequently used to create fertilizers and provided a thorough analysis of their value and limitations in agricultural activities. ZnO nanoparticles [167] were used on wheat fields during the harvest stage at different concentrations, which was much greater than that attained using standard chemical zinc in the form of zinc nitrate. Alternatively, copper oxide loaded chitosan/alginate nanoparticles may also affect such physiological and morphological changes [168]. As a result, a 300-nm spherically shaped shell is produced, encasing CuO and gradually releasing

Cu, which is essential for the growth of seedlings. Combining Zn and Cu [169] nanoparticles significantly changed the morphological aspects of the basil plant's biochemistry and demonstrated antioxidant activity. The Zn and Cu nanoparticles that originated from the basil extract are assumed to have been produced by reducing agents such vitamins, carbohydrates, phenolic compounds, or organic acid. The study promotes quality and quantity basil output, which has the added advantage of using less harmful fertilizer. Amorphous calcium phosphate that has been K- and N-doped has recently been mentioned as a potential strategy for successful nutrition regulation [170]. Nitrogen digested through urea and nitrate by altering reaction conditions provides a progressive release due to the two separate chemical forms. Because it impacts the kinetics of absorption, which regulates how quickly the roots can absorb it, the chemical configuration of the metal is a reliable system that can provide metal in the right level of oxidation [171]. Ingenious use of graphene oxide as a nanocarrier has revolutionized the administration of the pyrethroid insecticide [172] against the pest spider mite, a phytophagous mite. When using a chitosan-magnesium nanocomposite to increase rice production against the microbiological pests *Acidovorax oryzae* and *Rhizoctonia solani* [173] and a straightforward chitosan nanocomposite to combat *Spodoptera litura* [174] in the larval stage, similar positive outcomes were also noted. A natural *Bacillus* sp. strain RNT3 was used in the earlier nanocombination to create a spherically shaped nanocomposite with a size of 29 to 60 nm that was capable of destroying the pathogen's organelles.

In order to fulfil the requirements of the exploding global population and societal changes, the issue of declining soil fertility on farmlands must be taken seriously. By supplying or transporting macronutrients and micronutrients, nanotechnology has the capacity to enhance soil quality by fixing nutrient deficiency in this respect. They can be regulated by release thanks to their vast surface area to volume ratio, high adsorption capacity, and porous surface, and they serve as effective carriers of nutrients thanks to encapsulation.

One of the major issues is the cadmium absorption by plants, which is brought on by the high concentration of Cd^{2+} salts. Numerous instances where the huge specific surface area of nanoparticles aids in predicting the absorption of cadmium have been shown [175]. The impact on nutritional quality includes bio-absorption, where Cd (II) is changed into an interchangeable form that seedlings can digest, or through a molecular shift that boosts cadmium ion tolerance through a controlled metabolic pathway. Since soil moisture and metal uptake are tightly connected [176], most irrigational regimes consider soil moisture to be a key element [177].

Nanomaterials can enhance water absorption and retention capabilities when used in standard agricultural techniques. The creation of inexpensive superabsorbent nanocomposite that may be used to encase traditional fertilizers is a significant step in this approach [178].

Many biological systems, including microbes, enzymes, and other plant, have shown remarkable potential for the synthesis of nanomaterials (such as leaves, roots, fruits, shoots, and their extracts). Three primary advantages of employing biological systems as manufacturing hosts have been recognized by several researchers. In engineering processes, fewer hazardous materials can be used because of the following reasons: 1) biological systems can act as encapping, stabilizing, and reducing agents; 2) biosynthesis frequently uses ambient temperature and pressure as well as a neutral pH; this can reduce the need for energy resources and hazardous chemicals; and 3) the majority of biosynthesized nanomaterials are biocompatible which minute levels of toxicity. This section will provide a quick illustration of these three aspects of the biosynthetic application on creating and engineering nanomaterials.

Without the aid of chemical agents, biological systems alone are capable of capping, stabilizing, and reducing processes. It has been shown that several macro-biomolecules, such as proteins with functional amide and lipids with carboxyl groups, may adsorb on the surfaces of nanoparticles. This finding suggests that these molecules may contribute to the decrease of Au3+ and the stability of Au NPs [179, 180]. According to Jain *et al.* (2011)'s research, extracellular proteins' SDS-PAGE patterns are what stabilize silver nanoparticles throughout their formation [181]. However, there are downsides to biosynthesis of nanomaterials, such as less controlled engineering processes and low output rates, which can be avoided by employing nanomaterials that are inspired by biological systems. Designing new nanomaterials with comparable morphologies and functions to biological structures, such as mussel, cilia, insect tentacles, etc., is known as a "bioinspired" method. Due to the inherent character of the bioinspired method [185], it has been extensively investigated in biomedical investigations [182] as well as many other domains [183, 184].

Protein-polymer biomimetic membranes created for nanofiltration are one new way that nanobiomimetics are being used in agriculture. A relatively new membrane filtration technique called nanofiltration is utilized to filter out bacteria, parasites, and other particles from surface and fresh groundwater. The local brackish (saline) water was treated with nanofiltration membranes to provide high-quality desalinated irrigation water, which reduced the amount of water needed for irrigation by 25% [186, 187]. Smart biosensors for the detection of mycotoxines and biomimetic nanobiosensor designs based on immobilized tyrosinase enzyme for the detection of hazardous substances were disclosed. Without changing the shape of the enzyme molecule, biosensor design demonstrated high compatibility between membranes and enzymes, and binding always occurs outside the active regions of the enzyme [188]. Through the nanoformulation of fertilizers, nutritional quality, and yield barriers, nanotechnology holds the possibility of an innovation to enhance our existing low nutrient usage productivity. By employing clay minerals as storage spaces for nanoresources like nutrient ion receptors, soil fertility may be restored. Acidification can also be prevented, erosion-prone surfaces can be maintained, natural fiber can be strengthened, and the shelf life of fruits and vegetables can be increased [189].

15.5.5 Nanobiomimetics as the Future of Food Process Engineering

United Nation's World Population Prospects 2022 reported the present world population to be 7.9 billion and that it will reach 8 billion people by the end 2022 to the start of 2023 [190]. With such increase in population producing food for the masses seemingly is a challenge. However, reality is quite different. Back in 2018 the amount of food produced was enough to feed 10 billion people yet about 820 million people go hungry every day and malnutrition is rampant among kids. Regional conflict, the current economic crisis, political unrest, climate change, and COVID-19 are a few of the complex causes. Food loss, or the amount of food that does not make it to the consumer because of supply chain inefficiencies, is the cause of the discrepancy between the amount of food produced and the amount that is actually consumed. Pest infestation, rotting owing to improper temperature management and storage, and changes in the chemical makeup of the environment are just a few of the factors that contribute to this loss. Different businesses and start-ups are employing biomimicry to find answers to these issues in nature. This section describes many emerging bio-inspired technologies that are currently in the research stage but have the potential to completely transform the agro-food processing industries.

Each year, fungal spoilage results in the loss of around 25% of the entire fruit and vegetable production. Governments regulate or outright ban several synthetic fungicides because of their detrimental effect on public health and environment. Due to fungus developing resistance to them, many fungicides are useless. Additionally, a significant portion of fungicides washes off the surfaces of plants. Despite the fact that plants have their own built-in chemical defenses, UV radiation, oxidation, and heat cause these defenses to deteriorate quickly. Nanomik Biotechnology has strengthened the natural defense mechanisms utilized by plants to solve this issue. These safe, plant-based fungicides are made up of micro- and nanocapsules that release chemicals to protect the surface of the plant from fungus. Their invention was motivated by the plant's built-in defense mechanism against illness. To defend themselves against a variety of fungal infections, plants naturally create volatile chemicals. These molecules also encourage other plants in the area to activate their defensive mechanisms. As safer alternatives to dangerous synthetic pesticides, Pheronym has created products. Their solutions target nematodes, which are microorganisms that are naturally present in soil and are already widely employed in organic agriculture. Nematodes may either help plants flourish or harm them. Pheronym, a leader in biotechnology, instructs helpful nematodes to consume pest insects and to keep parasitic nematodes away from agricultural roots via pheromones, a chemical language. The exquisite chemical language that nematodes employ to communicate is the inspiration for the Pheronym team's biomimetic model. Rice Age has developed a rice seedling tray that mimics the hexagonal pattern seen in honeycomb in an effort to increase the efficiency and sustainability of rice production. This tray enables more rice to be planted in the same space. Recent advances in nanobiomimicry research and invention include [191]:

- A drainage system that mimics nature and preserves soil nutrients.
- A tool that promotes the growth of seedlings while repairing damaged soil.
- A mechanism inspired by nature that would allow people to eat insects.
- An autonomous system that enables individuals to grow wholesome food even without access to a green place.
- A modular, "groundless," growing system that allows people to grow more food on a smaller amount of land.
- A mangrove desalination still that is five times less expensive than conventional solar stills.
- A peer-to-peer networking tool for farmers that emulates how natural collectives work.
- A cheap aquaponics system driven by solar energy that is intended for subsistence farmers in developing countries.

Beyond only providing nutrients, plant-based meals can perform other functions that are beneficial to health in the intricate ecosystem of the human gastrointestinal (GI) tract. Natural plant foods have hierarchically self-assembling structures. With the creation of food structures that are inspired by the natural world, plant based biomimetic foods (BPFs) can provide answers for the future and promote better health and wellbeing. These structures are broken down during processing and digestion to allow for the release and absorption of nutrients and other bioactive compounds present in the food matrix. Plants have a hierarchy of structure that is related to assembly and disassembly. Within this hierarchy, interactions between several layers can affect nutrient bioavailability and digestion. BPFs can be designed to provide in-body functions and are inspired by nature. The emerging trends of biomimetics will potentially pioneer the path for the future of food [192].

15.6 Emerging Trends and Future Developments in Bioinspired Nanotechnology

Nature uses the smallest form of matter to display its structure by adopting a low-energy strategy analogous to self-assembly-type processes. Although the time scale is too long, innovations in nanoscience and nanotechnology have contributed to a better knowledge of the universe at the nanoscale, at which nature starts to form. Nature can easily produce multiple beneficial elements, such as a human finger with the ability to feel pressure, heat or cold, feel the wind blowing, report pain, move according to brain commands, be equipped with the ability to heal itself after a cut, be able to grasp objects, and leave a fingerprint. The nanoengineering field has not yet developed to the point where such sophisticated multiple advantages may be shown so easily and quickly. To solve nature's problem, more developments in materials, design, production, and sensing are needed.

A more effective control system must be developed because bionic materials, processes, and designs still is in need of a strong hierarchical network. Thus practical application of nature-inspired materials on a commercial scale (e.g. their scalable and affordable production) is an even bigger challenge. Materials advancements must thus be consistent with production advancements.

Manufacturing processes may be divided into the following categories: (i) subtractive methods - those based on material removal; (ii) additive methods - those involving material addition or deposition and (iii) procedures with neither material addition nor removal.

A number of factors, including the cost of manufacturing large-scale devices using biomaterials and the short life of biological substances like enzymes that may cause the degradation of these materials in devices, should be carefully considered before major commercial development of bioinspired materials for practical applications.

Future research in this area should concentrate on the development of cleaner, more sophisticated, yet cost-effective methods for designing them, despite the fact that biomimetic structures are fascinating and present new opportunities for creating novel materials with distinctive properties for a variety of applications. The creation of creative techniques to mimic the regulated self-assembly, adaptability, and self-healing properties of biological materials in synthetic materials inspired by biological materials will also be an interesting study area.

Because the lowest feature size for the bulk of production-scale equipment is in the hundreds of micron range, designing materials with natural influences can be challenging. However, in powder bed operations, the lowest size of voids is limited since the trapped powder needs to be released. After taking into account process restrictions, initial designs are created and parametric modeling is completed. To optimize the design for the function, simulations can be conducted. Any simulation cannot last as long as actual experiment, which might last anywhere from a few fraction of seconds to several weeks or even years. Hands-on sampling and testing using a strict design of experiments (DOE) technique can be used in some circumstances when the structure-function link is not completely known. It makes sense that prototype and testing, which need diverse efforts, are necessary following optimization, whether through simulations or sampling. A modern attempt in this area has partially begun to develop with the rise of artificial intelligence and machine learning.

In order to replicate complex hierarchical patterns and create materials with natural inspiration, these are the most important obstacles. Natural materials may therefore decompose on their own, and doing so has no negative effects on the environment [193].

15.7 Conclusion

Biomimetics is a new discipline that is already making strides in the technical and scientific worlds. It is obvious that nature has created a wide variety of substances and structured surfaces with very different properties. As we get more familiar with the underlying mechanics, we may start to use them for profitable applications. New nanomaterials, nanodevices, and processes can all be found in the commercial applications. Devices can be anything that provides aerodynamic lift, surfaces that are extremely hydrophobic, self-cleaning, or low drag, super-adhesives, robotics, fibers and materials with high mechanical strength, colored and antireflective surfaces, synthetic furs, textiles, foods, various biomedical implants, self-healing memory materials, and sensory-aid devices, to mention a few. In near future applications of nanobiomimetics will become indispensable in almost every sphere of scientific and technological innovations to develop a greener and sustainable future for the coming generations.

References

1. Bhushan, B., Biomimetics: Lessons from nature - an overview. *Philos. Trans. R. Soc., A*, 367, 1445–1486, 2009.
2. Lepora, N.F., Verschure, P., Prescott, T.J., The state of the art in biomimetics. *Bioinspiration Biomimetics*, 8, 013001, 2013.
3. Wegst, U.G.K., Bai, H., Saiz, E., Tomsia, A.P., Ritchie, R.O., Bioinspired structural materials. *Nat. Mater.*, 14, 23–36, 2015.
4. Katiyar, N.K., Goel, G., Hawi, S. et al., Nature-inspired materials: Emerging trends and prospects. *NPG Asia Mater.*, 13, 56, 2021.
5. Meyers, M.A., Chen, P.-Y., Lin, A.Y.-M., Seki, Y., Biological materials: Structure and mechanical properties. *Prog. Mater. Sci.*, 53, 1–206, 2008.
6. Wang, B., *Structural and functional design strategies of biological keratinous materials*, University of California, San Diego, 2016.
7. Naleway, S.E., Porter, M.M., McKittrick, J., Meyers, M.A., Structural design elements in biological materials: Application to bioinspiration. *Adv. Mater.*, 27, 5455–5476, 2015.
8. Antonietti, M. and Fratzl, P., Biomimetic principles in polymer and material science. *Macromol. Chem. Phys.*, 211, 166–170, 2010.
9. Xu, X., Rosi, N. L., Wang, Y., Huo, F., Mirkin, C. A., Asymmetric functionalization of gold nanoparticles with oligonucleotides. *Am. J. Chem. Soc.*, 128, 9286, 2006.
10. Mastroianni, A. J., Claridge, S. A., Alivisatos, A. P., Pyramidal and chiral groupings of gold nanocrystals assembled using DNA scaffolds. *Chem. J. Am. Soc.*, 131, 8455, 2009.
11. Park, S. Y., Lytton-Jean, A. K., Lee, B., Weigand, S., Schatz, G. C., Mirkin, C. A., DNA-programmable nanoparticle crystallization. *Nature*, 451, 7178, 553–6, 2008 Jan 31.
12. Zheng, J., Constantinou, P. E., Micheel, C., Alivisatos, A. P., Kiehl, R. A., Seeman, N. C., Two-dimensional nanoparticle arrays show the organizational power of robust DNA motifs. *Nano Lett.*, 6, 7, 1502–4, 2006 Jul.
13. Sharma, J., Chhabra, R., Cheng, A., Brownell, J., Liu, Y., Yan, H., Control of self-assembly of DNA tubules through integration of gold nanoparticles. *Science*, 323, 5910, 112–6, 2009 Jan 2.

14. Zan, G. and Wu, Q., Biomimetic and bioinspired synthesis of nanomaterials/nanostructures. *Adv. Mater.*, 28, 2099–2147, 2016.
15. Wang, Y., Naleway, S.E., Wang, B., Biological and bioinspired materials: Structure leading to functional and mechanical performance. *Bioact. Mater.*, 5, 745–757, 2020.
16. Dumitrescu, C., Nature works to maximum achievement at minimum effort. We have much to learn, Nano Technology Science Education (NTSE), Târgoviște, 2014.
17. Altun, D. A. and Örgülü, B., Towards a different architecture in cooperation with nanotechnology and genetic science: New approaches for the present and the future. *Architecture Res.*, 4, 1B, 1–12, 2014.
18. Arciszewski, T. and Cornell, J., Bio-inspiration: Learning creative design, in: *Intelligent Computing in Engineering and Architecture: 13th EG-ICE Workshop*, Virginia, 2006.
19. Jalil, W.D.A. and Kahachi, H.A.H., The implementation of nano-biomimicry for sustainability in architecture. *J. Eng. Sustain. Dev.*, 23, 03, 25–41, May 2019.
20. Bilek, M., Biofunctionalization of surfaces by energetic ion implantation: Review of progress on applications in implantable biomedical devices and antibody microarrays. *Appl. Surf. Sci.*, 310, 3–10, 2014.
21. Van Loo, S., Stoukatch, S., Axisa, F., Destiné, J., Van Overstraeten-Schlogel, N., Flandre, D., Lefevre, O., Mertens, P., Low temperature assembly method of microfluidic bio-molecules detection device, in: *Proceedings of the 2012 3rd IEEE International Workshop on Low Temperature Bonding for 3D Integration, LTB-3D 2012*, Tokyo, Japan, 22–23 May 2012, pp. 181–184.
22. Williams, E.H., Davydov, A.V., Oleshko, V.P., Lin, N.J., Steffens, K.L., Manocchi, A.K., Krylyuk, S., Rao, M.V., Schreifels, J.A., Biofunctionalization of Si nanowires using a solution based technique. *Nanoepitaxy Mater. Devices IV*, vol. 8467, p. 846702, 2012.
23. Jain, D., Athawale, R., Bajaj, A., Shrikhande, S., Goel, P.N., Gude, R.P., Studies on stabilization mechanism and stealth effect of poloxamer 188 onto PLGA nanoparticles. *Colloids Surf. B Biointerfaces*, 109, 59–67, 2013.
24. Vacchi, I.A., Guo, S., Raya, J., Bianco, A., Menard-Moyon, C., Strategies for the controlled covalent double functionalization of graphene oxide. *Chem. Eur. J.*, 26, 6591–6598, 2020.
25. Andrade, E.H.A., Maria das Graças, B.Z., Maia, J.G.S., Fabricius, H., Marx, F., Chemical characterization of the fruit of Annona squamosa L. occurring in the amazon. *J. Food Compos. Anal.*, 14, 227–232, 2001.
26. Noel, S., Liberelle, B., Robitaille, L., De Crescenzo, G., Quantification of primary amine groups available for subsequent biofunctionalization of polymer surfaces. *Bioconjug. Chem.*, 22, 1690–1699, 2011.
27. Samadarsi, R., Mishra, D., Dutta, D., Mangiferin nanoparticles fortified dairy beverage as a low glycemic food product: Its quality attributes and antioxidant properties. *Int. J. Food Sci. Technol.*, 55, 589–600, 2020.

28. Zarghami, V., Ghorbani, M., Bagheri, K.P., Shokrgozar, M.A., In vitro bactericidal and drug release properties of vancomycin-amino surface functionalized bioactive glass nanoparticles. *Mater. Chem. Phys.*, 241, 122423, 2020.
29. Brahmachari, G., Choo, C., Ambure, P., Roy, K., In vitro evaluation and in silico screening of synthetic acetylcholinesterase inhibitors bearing functionalized piperidine pharmacophores. *Bioorg. Med. Chem.*, 23, 4567–4575, 2015.
30. Vedani, A., Dobler, M., Hu, Z., Smieško, M., OpenVirtualToxLab-A platform for generating and exchanging in silico toxicity data. *Toxicol. Lett.*, 232, 519–532, 2015.
31. Guterres, S.S., Alves, M.P., Pohlmann, A.R., Polymeric nanoparticles, nanospheres and nanocapsules, for cutaneous applications. *Drug Target Insights*, 2, 147–157, 2007.
32. Razura-Carmona, F.F., Perez-Larios, A., Sáyago-Ayerdi, S.G., Herrera-Martínez, M., Sánchez-Burgos, J.A., Biofunctionalized nanomaterials: Alternative for encapsulation process enhancement. *Polysaccharides*, 3, 411–425, 2022.
33. Pelaz, B., Charron, G., Pfeiffer, C., Zhao, Y., de la Fuente, J.M., Liang, X.-J., Parak, W.J., Del Pino, P., Interfacing engineered nanoparticles with biological systems: Anticipating adverse nano-bio interactions. *Small*, 9, 1573–1584, 2013.
34. Sun, C., Du, K., Fang, C., Bhattarai, N., Veiseh, O., Kievit, F., Stephen, Z., Lee, D., Ellenbogen, R.G., Ratner, B. et al., PEG-mediated synthesis of highly dispersive multifunctional superparamagnetic nanoparticles: Their physicochemical properties and function in vivo. *ACS Nano*, 4, 2402–2410, 2010.
35. De la Fuente, J.M. and Penadés, S., Glyconanoparticles: Types, synthesis and applications in glycoscience, biomedicine and material science. *Biochim. Biophys. Acta*, 1760, 636–651, 2006.
36. Liu, J., Cao, Z., Lu, Y., Functional nucleic acid sensors. *Chem. Rev.*, 109, 1948–1998, 2009.
37. Jayagopal, A., Halfpenny, K.C., Perez, J.W., Wright, D.W., Hairpin DNA functionalized gold colloids for the imaging of mRNA in live cells. *J. Am. Chem. Soc.*, 132, 9789–9796, 2010.
38. Dhar, S., Daniel, W.L., Giljohann, D.A., Mirkin, C.A., Lippard, S.J., Polyvalent oligonucleotide gold nanoparticle conjugates as delivery vehicles for platinum(IV) warheads. *J. Am. Chem. Soc.*, 131, 14652–14653, 2009.
39. Prigodich, A.E., Alhasan, A.H., Mirkin, C.A., Selective enhancement of nucleases by polyvalent DNA-functionalized gold nanoparticles. *J. Am. Chem. Soc.*, 133, 2120–2123, 2011.
40. Jones, M.R., Macfarlane, R.J., Lee, B., Zhang, J., Young, K.L., Senesi, A.J., Mirkin, C.A., DNA-nanoparticle superlattices formed from anisotropic building blocks. *Nat. Mater.*, 9, 913–917, 2010.
41. Auyeung, E., Cutler, J.I., Macfarlane, R.J., Jones, M.R., Wu, J., Liu, G., Zhang, K., Osberg, K.D., Mirkin, C.A., Synthetically programmable nanoparticle

superlattices using a hollow three-dimensional spacer approach. *Nat. Nanotechnol.*, 7, 24–28, 2012.
42. Wang, J., Xu, J., Goodman, M.D., Chen, Y., Cai, M., Shinar, J. *et al.*, A simple biphasic route to water soluble dithiocarbamate functionalized quantum dots. *J. Mater. Chem.*, 18, 3270–3274, 2008.
43. Zhang, S.X., Wang, N., Yu, H.J., Niu, Y.M., Sun, C.Q., Covalent attachment of glucose oxidase to an Au electrode modified with gold nanoparticles for use as glucose biosensor. *Bioelectrochemistry*, 67, 15–22, 2005b.
44. Sahoo, B., Sahu, S.K., Pramanik, P., A novel method for the immobilization of urease on phosphonate grafted iron oxide nanoparticle. *J. Mol. Catal. B Enzym.*, 69, 95–102, 2011.
45. Conde, J., Dias, J.T., Grazú, V., Moros, M., Baptista, P.V., de la Fuente, J.M., Revisiting 30 years of biofunctionalization and surface chemistry of inorganic nanoparticles for nanomedicine. *Front. Chem.*, 2, Article 48, 1–27, 2014.
46. Basak, G., Hazra, C., Sen, R., Biofunctionalized nanomaterials for *in situ* clean-up of hydrocarbon contamination: A quantum jump in global bioremediation research. *J. Environ. Manage.*, 256, 109913, 2020.
47. Jalil, W.D.A. and Kahachi, H.A.H., The implementation of nano-biomimicry for sustainability in architecture. *J. Eng. Sustain. Dev.*, 23, 03, 25–41, May 2019.
48. Garg, P., Ghatmale, P., Tarwadi, K., Chavan, S., Influence of nanotechnology and the role of nanostructures in biomimetic studies and their potential applications. *Biomimetics*, 2, 2, 7, 1–25, 2017.
49. Barthelat, F., Biomimetics for next generation materials. *Philos. Trans. R. Soc.*, 365, 2907–2919, 2007.
50. Kim, P., Bio-inspired design of adaptive, dynamic, and multi-functional materials and architectures, Conference Proceedings, AIChE Annual Meeting, 2012.
51. Woon, H., Healthy Buildings in Singapore, Atelier ten, Sustainability-report-2015, pp. 28–29, 2015.
52. Huang, J., Lin, L., Sun, D., Chen, H., Yang, D., Li, Q., Bio-inspired synthesis of metal nanomaterials and applications. *Chem. Soc. Rev.*, 44, 6330–6374, 2015.
53. Madamsetty, V.S., Mukherjee, A., Mukherjee, S., Recent trends of the bio-inspired nanoparticles in cancer theranostics. *Front. Pharmacol.*, 10, 1264, 2019.
54. Kumar, A., Sharma, G., Naushad, M., Al-Muhtaseb, A.H., García-Peñas, A., Mola, G.T. *et al.*, Bio-inspired and biomaterials-based hybrid photocatalysts for environmental detoxification: A review. *Chem.*, 382, 122937, 2020.
55. Lai, Y., Li, F., Zou, Z., Saeed, M., Xu, Z., Yu, H., Bio-inspired amyloid polypeptides: From self assembly to nanostructure design and biotechnological applications. *Appl. Mater. Today*, 22, 100966, 2021.
56. Wu, W., Jiang, C.Z., Roy, V.A.L., Designed synthesis and surface engineering strategies of magnetic iron oxide nanoparticles for biomedical applications. *Nanoscale*, 8, 19421–19474, 2016.

57. Kankala, R.K., Han, Y.H., Na, J., Lee, C.H., Sun, Z., Wang, S.B. et al., Nanoarchitectured structure and surface biofunctionality of mesoporous silica nanoparticles. *Adv. Mater.*, 32, 1907035, 2020.
58. Reddy, L.H., Arias, J.L., Nicolas, J., Couvreur, P., Magnetic nanoparticles: Design and characterization, toxicity and biocompatibility, pharmaceutical and biomedical applications. *Chem. Rev.*, 112, 5818–5878, 2012.
59. Li, Y., Zheng, X., Chu, Q., Bio-based nanomaterials for cancer therapy. *Nano Today*, 38, 101134, 2021.
60. Ge, W., Wang, L., Zhang, J., Ou, C., Si, W., Wang, W. et al., Self-assembled nanoparticles as cancer therapeutic agents. *Adv. Mater. Interfaces*, 8, 2001602, 2021.
61. Ke, P.C. and Qiao, R., Carbon nanomaterials in biological systems. *J. Phys. Condens. Matter*, 19, 373101, 2007.
62. Nel, A.E., Mädler, L., Velegol, D., Xia, T., Hoek, E.M., Somasundaran, P., Klaessig, F., Castranova, V., Thompson, M., Understanding biophysicochemical interactions at the nano–bio interface. *Nat. Mater.*, 8, 543–557, 2009.
63. http://wyss.harvard.edu/
64. Zhou, H., Fan, T., Zhang, D., Biotemplated materials for sustainable energy and environment: Current status and challenges. *ChemSusChem*, 4, 1344–1387, 2011.
65. Sotiropoulou, S., Sierra-Sastre, Y., Mark, S.S., Batt, C.A., Biotemplated nanostructured materials. *Chem. Mater.*, 20, 821–834, 2008.
66. Wang, X., Cai, X., Hu, J., Shao, N., Wang, F., Zhang, Q., Xiao, J., Cheng, Y., Glutathione-triggered 'off-on' release of anticancer drugs from dendrimer-encapsulated gold nanoparticles. *J. Am. Chem. Soc.*, 135, 9805–9810, 2013.
67. Jeong, B., Akter, R., Han, O.H., Rhee, C.K., Rahman, M.A., Increased electrocatalyzed performance through dendrimer-encapsulated gold nanoparticles and carbon nanotube-assisted multiple bienzymatic labels: Highly sensitive electrochemical immunosensor for protein detection. *Anal. Chem.*, 85, 1784–1791, 2013.
68. Wang, X., Zhang, Y., Li, T., Tian, W., Zhang, Q., Cheng, Y., Generation 9 polyamidoamine dendrimer encapsulated platinum nanoparticle mimics catalase size, shape, and catalytic activity. *Langmuir*, 29, 5262–5270, 2013.
69. Helms, B. and Meijer, E.W., Dendrimers at work. *Science*, 313, 929–930, 2006.
70. Aime, S., Frullano, L., Geninatti Crich, S., Compartmentalization of a gadolinium complex in the apoferritin cavity: A route to obtain high relaxivity contrast agents for magnetic resonance imaging. *Angew. Chem. Int. Ed.*, 41, 1017–1019, 2002.
71. Kramer, R.M., Li, C., Carter, D.C., Stone, M.O., Naik, R.R., Engineered protein cages for nanomaterial synthesis. *J. Am. Chem. Soc.*, 126, 13 282–13 286, 2004.
72. Bhattacharya, P., Geitner, N.K., Sarupria, S., Ke, P.C., Exploiting the physicochemical properties of dendritic polymers for environmental and biological applications. *Phys. Chem. Chem. Phys.*, 15, 4477–4490, 2013.

73. Crooks, R.M., Zhao, M., Sun, L., Chechik, V., Yeung, L.K., Dendrimer-encapsulated metal nanoparticles: Synthesis, characterization, and applications to catalysis. *Acc. Chem. Res.*, 34, 181–190, 2001.
74. Myers, V.S., Weir, M.G., Carino, E.V., Yancey, D.F., Pande, S., Crooks, R.M., Dendrimer-encapsulated nanoparticles: New synthetic and characterization methods and catalytic applications. *Chem. Sci.*, 2, 1632–1646, 2011.
75. Wang, X., Cai, X., Hu, J., Shao, N., Wang, F., Zhang, Q., Xiao, J., Cheng, Y., Glutathione-triggered 'off–on' release of anticancer drugs from dendrimer-encapsulated gold nanoparticles. *J. Am. Chem. Soc.*, 135, 9805–9810, 2013.
76. Jeong, B., Akter, R., Han, O.H., Rhee, C.K., Rahman, M.A., Increased electrocatalyzed performance through dendrimer-encapsulated gold nanoparticles and carbon nanotube-assisted multiple bienzymatic labels: Highly sensitive electrochemical immunosensor for protein detection. *Anal. Chem.*, 85, 1784–1791, 2013.
77. Wang, X., Zhang, Y., Li, T., Tian, W., Zhang, Q., Cheng, Y., Generation 9 polyamidoamine dendrimer encapsulated platinum nanoparticle mimics catalase size, shape, and catalytic activity. *Langmuir*, 29, 5262–5270, 2013.
78. Helms, B. and Meijer, E.W., Dendrimers at work. *Science*, 313, 929–930, 2006.
79. Aime, S., Frullano, L., Geninatti Crich, S., Compartmentalization of a gadolinium complex in the apoferritin cavity: A route to obtain high relaxivity contrast agents for magnetic resonance imaging. *Angew. Chem. Int. Ed.*, 41, 1017–1019, 2002.
80. Uchida, M. *et al.*, Targeting of cancer cells with ferrimagnetic ferritin cage nanoparticles. *J. Am. Chem. Soc.*, 128, 16 626–16 633, 2006.
81. Bradshaw, T.D., Junor, M., Patanè, A., Clarke, P., Thomas, N.R., Li, M., Mann, S., Turyanska, L., Apoferritin-encapsulated PbS quantum dots significantly inhibit growth of colorectal carcinoma cells. *J. Mater. Chem. B*, 1, 6254–6260, 2013.
82. Wu, H., Engelhard, M.H., Wang, J., Fisher, D.R., Lin, Y., Synthesis of lutetium phosphate-apoferritin core-shell nanoparticles for potential applications in radioimmunoimaging and radioimmunotherapy of cancers. *J. Mater. Chem.*, 18, 1779–1783, 2008.
83. Wu, H., Wang, J., Wang, Z., Fisher, D.R., Lin, Y., Apoferritin-templated yttrium phosphate nanoparticle conjugates for radioimmunotherapy of cancers. *J. Nanosci. Nanotechnol.*, 8, 2316–2322, 2008.
84. Bode, S.A., Minten, I.J., Nolte, R.J.M., Cornelissen, J.J.L.M., Reactions inside nanoscale protein cages. *Nanoscale*, 3, 2376–2389, 2011. (10.1039/C0NR01013H).
85. Okuda, M., Iwahori, K., Yamashita, I., Yoshimura, H., Fabrication of nickel and chromium nanoparticles using the protein cage of apoferritin. *Biotechnol. Bioeng.*, 84, 187–194, 2003.

86. Allen, M., Willits, D., Young, M., Douglas, T., Constrained synthesis of cobalt oxide nanomaterials in the 12-subunit protein cage from listeria innocua. *Inorg. Chem.*, 42, 6300–6305, 2003.
87. Hoinville, J. et al., High density magnetic recording on protein-derived nanoparticles. *J. Appl. Phys.*, 93, 7187–7189, 2003.
88. Klem, M.T., Mosolf, J., Young, M., Douglas, T., Photochemical mineralization of europium, titanium, and iron oxyhydroxide nanoparticles in the ferritin protein cage. *Inorg. Chem.*, 47, 2237–2239, 2008, (10.1021/ic701740q).
89. Douglas, T., Strable, E., Willits, D., Aitouchen, A., Libera, M., Young, M., Protein engineering of a viral cage for constrained nanomaterials synthesis. *Adv. Mater.*, 14, 415–418, 2002.
90. Jaaskelainen, A., Harinen, R.-R., Soukka, T., Lamminmaki, U., Korpimaki, T., Virta, M., Biologically produced bifunctional recombinant protein nanoparticles for immunoassays. *Anal. Chem.*, 80, 583–587, 2008.
91. Yu, X., Wang, J., Du, Y., Ma, Z., He, W., Nanotechnology for tissue engineering applications. *J. Nanomater.*, 10, 3223–3230, 2011.
92. Hartgerink, J.D., Granja, J.R., Milligan, R.A., Ghadiri, M.R., Self-assembling peptide nanotubes. *J. Am. Chem. Soc.*, 118, 43–50, 1996.
93. Scanlon, S. and Aggeli, A., Self-assembling peptide nanotubes. *Nano Today*, 3, 22–30, 2008.
94. Zhang, S., Marini, D.M., Hwang, W., Santoso, S., Design of nanostructured biological materials through self assembly of peptides and proteins. *Curr. Opin. Chem. Biol.*, 6, 865–871, 2002.
95. Cui, H., Webber, M.J., Stupp, S.I., Self assembly of peptide amphiphiles: From molecules to nanostructures to biomaterials. *Biopolymers*, 94, 1–18, 2010.
96. Yemini, M., Reches, M., Gazit, E., Rishpon, J., Peptide nanotube-modified electrodes for enzyme–biosensor applications. *Anal. Chem.*, 77, 5155–5159, 2005.
97. Han, Y., Shen, X., Chen, S., Wang, X., Du, J., Zhu, T., A nanofiber mat with dual bioactive components and a biomimetic matrix structure for improving osteogenesis effect. *Front. Chem.*, 9, 740191, 2021.
98. Xing, Y., Qing, X., Xia, H., Hao, S., Zhu, H., He, Y., Mao, H., Gu, Z., Injectable hydrogel based on modified gelatin and sodium alginate for soft-tissue adhesive. *Front. Chem.*, 9, 744099, 2021.
99. Zhou, M., Luo, Y., Zeng, W., Yang, X., Chen, T., Zhang, L., He, X., Yi, X., Li, Y., Yi, X., A co-delivery system based on a dimeric prodrug and star-shaped polymeric prodrug micelles for drug delivery. *Front. Chem.*, 9, 765021, 2021.
100. Fang, D., Jin, H., Huang, X., Shi, Y., Liu, Z., Ben, S., PPy@Fe3O4 nanoparticles inhibit tumor growth and metastasis through chemodynamic and photothermal therapy in non-small cell lung cancer. *Front. Chem.*, 9, 789934, 2021.

101. Xia, J., Qing, X., Shen, J., Ding, M., Wang, Y., Yu, N., Li, J., Wang, X., Enzyme-loaded pH-sensitive photothermal hydrogels for mild-temperature mediated combinational cancer therapy. *Front. Chem.*, 9, 736468, 2021.
102. Yu, N., Ding, M., Li, J., Near-infrared photoactivatable immunomodulatory nanoparticles for combinational immunotherapy of cancer. *Front. Chem.*, 9, 701427, 2021.
103. Zhang, X., Ma, Y., Wan, J., Yuan, J., Wang, D., Wang, W., Sun, X., Meng, Q., Biomimetic nanomaterials triggered ferroptosis for cancer theranostics. *Front. Chem.*, 9, 768248, 2021.
104. Sharma, P., Pandey, V., Sharma, M.M.M. et al., A review on biosensors and nanosensors application in agroecosystems. *Nanoscale Res. Lett.*, 16, 136, 2021.
105. Aragay, G., Pino, F., Merkoci, A., Nanomaterials for sensing and destroying pesticides. *Chem. Rev.*, 112, 5317–38, 2012.
106. Sadik, O.A., Aluoch, A.O., Zhou, A., Status of biomolecular recognition using electrochemical techniques. *Biosens. Bioelectron.*, 24, 2749–65, 2009.
107. Bănică, F.-G., Nanomaterial applications in optical transduction, in: *Chemical Sensors and Biosensors*, pp. 454–72, Wiley, Chichester, 2012.
108. Grieshaber, D., MacKenzie, R., Vörös, J., Reimhult, E., Electrochemical biosensors—sensor principles and architectures. *Sensors*, 8, 1400–58, 2008.
109. Liu, G. and Lin, Y., Nanomaterial labels in electrochemical immunosensors and immunoassays. *Talanta*, 74, 308–17, 2007.
110. Kim, Y.A., Lee, E.-H., Kim, K.-O., Lee, Y.T., Hammock, B.D., Lee, H.-S., Competitive immunochromatographic assay for the detection of the organophosphorus pesticide chlorpyrifos. *Anal. Chim. Acta*, 693, 106–13, 2011.
111. Zhao, W.W., Ma, Z.Y., Yu, P.P., Dong, X.Y., Xu, J.J., Chen, H.Y., Highly sensitive photoelectrochemical immunoassay with enhanced amplification using horseradish peroxidase induced biocatalytic precipitation on a CdS quantum dots multilayer electrode. *Anal. Chem.*, 84, 917–23, 2012.
112. Trilling, A.K., Beekwilder, J., Zuilhof, H., Antibody orientation on biosensor surfaces: A minireview. *Analyst*, 138, 1619–27, 2013.
113. Jiajie, L., Hongwu, L., Caifeng, L., Qiangqiang, F., Caihong, H., Zhi, L., Tianjiu, J., Yong, T., Silver nanoparticle enhanced Raman scattering-based lateral flow immunoassays for ultra-sensitive detection of the heavy metal chromium. *Nanotechnology*, 25, 495501, 2014.
114. Liu, X., Xiang, J.J., Tang, Y., Zhang, X.L., Fu, Q.Q., Zou, J.H., Lin, Y., Colloidal gold nanoparticle probe-based immunochromatographic assay for the rapid detection of chromium ions in water and serum samples. *Anal. Chim. Acta*, 745, 99–105, 2012.
115. Li, M., Wang, Q., Shi, X., Hornak, L.A., Wu, N., Detection of mercury(II) by quantum dot/DNA/gold nanoparticle ensemble based nanosensor via nano-metal surface energy transfer. *Anal. Chem.*, 83, 7061–5, 2011.

116. Li, T., Li, B., Wang, E., Dong, S., G-quadruplex-based DNAzyme for sensitive mercury detection with the naked eye. *Chem. Commun.*, 24, 3551–3553, 2009.
117. Long, F., Zhu, A., Shi, H., Wang, H., Liu, J., Rapid on-site/in-situ detection of heavy metal ions in environmental water using a structure-switching DNA optical biosensor. *Sci. Rep.*, 3, 2308, 2013.
118. Liu, M., Wang, Z., Zong, S., Chen, H., Zhu, D., Wu, L., Hu, G., Cui, Y., SERS detection and removal of mercury(II)/silver(I) using oligonucleotide-functionalized core/shell magnetic silica Sphere@Au nanoparticles. *ACS Appl. Mater. Interfaces*, 6, 7371–9, 2014.
119. Ma, J., Chen, Y., Hou, Z., Jiang, W., Wang, L., Selective and sensitive mercuric(II) ion detection based on quantum dots and nicking endonuclease assisted signal amplification. *Biosens. Bioelectron.*, 43, 84–7, 2013.
120. Zhang, M., Yin, B.C., Tan, W., Ye, B.C., A versatile graphene-based fluorescence "on/off" switch for multiplex detection of various targets. *Biosens. Bioelectron.*, 26, 3260–5, 2011.
121. Evtugyn, G.A., Budnikov, H.C., Nikolskaya, E.B., Sensitivity and selectivity of electrochemical enzyme sensors for inhibitor determination. *Talanta*, 46, 465–84, 1998.
122. Bies, C., Lehr, C.M., Woodley, J.F., Lectin-mediated drug targeting: History and applications. *Adv. Drug Delivery Rev.*, 56, 425–35, 2004.
123. Willner, M.R. and Vikesland, P.J., Nanomaterial enabled sensors for environmental contaminants. *J. Nanobiotechnol.*, 16, 95, 2018.
124. Xu, Z. and Dong, J., Synthesis, characterization, and application of magnetic nanocomposites for the removal of heavy metals from industrial effluents, in: *Emerging Environmental Technologies*, V. Shah (Ed.), pp. 105–48, Springer, Netherlands: Dordrecht, 2008.
125. Li, M., Gou, H., Al-Ogaidi, I., Wu, N., Nanostructured sensors for detection of heavy metals: A review. *ACS Sustainable Chem. Eng.*, 1, 713–23, 2013.
126. Mohmood, I., Lopes, C., Lopes, I., Ahmad, I., Duarte, A., Pereira, E., Nanoscale materials and their use in water contaminants removal—A review. *Environ. Sci. Pollut. Res.*, 20, 1239–60, 2013.
127. Nagaraj, V.J., Jacobs, M., Vattipalli, K.M., Annam, V.P., Prasad, S., Nano channel based electrochemical sensor for the detection of pharmaceutical contaminants in water. *Environ. Sci. Process. Impacts*, 16, 135–40, 2014.
128. Sanvicens, N., Mannelli, I., Salvador, J.P., Valera, E., Marco, M.P., Biosensors for pharmaceuticals based on novel technology. *TrAC Trends Anal. Chem.*, 30, 541–53, 2011.
129. Sadik, O.A., Aluoch, A.O., Zhou, A., Status of biomolecular recognition using electrochemical techniques. *Biosens. Bioelectron.*, 24, 2749–65, 2009.
130. Koneswaran, M. and Narayanaswamy, R., l-Cysteine-capped ZnS quantum dots based fluorescence sensor for Cu2+ ion. *Sens. Actuators B Chem.*, 139, 104–9, 2009.

131. Algarra, M., Campos, B.B., Alonso, B., Miranda, M.S., Martínez, ÁM, Casado, C.M., Esteves da Silva, J.C.G., Thiolated DAB dendrimers and CdSe quantum dots nanocomposites for Cd(II) or Pb(II) sensing. *Talanta*, 88, 403–7, 2012.
132. Arriaza-Echanes, C., Campo-Giraldo, J.L., Quezada, C.P., Espinoza-González, R., Rivas-Álvarez, P., Pacheco, M., Bravo, D., Pérez-Donoso, J.M., Biomimetic synthesis of CuInS2 nanoparticles: Characterization, cytotoxicity, and application in quantum dots sensitized solar cells. *Arab. J. Chem.*, 14, 103176, 2021.
133. Romanholo, P.V.V., Razzino, C.A., Raymundo-Pereira, P.A., Prado, T.M., Machado, S.A.S., Sgobbi, L.F., Biomimetic electrochemical sensors: New horizons and challenges in biosensing applications. *Biosens. Bioelectron.*, 185, 113242, 2021.
134. Lowe, C.R., Chemoselective biosensors. *Curr. Opin. Chem. Biol.*, 3, 106–111, 1999.
135. Khan, S., Wong, A., Zanoni, M.V.B., Sotomayor, M.D.P.T., Electrochemical sensors based on biomimetic magnetic molecularly imprinted polymer for selective quantification of methyl green in environmental samples. *Mater. Sci. Eng. C*, 103, 109825, 2019.
136. Yang, S., Liu, M., Deng, F., Mao, L., Yuan, Y., Huang, H., Chen, J., Liu, L., Zhang, X., Wei, Y., Biomimetic modification of silica nanoparticles for highly sensitive and ultrafast detection of DNA and Ag+ ions. *Appl. Surf. Sci.*, 510, 145421, 2020.
137. Nasir, M., Rauf, S., Muhammad, N., Nawaz, M.H., Chaudhry, A.A., Malik, M.H., Shahid, S.A., Hayat, A., Biomimetic nitrogen doped titania nanoparticles as a colorimetric platform for hydrogen peroxide detection. *J. Colloid Interface Sci.*, 505, 1147–1157, 2017.
138. Cao, W., Ju, P., Wang, Z., Zhang, Y., Zhai, X., Jiang, F., Sun, C., Colorimetric detection of H2O2 based on the enhanced peroxidase mimetic activity of nanoparticles decorated Ce2(WO4)3 nanosheets. *Spectrochim. Acta A Mol. Biomol. Spectrosc.*, 239, 118499, 2020.
139. Vena, M.P., Jobbagy, M., Bilmes, S.A., Microorganism mediated biosynthesis of metal chalcogenides; a powerful tool to transform toxic effluents into functional nanomaterials. *Sci. Total Environ.*, 565, 804–810, 2016.
140. Queirós, R.B., Guedes, A., Marques, P., Noronha, J., Sales, M.G.F., Chemical, A.B., Recycling old screen-printed electrodes with newly designed plastic antibodies on the wall of carbon nanotubes as sensory element for *in situ* detection of bacterial toxins in water. *Sensors*, 189, 21–29, 2013.
141. Zhang, S.Q., Lin, F.F., Yuan, Q.P., Liu, J.W., Li, Y., Liang, H., Robust magnetic laccase-mimicking nanozyme for oxidizing ophenylenediamine and removing phenolic pollutants. *J. Environ. Sci.*, 88, 103–111, 2020.
142. Zhai, C., Miao, L., Zhang, Y., Zhang, L., Li, H., Zhang, S., An enzyme response-regulated colorimetric assay for pattern recognition sensing application using biomimetic inorganic-protein hybrid nanoflowers. *Chem. Eng. J.*, 431, 134107, 2022.

143. Chu, W., Zhang, Y., Li, D., Barrow, C.J., Wang, H., Yang, W., A biomimetic sensor for the detection of lead in water. *Biosens. Bioelectron.*, 67, 621–624, 2015.
144. Niu, X.H., He, Y.F., Li, X., Zhao, H.L., Pan, J.M., Qiu, F.X., Lan, M.B., A peroxidase-mimicking nanosensor with Hg2+-triggered enzymatic activity of cysteine-decorated ferromagnetic particles for ultrasensitive Hg2+ detection in environmental and biological fluids. *Sens. Actuators B Chem.*, 281, 445–452, 2019.
145. Wujcik, E.K., Londoño, N.J., Duirk, S.E., Monty, C.N., Masel, R.I., An acetylcholinesterase-inspired biomimetic toxicity sensor. *Chemosphere*, 91, 1176–1182, 2013, https://doi.org/10.1016/j.chemosphere.2013.01.027.
146. Zhang, D., Fan, Y., Li, G., Du, W., Li, R., Liu, Y., Cheng, Z., Xu, J., Biomimetic synthesis of zeolitic imidazolate frameworks and their application in high performance acetone gas sensors. *Sens. Actuators B Chem.*, 302, 127187, 2020.
147. Machini, W.B.S. and Teixeira, M.F.S., Application of oxo-manganese complex immobilized on ion-exchange polymeric film as biomimetic sensor for nitrite ions. *Sens. Actuators B Chem.*, 217, 58–64, 2015.
148. Cao, Q., Xiao, Y., Liu, N., Huang, R., Ye, C., Huang, C., Liu, H., Han, G., Wu, L., Synthesis of yolk/shell heterostructures MOF@MOF as biomimetic sensing platform for catechol detection. *Sens. Actuators B Chem.*, 329, 129133, 2021.
149. Cheng, Y., Chen, T., Fu, D., Liu, J., A molecularly imprinted nanoreactor based on biomimetic mineralization of bi-enzymes for specific detection of urea and its analogues. *Sens. Actuators B Chem.*, 350, 130909, 2022.
150. Wong, A., de Vasconcelos Lanza, M.R., Sotomayor, M.D.P.T., Sensor for diuron quantitation based on the P450 biomimetic catalyst nickel(II) 1,4,8,11,15,18,22,25-octabutoxy-29H,31H-phthalocyanine. *J. Electroanal. Chem.*, 690, 83–88, 2013.
151. Sgobbi, L.F. and Machado, S.A.S., Functionalized polyacrylamide as an acetylcholinesterase-inspired biomimetic device for electrochemical sensing of organophosphorus pesticides. *Biosens. Bioelectron.*, 100, 290–297, 2018.
152. Yan, X., Kong, D., Jin, R., Zhao, X., Li, H., Liu, F., Lin, Y., Lu, G.J.S., Chemical, A.B., Fluorometric and colorimetric analysis of carbamate pesticide via enzyme-triggered decomposition of gold nanoclusters-anchored MnO2 nanocomposite. *Sensors Actuators B Chem.*, 290, 640–647, 2019.
153. Jin, R., Kong, D.S., Zhao, X., Li, H.X., Yan, X., Liu, F.M., Sun, P., Du, D., Lin, Y.H., Lu, G.Y., Tandem catalysis driven by enzymes directed hybrid nanoflowers for on-site ultrasensitive detection of organophosphorus pesticide. *Biosens. Bioelectron.*, 141, 111473, 2019.
154. Gavrilaş, S., Ursachi, C.S., Perţa-Crişan, S., Munteanu, F.-D., Recent trends in biosensors for environmental quality monitoring. *Sensors*, 22, 1513, 2022.
155. Touloupakis, E., Giannoudi, L., Piletsky, S.A., Guzzella, L., Pozzoni, F., Giardi, M.T., A multi-biosensor based on immobilized photosystem II on

screen-printed electrodes for the detection of herbicides in river water. *Biosens. Bioelectron.*, 20, 1984–1992, 2005.
156. Naessens, M. and Leclerc JC Tran-Minh, C., Fiber optic biosensor using chlorella vulgaris for determination of toxic compounds. *Ecotoxicol. Environ. Saf.*, 46, 181–185, 2000.
157. Marty, J.-L., Garcia, D., Rouillon, R., Biosensors: Potential in pesticide detection. *Trends Anal. Chem.*, 14, 329–333, 1995.
158. Tibuzzi, A., Rea, G., Pezzotti, G., Esposito, D., Johanningmeier, U., Giardi, M.T., A new miniaturized multiarray biosensor system for fluorescence detection. *J. Phys.: Condens. Matter*, 19, 395006, 12 pp, 2007.
159. Scognamiglio, V., Raffi, D., Lambreva, M., Rea, G., Tibuzzi, A., Pezzotti, G., Johanningmeier, U., Giardi, M.T., Chlamydomonas reinhardtii genetic variants as probes for fluorescence sensing system in detection of pollutants. *Anal. Bioanal. Chem.*, 394, 1081–1087, 2009.
160. Giardi, M.T., Guzzella, L., Euzet, P., Rouillon, R., Esposito, D., Detection of herbicide subclasses by an optical multibiosensor based on an array of photosystem II mutants. *Environ. Sci. Technol.*, 39, 5378–5384, 2005.
161. Buonasera, K., Pezzotti, G., Pezzotti, I., Cano, J., Giardi, M., Biosensors: New frontiers for environmental analysis. *Rev. Politécnica.*, 13, 93–100, 2011.
162. Sugai, T., Yoshida, H., Shimada, T., Okazaki, T., Shinohara, H., Bandow, S., New synthesis of high-quality doublewalled carbon nanotubes by high-temperature pulsed arc discharge. *Nano Lett.*, 3, 6, 769–773, 2003.
163. Santoro, M., Gorelli, F.A., Bini, R., Haines, J., van der Lee, A., High-pressure synthesis of a polyethylene/zeolite nanocomposite material. *Nat. Commun.*, 4, 1, 1557, 2013.
164. Lou, X.W. and Zeng, H.C., Hydrothermal synthesis of α-MoO3nanorods via acidification of ammonium heptamolybdate tetrahydrate. *Chem. Mater.*, 14, 11, 4781–4789, 2002.
165. Khan, N.A., Kang, I.J., Seok, H.Y., Jhung, S.H., Facile synthesis of nano-sized metal-organic frameworks, chromium benzene dicarboxylate, MIL-101. *Chem. Eng. J.*, 166, 3, 1152–1157, 2011.
166. Takla, S.S., Shawky, E., Hammoda, H.M., Darwish, F.A., Green techniques in comparison to conventional ones in the extraction of Amaryllidaceae alkaloids: Best solvents selection and parameters optimization. *J. Chromatogr. A*, 1567, 99–110, 2018.
167. Guo, H., White, J.C., Wang, Z., Xing, B., Nano-enabled fertilizers to control the release and use efficiency of nutrients. *Curr. Opin. Environ. Sci. Health*, 6, 77–83, 2018.
168. Sheoran, P., Grewal, S., Kumari, S., Goel, S., Enhancement of growth and yield, leaching reduction in Triticum aestivum using biogenic synthesized zinc oxide nanofertilizer. *Biocatal. Agric. Biotechnol.*, 32, article 101938–101943, 2021.

169. Leonardi, M., Caruso, G.M., Carroccio, S.C. et al., Smart nanocomposites of chitosan/alginate nanoparticles loaded with copper oxide as alternative nanofertilizers. *Environ. Sci.: Nano*, 8, 174–187, 2021.
170. Abbasifar, Shahrabadi, F., Kaji, B.V., Effects of green synthesized zinc and copper nano-fertilizers on the morphological and biochemical attributes of basil plant. *J. Plant Nutr.*, 43, 8, 1104–1118, 2020.
171. Ramírez-Rodríguez, G.B., Dal Sasso, G., Carmona, F.J. et al., Engineering biomimetic calcium phosphate nanoparticles: A green synthesis of slow-release multinutrient (NPK) nanofertilizers. *ACS Appl. Bio Mater.*, 3, 3, 1344–1353, 2020.
172. Jahangirian, H., Rafiee-Moghaddam, R., Jahangirian, N. et al., Green synthesis of zeolite/Fe2O3 nanocomposites: Toxicity & cell proliferation assays and application as a smart iron nanofertilizer. *Int. J. Nanomedicine*, 15, 1005–1020, 2020.
173. Gao, X., Shi, F., Peng, F. et al., Formulation of nanopesticide with graphene oxide as the nanocarrier of pyrethroid pesticide and its application in spider mite control. *RSC Adv.*, 11, 57, 36089–36097, 2021.
174. Ahmed, T., Noman, M., Luo, J. et al., Bioengineered chitosanmagnesium nanocomposite: A novel agricultural antimicrobial agent against Acidovorax oryzae and Rhizoctonia solani for sustainable rice production. *Int. J. Biol. Macromol.*, 168, 834–845, 2021.
175. Namasivayam, S.K., Bharani, R.A., Karunamoorthy, K., Insecticidal fungal metabolites fabricated chitosan nanocomposite (IM-CNC) preparation for the enhanced larvicidal activity - an effective strategy for green pesticide against economic important insect pests. *Int. J. Biol. Macromol.*, 120, Part A, 921–944, 2018.
176. Gao, M., Yalei, X., Chang, X., Dong, Y., Song, Z., Effects of foliar application of graphene oxide on cadmium uptake by lettuce. *J. Hazard. Mater.*, 398, article 122859, 1–10, 2020.
177. Stafford, Jeyakumar, P., Hedley, M., Anderson, C., Influence of soil moisture status on soil cadmium phytoavailability and accumulation in plantain (Plantago lanceolata). *Soil Syst.*, 2, 1, 9, 2018.
178. Zhang, Z., Wu, X., Tu, C. et al., Relationships between soil properties and the accumulation of heavy metals in different Brassica campestris L. growth stages in a Karst mountainous area. *Ecotoxicol. Environ. Saf.*, 206, article 111150, 1–11, 2020.
179. Olad, Zebhi, H., Salari, D., Mirmohseni, A., Tabar, A.R., Slow-release NPK fertilizer encapsulated by carboxymethyl cellulose-based nanocomposite with the function of water retention in soil. *Mater. Sci. Eng.: C*, 90, 333–340, 2018.
180. Zhang, X., Qu, Y., Shen, W., Wang, J., Li, H., Zhang, Z. et al., Biogenic synthesis of gold nanoparticles by yeast Magnusiomyces ingens LH-F1 for catalytic reduction of nitrophenols. *Colloids Surf, A*, 497, 280e5, 2016.

181. Jain, N., Bhargava, A., Majumdar, S., Tarafdar, J., Panwar, J., Extracellular biosynthesis and characterization of silver nanoparticles using Aspergillus flavus NJP08: A mechanism perspective. *Nanoscale*, 3, 635e41, 2011.
182. Yoo, J.W., Irvine, D.J., Discher, D.E., Mitragotri, S., Bio-inspired, bioengineered and biomimetic drug delivery carriers. *Nat. Rev. Drug Discov.*, 10, 7, 521–35, 2011 Jul 1.
183. Zong, L., Li, M., Li, C., Bioinspired coupling of inorganic layered nanomaterials with marine polysaccharides for efficient aqueous exfoliation and smart actuating hybrids. *Adv. Mater.*, 29, 1604691, 2017.
184. Feng, Y., Zhu, W., Guo, W., Jiang, L., Bioinspired energy conversion in nanofluidics: A paradigm of material evolution. *Adv. Mater.*, 29, 1702773, 2017.
185. Zan, G. and Wu, Q., Biomimetic and bioinspired synthesis of nanomaterials/nanostructures. *Adv. Mater.*, 28, 2099e147, 2016.
186. Abid, M.F., Al-Naseri, S.K., Abdullah, S.N., Reuse of Iraqi agricultural drainage water using nanofiltration. *J. Membr. Sep. Technol.*, 013, 2, 53–62, 2013.
187. Hoek, E.M.V. and Ghosh, A.K., Chapter 4: Nanotechnology-based membranes for water purification, in: *Nanotechnology Applications for Clean Water*, N. Savage, M. Diallo, J. Duncan, A. Street, R. Sustich (Eds.), p. 47, William Andrew Inc., Norwich, NY, 2009.
188. Yotova, L., Yaneva, S., Marinkova, D., Biomimetic nanosensors for determination of toxic compounds in food and agricultural products (review). *J. Chem. Technol. Metall.*, 48, 3, 215–227, 2013.
189. Prasad, R., Bhattacharyya, A., Nguyen, Q.D., Nanotechnology in sustainable agriculture: Recent developments, challenges, and perspectives. *Front. Microbiol.*, 8, 1014, 2017.
190. United Nations Department of Economic and Social Affairs, Population Division, World population prospects 2022: Summary of results, UN DESA/POP/2022/TR/NO. United Nations, New York, USA, p. 3, 2022.
191. biomimicry.org
192. Do, D.T., Singh, J., Oey, I., Singh, H., Biomimetic plant foods: Structural design and functionality. *Trends Food Sci. Technol.*, 82, 46–59, 2018.
193. Katiyar, N.K., Goel, G., Hawi, S. et al., Nature-inspired materials: Emerging trends and prospects. *NPG Asia Mater.*, 13, 56, 2021.

Index

4,4′-bipyridine (4,4′-Bpy), 275

α-galactosidase enzymes, 299
α-helical structure, 268

Abiotic stress, 169–172, 174
Actinomycetes, 192
   unicellular filamentous bacteria, 192
Actuators, 221
Advantages of green nanoscience, 25, 33
   green nanoelectronics, 35
   in automobiles, 34
   in food and agriculture, 35
   in industries, 34
   in medicines, 35
Algae, 132–134, 192
   blue green algae, 192
   Fucus vesiculosus, 192
   Kappaphycus alvarezii, 192
   Sargassum wightii, 192
   Spirulina platensis, 192
Algal system-mediated nanomaterial synthesis, 143–145
Amino acid,
   3,4-dihydroxyphenylalanine (L-DOPA), 275, 279
   alanine, 268, 274
   arginine, 268
   cyteine, 268
   DL-alanine, 274
   glycine (Gly) (α,β,γ), 267, 270–275
   hydroxyproline (Hyp), 267, 274
   leucine, 268, 274
   N-acetyl alanine (AcA), 275

   phenylalanine, 275
   proline (Pro), 267, 274
   tyrosine, 275
Aminoacyl-tRNA synthetase (aaRS), 297
Animal architecture, 369
Animals & birds for bioinspiration,
   bats, 336
   eggshell membrane, 332, 339
   hummingbird, 331, 332, 338
   marsupials, 336, 340
   rodents, 336, 340
Anticancer drugs, 300
Application of bionanocomposites,
   dental applications, 248–250
   orthopedics, 246–248
   tissue engineering, 251
Applications of the green synthesized nanomaterials,
   anticancerous, 148
   antimicrobial agent, 148
   bioremediation, 149–150
   biosensing, 148–149
Aptamers, 45
Ascorbic acid, 267
Atomistic simulation, 267

Bacteria, 118–121
   Bacillus subtilis, 191, 193, 194
   Desulfovibrio desulfuricans, 191
   E. coli, 191
   Enterobacter cloacae, 191
   P. aerugenosa, 191
   Pseudomonas aeruginosa, 191

406 INDEX

Rhodospirillium rubrum, 191
T. thiooxidans, 191
Thiobacillus ferooxidans, 190–194, 196–200
Bioactivity-based material, 212
Bioavailability, 47, 53
Biocompatible, 265, 271, 274, 276–277
Bio-e-skin, 277–279
Biofunctionalization, 361–362
Biofunctionalized nanoparticle, 362
  fate, 362
Biogenic, 50, 52, 54, 56–58
Bioimaging, 44, 53, 54
Bioinspiration, 264, 325, 326, 355–356, 386
Bioinspired, 208, 209, 213–214
Bioinspired artificial materials, 264, 266
Bioinspired material properties,
  adhesion, 327
  dust resistance, 327
  lightweight, 327
  thermal insulation, 327, 329, 337
  water resistant, 327
Bioinspired materials, 325–327, 343
Bioinspired structures, 356
Biomass, 122–132
Biomaterials,
  bone, 264, 266–267, 279–280
  cellulose, 264, 279
  chitosan, 264, 279
  DNA, 264
  gelatin, 264, 269, 277
  hair, 264, 266
  skin, 264, 266, 268, 276–280
  tendon, 266
  wood, 266
Biomedical applications, 88, 370
  antidiabetic, 91
  antimicrobial activity, 88–90
  biomedication, 90
  diagnostic application, 91–92
  vaccines, 90

Biomedical applications of nanomaterial, 316
Biomedicine, 264, 266, 276–280
Biomimetic nanodrug delivery systems, 218, 214
Biomimetic synthesis, 214
  functional biomimetic synthesis (FBS), 214–215
  process biomimetic synthesis (PBS), 215
Biomimetics, 208, 355, 390
  sensor, 380
  synthesis, 358
  system, 358
Biomimicry, 355
  nano, 388
Bionics, 355
Biopiezoelectric platform, 265–266
Biopolymers,
  cellulose, 234
  chitin, 235
  chitosan, 234
  poly lactic acid (PLA), 235
  polyhydroxyalkanoates (PHA), 235
  starch, 234
Bioreduction, 116–118
Biosensor, 276–277, 377–383
Biosorption, 116–118
Biotemplates, 69
  microorganism-based biotemplates, 75
    algae, 82–84
    bacteria, 75–79
    fungi, 79
    yeast, 79–82
  plant-based biotemplates, 70–75
Bio-waste, 268
Body-friendly, 274
Bone regeneration, 266, 280
Bottom up approaches, 190

Calcination, 45
Capacitor, 276

Capping agent, 27
Carbon dots, 165, 172, 180
Carbon nanotubes, 165, 172, 178, 179
Carboxylate ions, 52
Carboxymethyl chitosan (CMCS), 213
Carotenoids, 267
Cell-free extract, 51
Chirality, 274, 276
Chlorella vulgaris, 133
Circular permutation, 296
Co-assembly, 275
Cobalt ferrite nanoparticles, 162
Coiled-coil peptide motifs, 297
Common mechanical compression,
 bending, 268–269
 foot tapping, 268–269
 sound vibrations, 268
 twisting, 268–269
 walking, 268–269
 wind flow, 268
Conductivity, 275, 279
Contaminant, 27, 34
Corona discharging method, 270
Cowpea chlorotic mottle virus
 (CCMV)-based VLP, 293–294
CRISPR, 158, 178, 182
Cryopreservation mechanisms,
 freezing preservation, 336, 340
 hypothermic preservation, 336, 340
 supercooling preservation, 336, 340
 vitrification, 336, 340
Cytochrome P450 (CYP), 299
Cytoplasm, 329
Cytotoxicity, 194, 198

Dendrimer, 372
DFT calculations, 264–274
Different types of bioinspired
 nanocomposites,
 clay-polymer nanocomposites, 238
 nanowhisker-based
 bionanocomposites, 237
 polymer-HAp nanoparticle
 composites, 236

Dimensions, 44, 45, 47
Dipole moment, 267, 269, 273
Directed evolution, 296
Drug delivery, 194, 198, 291, 292, 299, 300

Eco-friendly, 26, 27
Electric filed, 264, 271, 273, 280
Electrical stimulation, 265, 280
Electrical stimulation therapy, 265
Electrochemical, 52, 55
Electrospinning, 269–270, 277
Electrostatic interactions, 294–295
Energy applications,
 energy conversion, 331
 energy storage, 331, 332, 339
 power generation, 331, 332, 338
Energy harvester, 268–269, 271, 273–274, 277
Environmental application, 92–93
 agriculture, 94–96
 catalytic removal of textile dyes, 93–94
 environmental remediation, 93
 wastewater treatment, 94
Enzyme immobilization, 51
Enzyme replacement therapy (ERT), 301
Enzymes, 191, 193, 331, 338, 343
 anthraquinones, 190–192
 electron shuttle, 191
 mycelia cell wall enzymes, 191
 naphthoquinones, 191
Epidermis, 268
Evolution, 325, 326
Extracellular, 192
Extracellular matrix (ECM), 267
Extracellular synthesis, 315

Fabrication of bionanocomposites,
 3D printing, 242
 ball milling method, 243
 freeze drying, 242
 melt moulding, 241

microwave assisted method for bionanocomposite preparation, 244
solvent casting, 240
ultraviolet irradiation method, 245
Fabrication techniques,
  atomic layer deposition, 328, 341
  focused ion beam CVD, 341
  layer-by-layer assembly, 341
  sol-gel, 341
  spin-coating, 341
Factors affecting the green synthesis of nanomaterials,
  incubation period, 146–147
  light, 146
  pH, 147
  precursor concentration and bioactive catalyst, 147
  temperature, 146
Fatty acid, 267
Feather-light pressures, 279
Ferritins, 292, 294, 296, 299, 300, 371
Fish gelatin nanofibers (GNFs), 269, 277
Fish scale (FSC), 268
Fish swim bladder (FSB), 268
Fullerene, 45
Functional material, 211
Fundamental constructs of body, 265
Fungi, 122–127, 191–194
  Aperigillus flavus, 191
  fusarium oxysporum, 191
  Verticillium luteoalbum, 191
  Volvariella volvacea, 190, 191, 193, 196, 200

Gibbs free energy, 48
GlaxoSmithKline Biologicals' Cervarix®, 300
Glutathione (GSH) encapsulation, 299
Gold nanorods, 167, 180
Graphene, 165, 167
Green nanoscience, 25, 27, 33
Green synthesis, 29, 115–116, 308

Green technology, 27–28
GroEL chaperonin cage, 294
Guanine crystal, 329

H-aggregation, 275
Hard templates, 216
Health and environment, 35
Health monitoring, 276–277
Helical conformation, 267
HER2-positive breast cancer, 221
High-density lipoproteins (HDL), 210
Horizontally oriented nanotubes, 271–273
Human health, 26, 28, 35
Hydrogel scaffold, 271, 279–280
Hydrogen bonding, 267

Immunocompatibility, 214
Immunogenicity, 210
Immunotherapeutic, 220
Implantable medical devices, 265
Inductor-based biosensors, 276
Insects for bioinspiration,
  honeycomb, 332, 339
  insects, 336, 340
  male peacock spider, 337
  mosquito, 335, 336, 340
  moth eyes, 329
  papilo butterfly, 328, 330, 337
  silver ant, 329, 337
  white beetle *calothyrza margaritifera*, 329
  white beetle *cyphochilus* spp., 328, 330, 337
Intracellular, 191–193
Intracellular synthesis, 314
Inversion symmetry, 264
Iron oxide nanoparticles, 163, 173, 179, 180

J stacking, 275

Laser ablation, 45
Lignocellulosic molecules, 266

LiNbO$_3$, 270
Liposomes, 333, 334, 339
Localized surface plasmon resonance, 159, 165
Lumazine synthase protein cage (AaLS), 296

M. jannaschii, 297
Magnetic field, 270
Magnetic nanoparticles, 119
Marine creatures for bioinspiration,
　electric eel, 331, 338
　fishes, 336, 340
　Japanese koi, 328
　lookdown, 329, 330, 337
　sea cucumber, 335, 336, 340
　squid beak, 335, 336, 340
Mechanical force, 264, 268, 271–273
Mechanical modelling, 327
Mechanical rigidity, 270, 273, 275–276, 279
Medical applications,
　cryopreservation, 336, 343
　diagnosis, 334
　drug delivery, 333, 336, 342
　neural implants, 336
　therapy, 333
　tissue engineering, 336
　wound healing, 333
Melanin pigment, 210
Merck's Gardasi, 301
Metal nanoparticles, 27, 32, 117–118, 191–194, 200
　copper, 31, 32
　gold, 30
　lead, 33
　palladium, 31, 32
　platinum, 32, 35
　selenium, 31, 33
　silver, 30, 32
Metal oxide nanoparticles, 190
Metal oxides, 32
　copper oxide, 31, 32
　iron oxide, 31, 32
　titanium dioxide, 31, 32

Metallization, 129
Microbe-based synthesis, 314
　actinomycete, 317
　bacteria-mediated, 315
　fungus-mediated, 316
Microbubbles, 45
Microorganism, 136
Microorganism-induced synthesis, 190
Molecular dynamics simulation, 267
Monoclinic, 269
Moving toward green nanostructures, 309
MS2 bacteriophage, 296
Multidrug-resistant, 50

Nanobiomimetics, 360–361, 387
Nanobiomimicry, 359–360
Nanobiosensor, 377
Nanoceria, 170
Nano-emulsion, 50
Nanofiber mat, 375
Nanomaterial-based biosensors, 376
Nanomedicine platform, 374
Nanoparticle synthesis, 309
　microbe-based, 314–318
　plant-mediated, 310–314
Nanoparticles, 26–30
　lipid, 30
　liposomes, 30
　nanocapsules, 30
　nanoemulsion, 30
　nanosphere, 30
　polymeric, 30
Nanoprisms, 48
Nanoscale metal organic frameworks, 165
Nanosensors, 158, 167
　design, 378–380
　operation, 380
Nanostructured materials, 26, 33
Nanotechnology, 26–28, 325, 326, 336, 355
Nanotheranostics, 53

Natural materials, 264–266
Natural resources, 35
Non-centrosymmetric crystal, 274, 275
Non-covalent interaction, 267
Non-toxic, 265
Nuclear quadruple resonance, 274
Nucleation, 48, 55

Oligodeoxynucleotides (CpG), 300
Oligosaccharides, 49
Open-circuit voltages, 268–271, 273–277
Optical mechanisms,
 bioluminescence, 328
 pigments, 328, 341
 structural colours, 328, 341
Optics applications,
 antireflector, 329, 337, 341
 broad band light absorption, 329, 338
 camouflage, 328, 329, 337
 colour perception, 337
 fibers, 329
 paint, 329
 photocatalysis, 329
 photothermal absorption, 329
Organic nanoparticles,
 dendrimers, 10
 liposomes, 10
 micelles, 10
Orientation of structure, 266, 269, 271, 277
Other sources for bioinspiration,
 amphibians, 336, 340
 mollusk, 336, 340
 reptiles, 336, 340
 West African Gaboon viper, 337
 yeast, 332

P. horikoshii, 297
Peptide, 265
 cyclo-Dip-Dip, 273
 cyclo-Gly-Trp, 273
 cyclo-Phe-Trp, 273
 Dip-Dip, 273

diphenylalanine (FF), 271
Fmoc-Gly-Pro-Hyp, 271
Hyp-Phe-Phe, 270–271
poly(l-lactic acid) (PLLA), 280
poly-γ-benzyl-L-glutamate (PBLG), 270
poly-γ-methyl-L-glutamate (PMLG), 270
Pro-Phe-Phe, 270
Photoacoustic, 54
Photonic crystals, 328
Physical functions of body,
 blood circulation, 265, 279
 body movement, 265, 268–279
 heartbeat, 265, 279
 muscle construct, 265, 279
Physicochemical, 44, 45, 56
Physiological behaviors of cells,
 apoptosis, 265
 differentiation, 265, 276
 migration, 276
 proliferation, 265, 276
Physiological signals (PS), 277, 279
Phytochemistry, 48
Phytomining, 46
Piezoelectric coefficient, 267, 269–271, 273–277
Piezoelectric green energy harvester,
 anti-fatigue, 269
 biopiezoelectric nanogenerator (BPNG), 268–277, 279
 copper wires, 269–270
 device fabrication, 268–273
 durable, 268
 efficiency, 274, 278
 electrode, 268–273
 finger tapping, 279
 functional steadiness, 269
 Kapton tape, 270
 lead wires, 271
 pressure sensitivity, 277
 self-powered, 264–280
 sensitivity, 277, 279
 stability, 271, 279

Piezoelectric sensors, 276–278
Piezoelectricity,
   fine tune, 274
   longitudinal piezoelectricity, 264–280
   shear piezoelectricity, 270–271
Piezoresponse force microscopy (PFM), 267, 269–271, 278
Plant extracts, 27, 33, 190, 193
   Aloe vera, 193
   Azadirachta indica, 193, 194
   Cycas, 193
   Geranium, 193
   lucerne, 193
   Magnolia kobus, 193
   Pelargonium graveolens, 193
   Stevia rebaudiana, 193
   Sunflower Asteraceae, 193
Plant-mediated synthesis,
   gold, 311
   selenium, 314
   silver, 310
   zinc oxide, 313
Plants for bioinspiration,
   leaf venation system, 331, 338
   plants, 336, 340
Plants, 134–135
Plasma arcing, 44
Polarization, 266
Polyamidoamine (PAMAM) dendrimers, 179
Polydimethylsiloxane (PDMS), 268–269, 279
Polymeric nanoparticles, 210, 213
Polymersomes, 333, 334, 339
Polyproline II-structure, 271
Polypropylene film, 269
Pomegranate peel, 269
Portable electronics, 277
Properties of nanoparticles,
   bioavailability, 30
   controllable size, 30
   solubility, 30
   stability, 30

Protein,
   collagen, 264, 266–270, 274, 277, 279
   keratin, 268–269
   lysozyme, 268–269
   M13 bacteriophase, 269–270
   prestin, 268
   silk, 267, 278–279
Protein cages, 371
Protein cages and their potential application in therapeutics,
applications,
   protein cage as targeted delivery vehicles for therapeutic protein, 298–299
   protein cage-based encapsulation and targeting of anticancer drugs, 299–300
   protein cage-based immune-therapy, 300–301
   cage modifications and cargo loading, 295–298
   future perspective, 301
   introduction, 292–295
   main sources of, 292–293
Proteins & peptides, 336, 340
PVDF, 277, 280

Quantum dots, 165, 166
Quartz, 266

Rat model, 280
Real-time physiological signals, 277
Recemic mixtures, 274
Reducing agent, 189, 190, 194, 199
   Ananas comosus, 194
   banana peel, 193
   Berberis vulgaris leaf, 193
   black tea leave (*Camellia sinensis*), 195
   Carica papaya leaf extract, 197
   Carob pod (*Ceratonia siliqua*), 199
   Chickpea (*Cicer arietinum*), 194
   mango peel extract, 194

Moong Bean (*Vigna radiata*), 194
Ocimum tenuiflorum (black tulsi), 194
orange peel extract, 197
Removal of metal ions, 191
Response surface methodology (RSM), 57
Rigid assembly, 275
RNA virus, 130

Salmonella typhimurium, 296
Seeding platform, 271
Selenium nanoparticles, 169
Short-circuit current, 268–271, 273–277
Signal-generation tags, 51
Silica nanoparticles, 176, 177, 179
Silver nanoparticles, 162, 163, 170
Soft templates, 216
Solvothermal, 45
Sonchus asper, 267
SpyTag/SpyCatcher system, 297–298
Stimulus-responsive materials, 212
Stokes-Einstein, 58
Structural material, 211
Super critical water oxidation, 366
Supercharging method, 295–296
Surface enhanced Raman spectroscopy, 165
Surface modification, 46, 51, 55
Surface plasmon, 50, 57, 58
Surfactants, 48, 51
Sustainability, 326
Synthesis routes, 82
  effect of biomolecules, 86
    microorganism-based, 87
    plant-based, 86
  effect of pH, 84–85
  effect of temperature, 85

Target triggered nanomaterials, 375
Tautomerism, 49
Tetragonal, 269
Thermal stability, 275–276
Tissue function,
  healing, 264, 266
  regeneration, 264, 266
  remodeling, 264
  tissue growth, 264, 266
Tissue regeneration,
  hydroxyapatite, 279
  intracellular ion level, 279
  voltage gated ion channel, 279
Tobacco mosaic virus (TMV)-based therapeutic cage, 299
Toll-like receptors, 300
Top down approaches, 190
Toxicity, 30
Toxicity of nanostructures, 318
Triple helical conformation, 267
Types of nanoparticles,
  green synthesis of copper (Cu) nanoparticles, 8–9
  green synthesis of gold (Au) nanoparticles, 7–8
  green synthesis of silver (Ag) nanoparticles, 4–6
  inorganic nanoparticle, 4–9
  iron oxide nanoparticles, 9

Unnatural amino acids (uAAs), 297
Upconversion nanophosphors, 165

Vaccine development, 292, 298, 300
Vertically oriented rods, 271
Virus, 129–132
Virus-like-particles (VLPs), 292–294, 299
Voice recognition, 279

Wet air oxidation, 366
What are nanoparticles?, 2–3
Wolff's law, 264
Wound healing,
    inflammation, 280
    proliferation, 280
    remodeling, 280
Wrist pulse, 279

X stacking, 275

Yeast, 127–129

Zein nanoparticles, 161
Zeta potential, 56
ZnO, 270